Frontiers in International Relations

This book series offers an outlet for cutting-edge research on all areas of international relations. *Frontiers in International Relations* (FIR) welcomes theoretically sound and empirically robust monographs, edited volumes and handbooks from various disciplines and approaches on topics such as IR-theory, international security studies, foreign policy, peace and conflict studies, international organization, global governance, international political economy, the history of international relations and related fields. All books published in this series are peer-reviewed.

More information about this series at http://www.springer.com/series/16555

Joana Castro Pereira · André Saramago
Editors

Non-Human Nature in World Politics

Theory and Practice

Springer

Editors
Joana Castro Pereira
Portuguese Institute of International Relations
NOVA University of Lisbon
Lisbon, Portugal

André Saramago
Faculty of Economics
University of Coimbra
Coimbra, Portugal

ISSN 2662-9429 ISSN 2662-9437 (electronic)
Frontiers in International Relations
ISBN 978-3-030-49495-7 ISBN 978-3-030-49496-4 (eBook)
https://doi.org/10.1007/978-3-030-49496-4

This Springer imprint is published by the registered company Springer Nature Switzerland AG
The registered company address is: Gewerbestrasse 11, 6330 Cham, Switzerland

Contents

Empirical Investigations

Introduction: Embracing Non-Human Nature in World Politics

Joana Castro Pereira and André Saramago

Abstract This book theorizes and empirically analyses the intertwinement between world politics and non-human nature breaking the anthropocentric boundaries that have characterized the field of International Relations thus far. Gathering contributions from several different perspectives ranging from post-humanism and ecomodernism, to new materialism and post-colonialism, the book is divided into two main parts. The first is dedicated to new ways of theoretically conceiving the relationship between non-human nature and world politics. The second presents empirical investigations focusing on specific case studies. It includes chapters on state actors and international organizations and bodies. This edited volume is a distinctive intervention in the field, offering new perspectives on how to conceptualize the embeddedness of world politics in non-human nature and how to reorient political practice to better address the challenges posed by climate change and the degradation of the Earth's ecosystems.

Introduction

Humans have become a planetary force—anthropogenic activities are transforming the functioning and risking the resilience of the Earth system. Humanity is dangerously changing the climate and eroding the planet's life-support systems upon which it depends (Dasgupta et al. 2019; Gaffney and Steffen 2017; IPBES 2019; IPCC

Joana Castro Pereira would like to acknowledge the support of the Portuguese Foundation for Science and Technology—FCT (CEECIND/00065/2017). The authors would also like to thank all the contributors to this volume as well as all of those that, in one way or another, helped make it possible.

J. C. Pereira (✉)
Portuguese Institute of International Relations, NOVA University of Lisbon, Lisbon, Portugal
e-mail: joanacastropereira@fcsh.unl.pt

A. Saramago
Faculty of Economics, University of Coimbra, Coimbra, Portugal
e-mail: asaramago@fe.uc.pt

J. C. Pereira and A. Saramago (eds.), *Non-Human Nature in World Politics*,
Frontiers in International Relations, https://doi.org/10.1007/978-3-030-49496-4_1

2018; Ripple et al. 2019; Steffen et al. 2011, 2015). As a result, climate change and biodiversity loss have emerged as two of the most pressing threats to security and survival on a global scale. However, notwithstanding decades of multilateral environmental negotiations, both crises have aggravated significantly, and political efforts are still highly insufficient to ensure "a safe operating space for humanity" (Rockström et al. 2009).

Global temperatures are rising at a rapid rate and climate change is accelerating. 2015–2019 was the warmest 5-year period ever recorded (WMO 2019). According to the latest estimates, current policies are consistent with a global temperature increase of >1.5 °C by the mid-2030s, >2 °C by the early 2050s, and >3.2 °C by 2100; if parties fully achieve the emissions reduction targets they have committed to under the Paris Climate Agreement, the global average temperature is likely to rise to 2.9 °C by the end of the century (CAT 2019). This would place the Earth at the edge of a climate catastrophe (Xu and Ramanathan 2017; Pereira and Viola 2018). As nearly two-thirds of the major greenhouse gas emitters are not even on course to meet their climate pledges (Kuramochi et al. 2018), there are strong reasons for concern. In addition, anthropogenic pressure on the biosphere is pushing biodiversity into a new mass extinction event (Ceballos et al. 2017)—current species extinction rates have no parallel in human history (Laybourn-Langton et al. 2019). Representing no more than 0.01% of all life on the planet, humanity has caused the loss of approximately 80% of wild and marine mammals, nearly half of all plant species, and roughly 15% of fish (Bar-On et al. 2018); one million species are currently threatened with extinction, risking ecosystem collapse (IPBES 2019). The loss of those organisms, which are integral working parts of the natural systems that support life on Earth, is permanent and has potentially catastrophic consequences for the human species, whose very existence is owed to the activities of past and current non-human life forms. This fact remains critically under-recognized in political discussions and within the policy domain (Lenton 2019). As in the case of climate change, there is a massive gap between scientific evidence and political action—most of the Aichi Biodiversity Targets for 2020 will not be met (IPBES 2019).

Climate change and biodiversity loss, and the failure to mitigate them, can be understood as the most visible manifestations of the anthropocentrism underlying academic thought and environmental politics and policy at all levels of governance; in other words, of a viewpoint according to which human beings are separate from, and superior to, nature; a belief that is profoundly contradicted by the essence of the ecological crisis facing the world (Youatt 2014; Cudworth and Hobden 2013). An understanding of the crisis and the development of an adequate and effective response to its challenges is thus inseparable from a post-anthropocentric perspective that breaks with conventional thinking and practice. This perspective would recognize humanity's (a) embeddedness in the Earth system, (b) predominant role in driving ecological change at the planetary level, and (c) dependence on, and "shared ties of vulnerability" (Braidotti 2013, p. 69) with, non-human living beings. Two basic assumptions should thus inform knowledge production and policymaking. First, nature is not external to human politics. Humanity is reshaping the planet and remaking its own context at a pace and with an impact hitherto unknown in its history;

the ecological crisis is the result of political choices, and only new ones can mitigate its effects and prevent the worst-case scenarios from becoming a reality (Dalby 2017). Second, harm and violence inflicted on non-human nature compromises human security and the very conditions that enable life (human and non-human) on Earth to flourish (Mitchell 2014; Fishel 2017). Accordingly, if humanity is to create "a safe operating space" for itself, policies must be guided by ontological, ethical, and moral principles that ensure a fair sharing of the Earth between humans and non-humans (Gupta 2016; Pereira and Freitas 2017).

The fact that life can radically transform the Earth is not new. It is an inherent feature of life itself (Lovelock 1979). However, "it is the first time that a single animal species has wielded such world-changing power" (Lenton 2019, p. 34)—a power that, ironically, is risking that same species' well-being and survival. International Relations (IR), as both a field of knowledge and institutional practice, is yet to understand such power, its implications, and the multiple ways in which humanity depends on, and is enmeshed in, nature. Chained to an anthropocentric worldview, IR is failing to acknowledge and address "the collective human interaction with the biosphere" (Burke et al. 2016, p. 501), finding itself voiceless before the potentially catastrophic or existential risks arising from anthropogenic interference with the Earth system (Mitchell 2017). In fact, despite the gravity of the planetary ecological crisis, environmental issues, in general, are still far from being a priority in IR's research agenda (Green and Hale 2017; Pereira 2017). This is particularly odd for a discipline "explicitly and profoundly occupied with the issue of survival" (Mitchell 2017, p. 2) and whose "*raison d'etre*" is to grapple with problems of coexistence (Corry and Stevenson 2018, p. 1).

The idea for this book was born out of the editors' wish to contribute to overcome IR's ecological blindness and anthropocentrism and enhance the field's capacity to both deal with the challenges posed by a rapidly changing planet and have a positive and constructive impact on policy and practice. It was inspired by an emerging body of the literature that is seeking to reorient research towards novel ontological, epistemological, methodological, ethical, moral, political, and governance approaches aligned with the planetary real (see, for instance, Arias-Maldonado and Trachtenberg 2019; Biermann and Lövbrand 2019; Burke et al. 2016; Cudworth and Hobden 2018; Dryzek and Pickering 2018; Eckersley 2017; Eroukhmanoff and Harker 2017; Fagan 2017; Harrington and Shearing 2017; Inoue 2018; Leep 2017; McDonald 2018; Nicholson and Jinnah 2016; Pereira 2017; Pereira and Freitas 2017; Youatt 2014).

The notion of "non-human nature", presented in the book's title, acts as the conceptual framework for this edited volume. In this context, "non-human nature" is understood as an attempt to overcome the dualism between humans and nature that is so prevalent in the literature. On the one hand, the use of separate terms for "humans" and "nature" reinforces anthropocentric ontologies by inherently assuming that humans are separate from "nature", broadly conceived; something that, consequently, frequently reifies "nature" as a separate ontological entity that lacks agency and is acted upon by humans. On the other hand, while humans are a part of nature, they are also a species that, as a result of a long-term process of natural evolution, has developed a capacity to transform itself and the rest of nature to a far greater extent than other species on the planet—the interdependent development of industrial civilization and the current ecological crisis being one example of just such a capacity (see Damásio 2018; Elias 2011). Humans are, therefore, uniquely placed in the Earth system to influence its future development, for better or for worse. Accordingly, the concept of non-human nature opens up discussions centred on how humans relate to the "non-human part of the universe"—with which humans are inherently and deeply intertwined, being themselves a part of "nature"—while recognizing that not all of nature has the same level of responsibility towards the future of earthly life, given wide variations between different species' evolved capacity to influence its development (Goudsblom et al. 1996; Vries and Goudsblom 2003).

The book assembles scholars working in IR and associated fields, such as Political Science, Sociology, Law, and Philosophy, and practitioners, to consider a broad range of questions concerning the intertwinement between world politics and non-human nature. It is divided into two main parts—the first is focussed on theoretical investigations and the second on empirical studies. Part I is formed of six chapters discussing news ways of theoretically conceiving the relationship between non-human nature and world politics. While expressive of a wide range of theoretical approaches, from post-humanism and new materialism to ecomodernism, post-colonialism, process sociology, and post-structuralism, these chapters are connected by a collective concern with thinking beyond anthropocentric ways of conceiving human/non-human nature relations. Part II gathers eight empirical chapters that examine both the destruction and protection of non-human nature across different contexts in world politics. Specific case studies are addressed—these cover some of the key state actors and international organizations and bodies in mitigating climate change and halting biodiversity loss.

The broad argument of this edited volume is that the development of a truly post-anthropocentric IR should not be conceived of as another "turn" in the discipline that shifts its focus to the study of non-human nature, nor as an add-on for analyses of world politics or undergraduate courses. Rather, post-anthropocentric IR must integrate human/non-human nature relations in its study of the "international", seeking to understand the inherent intertwinement between world politics and the Earth system. Embracing non-human nature in world politics thus means recognizing, theorizing, and analysing the fundamental embeddedness of the "human" in the "non-human part of the universe", and how every manifestation of world politics is necessarily entangled with non-human processes and species. The chapters in this book try to contribute to the development of such a post-anthropocentric IR by theoretically and empirically conceiving human/non-human nature relations in world politics as an entangled whole.

Chapters 2 and 3 approach the topic of non-human agency seeking to understand how it can be recognized and taken into account in analyses of world politics, and how human institutions can integrate non-human points of view. In Chap. 2, Bob Carter and Oliver J. T. Harris contest "normal politics" and their inherent anthropocentrism by proposing the alternative of "assemblage politics" that emphasize relational entanglements and dissolve conceptual borders between the human and the non-human; an argument that is illustrated by the examples of car use, Brexit, and the climate crisis. In Chap. 3, Anthony Burke and Stephanie Fishel develop a new materialist and post-humanist approach to the material agency of ecosystems in a way that informs the conception of ecological democratic institutions that include non-human communicative systems and permit the development of "interspecies politics".

Chapters 4 and 5 share a concern with the ontological dimension of IR seeking, in different ways, to problematize and/or move beyond the dominant ontologies via which human/non-human nature relations have been understood in the discipline. In Chap. 4, Cameron Harrington uses quantum social theory to analyse different emerging discourses on the Anthropocene, contrasting those embedded in the "classic" world view, which tend to conceive of the Anthropocene as a radical break with the past, with discourses that, in various ways, are closer to a "quantum" world view that emphasizes how the Anthropocene makes apparent a fundamental entanglement between the human and the non-human. In Chap. 5, Rafi Youatt problematizes what he considers to be the reductionist way in which mountains are produced as political objects by secular environmental governance regimes, instead proposing an understanding of mountains via an analytic and political focus on routes and routing processes that permits conceiving them as multiple rather than singular natural objects.

Chapters 6 and 7 present different arguments for what might be considered the development of humanist approaches to the analysis of human/non-human nature relations in world politics. In Chap. 6, Rasmus Karlsson contrasts ecomodernism with traditional environmentalism emphasizing the shortcomings of the latter's aspiration to an ecologically sustainable, less integrated, low-energy world, while recognizing the tension between ecomodernism's vision of an ecological future in a fully integrated, high-energy planet and present-day science. In Chap. 7, Stephen Quilley assesses the de-growth and ecomodernist agendas from the perspective of process sociology and considers the latter's contribution to the development of a deeper understanding of the sociological processes involved in ecological consciousness and community formation in the context of the Anthropocene.

Chapters 8 to 10 shed light on the politics and governance of non-human nature in Indonesia, China, and the United States. In Chap. 8, Shofwan Al Banna Choiruzzad analyses the Indonesian case, discussing the role that historical colonization and state-building processes had in both eroding the social–ecological governance systems of indigenous communities, and establishing anthropocentric ontological assumptions and institutional practices in the country. In Chap. 9, Justin Heinzekehr considers the importance of sustainability narratives to environmental policy by critically exploring the Chinese concept of "ecological civilization"—both the ideological framework for China's environmental policies and a cornerstone of the Chinese vision for sustainable development and humanity's common future. In Chap. 10, Sheila D. Collins addresses the right-wing populist–nationalist challenge to climate change science and policy by examining the Donald Trump administration's efforts to reverse climate regulations in the United States. In addition, she briefly describes the role that sub-state and civil society actors have been playing in countering climate change denial in the country.

Chapters 11 and 12 look at the European Union's (EU) internal and external environmental action. First, Hendrik Schoukens discusses the significance of the rights of nature paradigm to ecological governance; sheds light on the potential overlaps between a rights-based approach to nature and the EU's environmental laws; and elaborates on the challenges facing full operationalization of the concept within the existing EU legal order. In the next chapter, Simon Schunz, Bram De Botselier, and Sofía López Piqueres critically analyse the EU's environmental diplomacy. Reviewing the union's positions and behaviour regarding global political efforts to mitigate climate change, halt biodiversity loss, and protect the Arctic region's stability and fragile ecosystems, they assess the ecological effectiveness of the EU's external action.

Chapters 13 and 14 are devoted to the work developed by environmental non-governmental organizations (ENGOs). First, Danita Catherine Burke addresses the case of the World Wide Fund for Nature/World Wildlife Fund (WWF) and its work in the Arctic Council, and scrutinizes the importance of cultivating trust and considering the social dimension and impacts of environmental protection actions for the organizational effectiveness and success of ENGOs. In the next chapter, Teale N. Phelps Bondaroff furthers our understanding of direct action strategies and their application by activists on transnational issues and campaigns by investigating the operations of a wide range of marine conservation organizations. He categorizes the different strategies into a typology and analyses activist–state relationships.

Finally, Chap. 15 considers the biodiversity knowledge–policy interface. Lavínia Pereira and Olivia Bina reflect on the need for effective dialogue across different paradigms and knowledge and value systems to better capture the complex interactions between the natural and the social worlds, and to inform transformative political change, by critically reviewing the conceptual framework of the Intergovernmental Science-Policy Platform on Biodiversity and Ecosystem Services (IPBES).

The book concludes in Chap. 16 with a discussion by Olaf Corry that assesses the volume's various contributions and explores some of the possible pathways for, and implications of, bringing non-human nature into world politics analyses. Corry traces the origins of anthropocentric IR; draws attention to the need of considering the international dimension of world politics in post-anthropocentric theorizations and empirical examinations; and proposes a materialist understanding of societal multiplicity as a means of including the "international" within analyses of the human/non-human nature dynamics.

References

Arias-Maldonado, M., & Trachtenberg, Z. (Eds.). (2019). *Rethinking the environment for the Anthropocene: Political theory and socionatural relations in the new geological epoch*. Abingdon and New York: Routledge.

Bar-On, Y. M., Phillips, R., & Milo, R. (2018). The biomass distribution on earth. *Proceedings of the National Academy of Sciences of the United States of America, 115*(25), 6506–6511.

Biermann, F., & Lövbrand, E. (Eds.). (2019). *Anthropocene encounters: New directions in green political thinking*. Cambridge: Cambridge University Press.

Braidotti, R. (2013). *The posthuman*. Cambridge and Malden: Polity Press.

Burke, A., Fishel, S., Mitchell, A., Dalby, S., & Levine, D. J. (2016). Planet politics: A manifesto from the end of IR. *Millennium: Journal of International Studies, 44*(3), 499–523.

CAT (2019). Climate Action Tracker Warming Projections Global Update: September 2019. Retrieved from https://climateactiontracker.org/documents/644/CAT_2019-09-19_BriefingUNSG_WarmingProjectionsGlobalUpdate_Sept2019.pdf.

Ceballos, G., Ehrlich, P. R., & Dirzo, R. (2017). Biological annihilation via the ongoing sixth mass extinction signaled by vertebrate population losses and declines. *Proceedings of the National Academy of Sciences of the United States of America, 114*(30), E6089–E6096.

Corry, O., & Stevenson, H. (2018). IR and the earth: Societal multiplicity and planetary singularity. In O. Corry & H. Stevenson (Eds.), *Traditions and trends in global environmental politics: International relations and the earth* (pp. 1–25). Abingdon and New York: Routledge.

Cudworth, E., & Hobden, S. (2013). Of parts and wholes: International relations beyond the human. *Millennium: Journal of International Studies, 41*(3), 430–450.

Cudworth, E., & Hobden, S. (2018). *The emancipatory project of posthumanism*. Abingdon and New York: Routledge.

Dalby, S. (2017). Anthropocene formations: Environmental security, geopolitics and disaster. *Theory, Culture & Society, 34*(2–3), 233–252.

Damásio, A. (2018). *The strange order of things: Life, feeling, and the making of cultures*. New York: Pantheon Books.

Dasgupta, P., Raven, P. H., & McIvor, A. L. (Eds.). (2019). *Biological extinction: New perspectives*. Cambridge: Cambridge University Press.

Dryzek, J. S., & Pickering, J. (2018). *The politics of the Anthropocene*. Oxford: Oxford University Press.

Eckersley, R. (2017). Geopolitan democracy in the Anthropocene. *Political Studies, 65*(4), 983–999.

Elias, N. (2011). *The symbol theory*. (The collected works of Norbert Elias: Vol. 13). Dublin: University College Dublin Press.

Eroukhmanoff, C., & Harker, M. (Eds.). (2017). *Reflections on the posthuman in international relations: The Anthropocene, security and ecology*. Bristol: E-International Relations Publishing.

Fagan, M. (2017). Security in the Anthropocene: Environment, ecology, escape. *European Journal of International Relations, 23*(2), 292–314.

Fishel, S. (2017). Performing the posthuman: An essay in three acts. In C. Eroukhmanoff & M. Harker (Eds.), *Reflections on the posthuman in international relations: The Anthropocene, security and ecology* (pp. 50–60). Bristol: E-International Relations Publishing.

Gaffney, O., & Steffen, W. (2017). The Anthropocene equation. *The Anthropocene Review, 4*(1), 53–61.

Green, J. F. & Hale, T. N. (2017). Reversing the marginalization of global environmental politics in international relations: An opportunity for the discipline. *PS: Political Science & Politics, 50*(2), 473–479.

Goudsblom, J., Jones, E., & Mennell, S. (1996). *The course of human history: Economic growth, social process and civilization.* Abingdon: Routledge.
Gupta, J. (2016). Toward sharing our ecospace. In S. Nicholson & S. Jinnah (Eds.), *New earth politics: Essays from the Anthropocene* (pp. 271–292). Cambridge and London: The MIT Press.
Harrington, C., & Shearing, C. (2017). *Security in the Anthropocene: Reflections on safety and care.* Bielefeld: Transcript Verlag.
Inoue, C. Y. A. (2018). Wordling the study of global environmental politics in the Anthropocene: Indigenous voices from the Amazon. *Global Environmental Politics, 18*(4), 25–42.
IPBES (2019). Summary for Policymakers. In S. Díaz, J. Settele, E. S. Brondizio, H. T. Ngo, M. Guèze, J. Agard, & C. N. Zayas (Eds.), *Global assessment report on biodiversity and ecosystem services of the intergovernmental science-policy platform on biodiversity and ecosystem services.* IPBES Secretariat: Bonn.
IPCC (2018). Summary for policymakers. In V. Masson-Delmotte, P. Zhai, H.-O. Pörtner, D. Roberts, J. Skea, P. R. Shukla, T. Waterfield (Eds.), *Global Warming of 1.5°C. An IPCC Special Report on the impacts of global warming of 1.5°C above pre-industrial levels and related global greenhouse gas emission pathways, in the context of strengthening the global response to the threat of climate change, sustainable development, and efforts to eradicate poverty.* Geneva: WMO.
Kuramochi, T., Fekete, H., Luna, L., de Villafranca Casas, M. J., Nascimento, L., Hans, F., et al. (2018). *Greenhouse gas mitigation scenarios for major emitting countries. Analysis of current climate policies and mitigation commitments: 2018 update.* Cologne: New Climate Institute.
Laybourn-Langton, L., Rankin, L., & Baxter, D. (2019). *This is a crisis: Facing up to the age of environmental breakdown. Initial report.* London: Institute for Public Policy Research.
Leep, M. (2017). Cosmopolitanism in a carnivorous world. *Politics and Animals, 3,* 16–30.
Lenton, T. M. (2019). Biodiversity and global change: From creator to victim. In P. Dasgupta, P. H. Raven, & A. L. McIvor (Eds.), *Biological extinction: New perspectives* (pp. 34–79). Cambridge: Cambridge University Press.
Lovelock, J. (1979). *Gaia: A new look at life on earth.* Oxford: Oxford University Press.
McDonald, M. (2018). Climate change and security: Towards ecological security? *International Theory, 10*(2), 153–180.
Mitchell, A. (2014). Only human? A wordly approach to security. *Security Dialogue, 45*(1), 5–21.
Mitchell, A. (2017). Is IR going extinct? *European Journal of International Relations, 23*(1), 3–25.
Nicholson, S., & Jinnah, S. (2016). *New earth politics: Essays from the Anthropocene.* Cambridge and London: The MIT Press.
Pereira, J. C. (2017). The limitations of IR theory regarding the environment: Lessons from the Anthropocene. *Revista Brasileira de Política Internacional, 60*(1), e018.
Pereira, J. C., & Freitas, M. R. (2017). Cities and water security in the Anthropocene: Challenges and opportunities for international relations. *Contexto Internacional, 39*(3), 521–544.
Pereira, J. C., & Viola, E. (2018). Catastrophic climate change and forest tipping points: Blind spots in international politics and policy. *Global Policy, 9*(4), 513–524.
Ripple, W. J., Wolf, C., Newsome, T. M., Barnard, P., & Moomaw, W. R. (2019). World scientists' warning of a climate emergency. *BioScience*, biz088.
Rockström, J., Steffen, W., Noone, K., Persson, Å., Chapin, F. S., Lambin, E., et al. (2009). Planetary boundaries: exploring the safe operating space for humanity. *Ecology and Society, 14*(2), 32.
Steffen, W., Persson, A., Deutsch, L., Zalasiewicz, J., Williams, M., Richardson, K., et al. (2011). The Anthropocene: From global change to planetary stewardship. *Ambio, 40*(7), 739–761.
Steffen, W., Richardson, K., Rockström, J., Cornell, S. E., Fetzer, I., Bennett, E. M., et al. (2015). Planetary Boundaries: Guiding human development on a changing planet. *Science, 347*(6223), 1259855.

Vries, B., & Goudsblom, J. (2003). *Mappae mundi: Humans and their habitats in a long-term socio-ecological perspective.* Amsterdam: Amsterdam University Press.
WMO (2019). *The global climate in 2015-2019.* Geneva: World Meteorological Organization.
Xu, Y., & Ramanathan, V. (2017). Well below 2°C: Mitigation strategies for avoiding dangerous to catastrophic climate changes. *Proceedings of the National Academy of Sciences of the United States of America, 114*(39), 10315–10323.
Youatt, R. (2014). Interspecies relations, international relations: Rethinking Anthropocentric politics. *Millennium: Journal of International Studies, 43*(1), 207–223.

Theoretical Investigations

The End of Normal Politics: Assemblages, Non-Humans and International Relations

Bob Carter and Oliver J. T. Harris

Abstract Over the past decade, notions of the non-human have become established in a range of disciplines—for example, archaeology, human geography, anthropology, and architecture. However, their impact on politics and international relations has been restricted. In this chapter, we consider the reasons for this, identifying what we term 'normal politics' as the dominant form of anthropocentric discourse that prevents non-humans from taking their proper place in the political realm. In contrast, we propose a form of assemblage politics that is better placed to accommodate notions of the non-human, whilst also offering fresh perspectives on politics and international relations. We develop and illustrate our notion of assemblage politics using three examples: car use, Brexit and the climate crisis.

Introduction

We are aware in writing this chapter that we are hardly well qualified to contribute to a text on International Relations. One of us is a sociologist, the other an archaeologist. Yet in both our disciplines, as in the discipline of International Relations (hereafter IR), there has been a growing concern at the uneven response to the emergence of what Burke et al. (2016) has termed 'planet politics'. This concern has expressed itself in various ways: in calls for a posthumanist approach within the social sciences (for example, Braidotti 2013; Cudworth and Hobden 2011); in calls for a robust reassertion of the 'positive' promise of the social sciences (Fuller and Lipinska 2014; Chandler 2013); in the advocacy of a 'new materialism' (Barad 2007; Haraway 2016; Coole and Frost 2010); and in the development of a wide range of interdisciplinary theories associated with notions of complexity (see for example Walby 2007). These responses in the main share two common commitments: to

B. Carter (✉)
Department of Media, Communications and Sociology, University of Leicester, Leicester LE1 7RH, UK
e-mail: rc300@le.ac.uk

O. J. T. Harris
School of Archaeology and Ancient History, University of Leicester, Leicester LE1 7RH, UK

J. C. Pereira and A. Saramago (eds.), *Non-Human Nature in World Politics*, Frontiers in International Relations, https://doi.org/10.1007/978-3-030-49496-4_2

moving beyond conventional disciplinary boundaries and constructing accounts of the world capable of addressing the overwhelming threat of human extinction carried by the current climate crisis; and an acknowledgement that to do so requires the inclusion of the non-human, animate and inanimate, in our theories. There are resources from within our own disciplines that share these commitments. It is in this spirit of contributing to a common enterprise that we offer perspectives from what would appear to be outlier disciplines for IR. It is also why our notion of what counts as a concern for IR is considerably broader than is usually found in the discipline's literature; analogously, many sociologists and archaeologists would find our view of what should concern their subjects unconventionally wide.

In this chapter, we argue that a political language that is predominantly human in focus cannot hope to meet the challenges that face IR in the twenty-first century. We are not alone in making this claim, of course. Writers such as Latour (2018), Harman (2018), Bennett (2010), Moore (2015) and Tsing (2015), amongst others, have all argued for a non-anthropocentric approach to the world and suggested theoretical resources for such an endeavour. These arguments have found echoes in the work of some of those writing within IR, such as, amongst others, Sassen (2006), Ong and Collier (2005), Eroukhmanoff and Harker (2017), Cudworth and Hobden (2018; cf. Cudworth et al. 2018), Mitchell (2014), Pereira (2017), Salter (2015) and McDonald (2018). Yet, as Acuto and Curtis point out, '…a pressing need has developed for theoretical perspectives and methodologies that can enable us to understand the impact of the changing configurations of the natural and social worlds' (2014, p. 2). Whilst IR, and the social sciences more generally, may not be as 'fully human centred' in their approach as Cudworth and Hobden claimed in 2011, we would insist that human-centred political analysis remains a persistent force, despite proving less and less adequate both to diagnose and to offer a response to the challenges—of climate crisis, of international migration, of the threats posed by the rise of authoritarian populisms, of growing inequalities—facing the contemporary world.

There are a number of reasons for the persistence of human-centric perspectives in the social sciences. These include the privileging of what we term, following the anthropologist Anna Tsing (2015), 'kinds thinking', which artificially separates and delineates the world into discrete, hermetic types; the casting of these 'kinds' as transcendent, that is, outside of history and the local conditions from which they are emergent; and the 'scaling up' of local conditions into forms of kinds thinking, producing a radical reductionism in which the singularity of the local vanishes in favour of generic mechanisms of change. Kinds thinking of this sort not only denies a political voice to many of the non-humans to whom we need to listen, it also masks the powerful political agency possessed by some of them. Political visions that rely on kinds thinking of one sort or another define what we will identify as Normal Politics. All forms of Normal Politics are anthropocentric, regardless of their political alignment on 'the Left' or 'the Right'. We might refer to the former as 'critical' Normal Politics, and the latter as 'conventional', but both adopt, as we set out in detail below, the same fundamental ontological position: that nature and culture are opposed; that politics is the provenance of human beings; that agency[1] is the possession of actors either individual, state or governmental (cf. Latour 1993). Each of these claims rests on kinds thinking. Even environmental politics in its Normal Politics mode does not escape a fundamental opposition between human beings, on the one hand, and everything non-human on the other, especially those elements of the world that would conventionally be classified as nature.

The dominance of Normal Politics, and its anthropocentric concerns, is a major obstacle to imaginative political responses to the crises of the contemporary world, as writers within IR have pointed out (e.g. Cudworth and Hobden 2011). Here, we wish to adopt a different starting point to Normal Politics. In particular, we draw upon the assemblage theory of the philosophers Manual DeLanda (2016), Gilles Deleuze and Félix Guattari (2004; cf. Acuto and Curtis 2014), and add to it the notion of 'patch politics', a concept we develop through our reading of the work of Anna Tsing (2015). Assemblage theory is an approach that treats entities not as homogeneous bounded wholes, but as relational entanglements that are always changing and in production (DeLanda 2016). Assemblages operate at multiple scales from individual human beings, an assemblage of DNA, proteins, muscles, skeleton, bacteria, memories, family members and more, via cities, assemblages of buildings, laws, people, animals, birds, cars, sewers, water, roads and much more, to planets themselves and beyond. Each assemblage gathers together heterogeneous elements and is always only temporarily stabilized, however long it lasts. Tsing's (2015) notion of patches provides an essential additional element to this. By focusing on patches of livability, Tsing's work mapping the worlds of Matsutake mushrooms, their pickers, their histories, their lives, reveals how even in a seemingly globalized world, patches of difference continue to emerge. Amidst the destruction caused by global capitalism, Tsing's patches capture the temporally emergent forms of living and politics that come into being, exist and then fade away. In contrast to calls for a 'planet politics' (Burke et al. 2016) as the future for IR, we argue that any approach must necessarily take a more specific 'patch politics' approach.

Using these conceptual resources we put forward an alternative view of politics, political language and IR which emphasizes: the pre-eminence of relations over 'kinds'; the irreducibility and heterogeneity of sets of relations, and therefore their resistance to transcendence; and, through this emphasis on relations, we offer a

[1]The question of agency is one of long standing in the social sciences and humanities. It is often conceptualized as an attribute of rational thinking human beings, and, in so far as this is the case, such a view of agency is inescapably anthropocentric. Against this view, in this paper we advocate a view of agency as: *assemblage derived*, that is relational; *emergent*, because agency emerges through the relations that form an assemblage; and *contingent*, because all assemblages are only temporarily stabilized. On this account, agency is not a possession or an attribute of individual entities. Rather, our approach recognizes and emphasizes the historical specificity of agency, and requires that we attend empirically to its specific forms. We cannot know in advance the agency that assemblages such as 'a mountain' or 'a human' might have, or the differences between them—these have to be mapped in their specific, immanent, singularity. Whilst this complicates notions of 'human responsibility' because it also complicates any notion of simple linear causation, this does not, as we will demonstrate, mean we are unable to identify causally relevant actors, whether they are human or not.

critical reassessment of the place of non-humans in politics. This reassessment has significant implications for the study of IR, not least in how the purview of IR is defined (cf. Burke et al. 2016).

We develop our argument through a series of examples (cf. Tsing 2015, p. 23). Only one of these might readily be considered as pertinent to a discussion of IR (the efforts of the UK government to secure its withdrawal from the EU), but in our view all three examples point to important features of an assemblage approach to IR. In particular, they remind us that politics is multi-scalar in a sense not captured by orthodox notions of the local versus the global (which in our view is itself a form of kinds thinking). To begin with, we consider the seemingly everyday practice of driving a car. In particular, we ask why cars remain central to modern life, given the enormous environmental damage they inflict, the large numbers of people killed and injured in vehicle collisions and accidents, and the political dependency on oil that they sustain. Although not immediately obvious as an example of IR, car use, we contend, demonstrates clearly why an approach that operates at a variety of complex scales, and that embraces the power of the non-human, is essential.

Our second example takes us to a more obvious political place, the trauma of Brexit, and in particular, the UK government's attempt to extricate itself from membership of the European Union. We argue that the reliance on kinds thinking and the refusal to give political voice to the non-human are key elements in the seemingly intractable nature of this predicament. Finally, we turn to an example on the largest scale, one that clearly transcends international boundaries and refuses to be contained by the type of 'kinds thinking' that Normal Politics delights in: climate crisis. Here the need for an approach that focuses on relations, one that denies that politics is solely a matter for rational actors, and one that embraces the non-human, is central to any attempt to ameliorate the worst excesses of the damage to come.

Autogeddon: The Price of Automobility

The scientific evidence for the damage that car-human assemblages do to living beings and the larger environment, what Williams (1991) refers to as 'autogeddon', is incontrovertible. (We use the term car-human assemblage to reinstate the central symbiotic relationship between the human and the non-human; after all, cars do not kill people, they need a driver to do that). To begin with, internal combustion engines are fossil fuel driven devices and so their production and use is an integral part of the infrastructure necessary for the extraction of oil. More specifically, the car-human assemblage is lethal in particular ways. According to the World Health Organisation (WHO) (2018), globally there were 1.35 million road traffic deaths in 2018, with the highest fatality rates occurring in low-income countries. According to their report, 'Road traffic injuries are now the leading killer of people aged 5–29 years', and they urge 'that the price paid for mobility is too high' (see also Monbiot 2019; Bohm et al. 2006).

Not only does the car-human assemblage kill directly, it is also lethal in other ways, principally through ambient air pollution and its contribution to global warming, a topic we return to below. Vehicle engines, especially those using diesel fuel, emit particulate matter, which penetrates the lungs and is a serious threat to human health. All cars also emit nitrogen oxide, which forms a ground-level ozone layer and constitutes secondary particulate matter, and carbon monoxide, a key contributor to global warming. (Again, cars in themselves do not pollute; someone has to turn the ignition for a vehicle to emit toxic gases). The WHO (2018) estimates that annually there are 4.2 million premature deaths worldwide linked to ambient air pollution, mainly from heart disease, stroke, chronic obstructive pulmonary disease, lung cancer, and acute respiratory infections in children (see also Gardiner 2019). These deaths are not distributed equally but fall disproportionately on those with low incomes. It has been estimated that each car in London costs the National Health Service £8000 because of the effects of air pollution, which is linked to about 40,000 deaths annually in the UK (Gabatiss 2018). The car-human assemblage is also strikingly self-destructive, since exposure to toxic air is often far higher inside than outside vehicles.

Why then do people remain wedded to these fossil fuel based toxic emission devices? Car ownership remains widespread throughout the world, enormous resources are devoted to the roads they need, the oil required to power them, the infrastructure necessary to build and repair them. There are conventional, Normal Politics accounts of the persistent use of cars in the face of looming environmental disaster. Many of these stress, for example, the exercise of rights to personal mobility that cars enable (a feature invariably stressed by car advertising where happy drivers enjoy themselves on roads mysteriously free of other road users), and the freedom of movement they make possible. Cars are also, in the accounts of Normal Politics, crucial to the 'national interest' through the jobs created by their production, the components manufacturing infrastructure entailed by this, and the vast programme of road building that car use both requires and encourages (Spelman and Dromey 2019). Cars, in this view, are a good thing.

The critical version of Normal Politics, on the other hand, seeks to explain continued car use in the face of overwhelming evidence of its destructive environmental effects by the key role played by car companies and their ad agency hirelings in selling cars as affective objects: from this perspective, we regard cars not simply as aids to personal mobility, but as expressive of core features of our identities (a link especially strong, for example, in the case of masculinity and sexuality; see, for instance, Balkmar 2012; Redshaw 2017; Horowitz 2013; Holt and Thompson 2004).

These distinct accounts offered by the conventional and critical forms of Normal Politics propose different responses to concerns about car use, but ontologically they occupy similar positions: they are anthropocentric explanations. In the case of the 'cars are a form of personal freedom' argument, cars are simply an economic product. In the case of the 'cars are a means of consumer conformity' argument, cars are a means of securing capitalist hegemony. In both perspectives, cars themselves are passive material entities, without agency.

A non-anthropocentric approach would reject this distinction between human agent and inert material object and allow an exploration of the ways in which desire and entanglement with non-humans generate forms of reliance and alliance between them and humans. Such forms of alliance point towards alternative understandings of why people continue routinely to destroy the conditions of their own flourishing by poisoning the air they breathe. For example, one significant assemblage that arises from the entanglement of vehicles and people is what we shall call autonormativity, the quotidian, unreflected-upon use of cars as a mobility prosthetic—the daily commute, the school run, the road trip—and the political priority given to car use and ownership by governments. By thoroughly normalizing automotive mobility—which, as we have argued, is an assemblage of human and vehicle together—autonormativity conceals the key role of the non-human by incorporating car use into the practices and discourses of freedom and personal choice; it privatizes and individualizes mobility. The non-human itself here plays a critical role, since this privatization would be impossible without the vehicle itself whose design and aesthetics enclose its occupants off from their surroundings.

Recognizing autonormativity allows us to acknowledge that the decision to drive a car is not simply a matter of exercising an abstract choice. Nor is it merely a choice made within a socio-historical context where individual actions are reduced to the outcome of social pressures. Rather it is a choice made within, to use Tsing's (2015, 63) term, conditions of 'salvage'. In these conditions, choices are neither mechanical responses to the pressures of capitalist advertising, nor are they the outcome of purportedly rational, cognitive appraisals of consumer options. Both of these interpretations of people's choices share profoundly anthropocentric assumptions. Car driving as a practice of freedom is not only an emergent product of the car-human assemblage, but is also a political response to a world where the practice of freedom is profoundly shaped and restricted. In this sense, driving a car is a *comparatively* democratic expression of a particular, and limited, form of freedom. Car use can only be seen as a practice of freedom in this context (see, for example, Lumsden 2010). This is why efforts to argue against car use by pointing to its catastrophic consequences for air toxicity, global warming and the mortality of other users of public space (non-human as well as human) are often unpersuasive. Driving a car is a practice of freedom; in a world where more substantive freedoms are no longer possible, the practice of unrestricted personal mobility appears realizable in ways that a desire for an equal or a just world, say, or one not threatened by climate crisis, are not. People do not drive cars because they cannot have substantive freedoms, but a certain kind of freedom is realizable through the car-human assemblage.

Autonormativity conceals the role of the non-human in other ways that make car use seem a matter of human choice. A first step in recovering the key role of the non-human in autonormativity is to grasp the power that cars have: not the car companies, or the car owners or the economic and cultural processes that sustain car production and use, but the cars themselves. Cars are affective, phenomenological, experiential (and toxic) material objects in addition to being technical devices. A second step is to identify the car-human assemblage and its mundanity, the routine everyday practice of aided mobility. This is a major source of the current climate crisis, just as much as the egregious breaches of environmental laws by large companies.

Moreover, it is difficult to address autonormativity politically whilst these other emergent dimensions of the human-car assemblage are ignored. An anthropocentric Normal Politics is unable to grasp these dimensions. For example, the technical boundedness of the car creates an illusion of safe private space, but once cars are viewed relationally it is clear that they rely on and exploit (and render less safe) a commonly shared public space, a public resource, made up of humans and non-humans. The car-human assemblage appropriates public space physically (roads, car parks, parking on pavements), aurally (engine noise, music) and atmospherically (poisoning the air and heating up the planet). Its demand for oil makes the landscape look the way it does—it covers the countryside and smothers the water table with roads while weeds grow through the railway tracks. It is possible to acknowledge here how the car-human assemblage and autonormativity become aspects of international relations: driving cars creates a demand for oil, a limited natural resource controlled by particular geopolitical entities, whose manipulation of this demand, leads to conflicts as well as to climate emergency.

Normal Politics views car use as a matter of private decisions, an exercise of consumer choice and freedom. In so far as environmental politics posits the production of pollution as a consequence of individual choices—for example, to drive or not to drive—it equally ignores the role of the car-human assemblage. In short, an environmental politics based on the management of consumer choice—recycle properly, use a greener car and so on—rests upon the same ontological assumptions as Normal Politics. (This is not to say that efforts to live a greener life are worthless, merely that they cannot provide a basis for a systemic response to the dominance of the car-human assemblage).

How does an analysis based on the notion of autonormativity lead to a different politics? Car use is an example of acting politically, that is, car use has political effects, although it is seen as apolitical because of autonormativity. If we want people to stop using cars, we need to do more than simply tell them not to use them or present them with facts about the deleterious environmental and other consequences of car use. Banning cars is simply impracticable, but in any case, it would not follow from the analysis developed here. Instead, we need to make other forms of attachment possible and desirable; in other words, by making a different view of freedom viable, the freedom *from* cars. There are parallels here with the UK government response to the public health risks posed by smoking. Governments, under pressure from medical and health professionals, were aware that people became addicted to cigarettes because of the nicotine in tobacco. For as long as smoking was a freely undertaken choice, it was regarded as a private act, and therefore outside the realm of political intervention. Yet the UK government was able to ban smoking in public spaces and to reduce significantly the number of people smoking (Ash Report 2017). This required the reclaiming of a public space for non-smokers—the air belongs to all—and the redesignation of public spaces as a shared resource. This modified assemblages for both smokers and non-smokers, altering, for example, their relationships with different buildings (restaurants, bars) and with the topography surrounding smoking: it became less of an expression of coolness and control and more of an undesirable, anti-social and unhealthy minority activity. As an affective resource, smoking was much altered and diminished.

There are three key points from this account of autonormativity that we would like to carry forward into the rest of the chapter. Firstly, that the practice of Normal Politics obscures the role of the non-human (although the non-human and its effects are, to an important degree, independent of people's understanding or recognition of them. So affective desire is real, but so are its often-hidden non-human effects—the cumulative effects of carbon emissions on biodiversity for example). Secondly, the car-human assemblage is both a product and reproducer of a privatized world view or cosmology, one in which individual rights are regarded as a fundamental existential given, and appropriation of the commons (clean air, safely built environment and so on) is occluded. Thirdly, from our perspective, driving a car is acting politically, and challenging the car politically means generating an affective response to autonormativity, striking at the ways in which the car is at the core of cosmological assemblages to do with freedom and individual (human) rights. These are concerns at the heart of IR.

Brexit

Our discussion of autonormativity focused primarily on the relationships that constitute the car-human assemblage and that transform the act of driving a car into a privatized matter of individual choice, rather than an active abuse of a common environment. The politics that flow from autonormativity are thus concerned with the management of individual choices through persuasion, argument and manipulation. Most phenomena regarded as politics by Normal Politics, however, have to do with collective choices expressed through elections and referenda. These tend not to be a matter so much of individual choices but of 'political decision-making'. It is unsurprising, then, that driving a car (like smoking a cigarette) is not seen as a political matter, but as a matter of consumer choice and personal freedom (see Douglas et al. 2011 for an account of 'cars as the new tobacco'). However, we have argued that including non-humans in the scale of political analysis prompts a different understanding of 'the political' and political decision-making.

Writing this chapter in 2019, it is impossible to think of political decision-making without evoking the spectre of Brexit, the UK government's attempt to extricate itself from its membership of the European Union, a classic example of the complexity of international relations. Following the narrow victory for the Leave campaign in a referendum held in 2016, UK politicians have remained deadlocked as to how to implement this result.

Parliamentary politicians primarily work within conventional approaches to politics, characterized broadly by Graham Harman (2014) in his book on Bruno Latour's politics, as either 'Truth Politics' (where political aims are shaped by assumptions about what is true and what is not) or 'Power Politics' (where political ambitions are driven by a desire for attaining and retaining power). In practice, most politicians are motivated by an unstable combination of the two. In the case both of 'Truth Politics' and 'Power Politics', though, politics rests on kinds thinking and the pursuit of singular, universal solidarities—whether these be of the Right, with its concern for the nation, the will of the people, the interests of the British economy, the faith and so on, or of the Left, with its pursuit of human emancipation, socialism, freedom and so on. This is not to suggest that these 'kinds' of solidarities are somehow equivalents in terms of their content, but rather to point out that both 'Left', or critical, and 'Right', or conventional, modes of Normal Politics have an inexorable momentum for scalability, that is, the capacity to move from the small scale—how this or that voter expresses a preference on a ballot paper—to the grand scale—'the nation has decided'. Thus, the Brexit referendum, for the Right, expressed the 'will of the people' and so 'Brexit means Brexit'. Many on the Left, though, have responded to this with a call for a second referendum on the grounds that this could result in a different outcome that would also be legitimated as 'the will of the people'. Tellingly, the major political pressure for a second referendum has come from the 'Campaign for a People's Vote'.

Parliamentary democracy rests on the notion that 'the people' are 'represented' in Parliament, again a view that is not peculiar to either the Left or the Right of politics. One way in which we might understand the current political impasse over Brexit is that politicians, Left and Right, find it difficult to overcome the apparently insurmountable certainty, embedded in the 2016 referendum, that 'the people' have expressed a wish to leave the EU. Once the referendum result, with a narrow majority (51.89%, or 17,410,742 people, in favour of Leave and 48.11%, or 16,141,241 people, in favour of Remain; the turnout was 72.3%) supporting leaving the EU, is viewed as a definitive expression of the people's will, then it is impossible to reconcile it with a notion of representative democracy: any decision for Leave or Remain will necessarily exclude roughly 50% of the electorate. This is one reason why Normal Politics finds Brexit so difficult—difficult to explain, difficult to carry out, and in the case of the major political parties in Britain, difficult perhaps even to survive. There are three aspects to this difficulty.

Firstly, the rise of populism has created assemblages of ideas and technologies that have enabled people to find new methods of acting politically, that is, specifically in ways that are expressive of concerns central to their affective economies, concerns usually viewed by Normal Politics as 'non-political'. These forms of acting politically—through random violence, say, or protest voting, or social media trolling—are rarely able to be expressed in modes that Normal Politics recognizes. The emergence of new forms of acting politically has resulted in a radical democratization of politics through access to social media. This democratization, though, has been accompanied by a deep distrust of expertise, now seen as inescapably tied to sectional interests such as 'the elite' or 'the Establishment'. The 'public', the 'radicalised'—do not respond to 'facts' about the future of the economy, the extent of the climate crisis or levels of immigration because these come from the authorized voices of Normal Politics that have traditionally identified 'what really matters', voices that are no longer trusted.

Normal Politics experiences a crisis at this point, since it is no longer able to translate what is of concern to people into the 'matters of fact' about the economy, migration and so on as understood by politicians (cf. Latour 2004). The reverse is also true: people's matters of concern—about jobs, security, their quality of life, their sense of identity, for example—are difficult to translate into the politician's 'matters of fact'. To an extent this is a matter of social class: as Friedman and Laurison (2019) have recently demonstrated, the UK is still governed by the white, upper-middle class, mainly male products of public schools and Oxbridge. But other factors continue to obstruct the passages of translation from matters of concern to matters of fact. One of the most important of these is the relentlessly privatized forms of entrepreneurial self-hood that are characteristic of much of the contemporary world (Han 2017). Such forms of self-hood require material affluence to flourish as they are largely expressed through consumption; as Han notes, 'Free choice is eliminated to make way for a free selection from among the items of offer' (Han 2017, p. 15). As we have seen with autonormativity, these choices are underpinned by an ideational emphasis on individual rights, and their corresponding denigration of responsibilities for those features of the world—the air, the oceans, other beings and so forth—that are common to all species. Private appropriation of the commons, through oil, gas and coal extraction, for example, or pollution of the air or the oceans, is a political act.

The question of how one person's matter of concern might be transformed into a political fact thus becomes a key political question. The vote for Brexit, led by a Leave campaign under the slogan of 'take back control', offered people an opportunity to act politically, to place their matters of concern into the political mainstream, but with no way of translating them into a language Normal Politics can understand. In this way, the Brexit vote acted as an assemblage converter (cf. Deleuze and Guattari 2004, 358)—it took multiple, already existing assemblages (austerity, racism, Euroscepticism, anger at Parliamentary systems and so on), and allowed them to come together in a way not previously possible (and certainly not one envisioned by politicians). Thus 'Brexit' was not *simply* a political decision, and so could not be addressed in the terms of Normal Politics: it was always a complex entanglement of intersecting assemblages. Since parliamentary democracy has worked by linking the powers of legitimacy to those who decide 'what really matters', Brexit has thrown Normal Politics into crisis. Under Normal Politics, elections lend one party the right to define what really matters for everybody else. A referendum subverts this. Suddenly it is not the compromises and complexities of parliament that take priority, but rather 'the will of the people'. Intended to be an act of Normal Politics, and to operate under those rules, the Brexit referendum opened up a form of engagement that presented people routinely excluded from Normal Politics with a chance to act politically. Many of them took it, prompting a radical loss of control by Normal Politics that has resulted in the current impasse. Politicians' attempts to negotiate departure from a major customs union and political entity represents a Normal Politics response. But this can never meet the demands of those who 'just want us to leave' or to 'get with on with it' because their demands are not those of Normal Politics. Whole populations now act politically. They demand Britain leaves the EU, and do not understand why politicians, who operate within Normal Politics, cannot deliver this. Populism invariably demands solutions that Normal Politics cannot deliver.

The second difficulty for Normal Politics is that its engagement with Brexit ignores the role of the non-human. The movement of goods, for instance, into the UK after Britain's departure from the EU is clearly of critical importance and the regulation of this movement has been a key feature of the rancorous bargaining over the Irish back-stop (the EU's proposed scheme for dealing with the fact that, once the UK leaves the EU, Northern Ireland will leave whilst Ireland will remain in the EU. Thus a part of the UK will border directly the EU and will thus require border management of goods and people. The EU, and those wishing to leave it, have very different views about what this means precisely). Trade deals are fundamentally about non-humans, about the way in which non-humans flow along supply chains that link multiple sites and locations across the world. However, to manage these flows requires the use of complex modern technologies; this in turn depends on access to materials across the world and their movement from one place to another. This movement relies on planes, trains and, as Chris Grayling, the Secretary of State for Transport, failed to learn in 2018, ferries. Making things is never a human enterprise solely; it requires us to attend to the location of materials in the world and to respect and respond to this. 'Making things', manufacture, in Normal Politics, rests on a radical distinction between the human and the non-human, a distinction enacted partly through the insistence that whilst the movement of goods and things should be, in the main, unrestricted, the movement of people must be severely limited. Yet this distinction obscures the fact that the success of the EU's single market rests on the movement of both people and things, and not simply co-incidentally—it is the movement of both together: people driving trucks, moving for work, moving with new skills and technologies. An anthropocentric approach treats 'migration' as something humans do, but if you walk down any high street you will see that humans come with new foods, new clothes, all sorts of new non-humans. Neither human nor non-human ever travel alone.

The third aspect of Brexit that is problematic for Normal Politics is its vision of the EU as merely a political arrangement, one that can be changed through the conventional modes of Normal Politics (laws and treaties, lobbying, appeals to sectional interests, use of trade threats and tariffs and so on, a view captured well in Robert Menasse's (2019) novel *The Capital*). However, the UK's relationship with the EU, like people's relationship with cars, is stratified. This stratification happens in all sorts of ways—through personal relationships (e.g. marriage to an EU citizen); the ability to buy imported goods; the ability to travel for business and holidays; the flow of money and materials that sustain countless jobs and small businesses. The assumption of Normal Politics that one can simply withdraw from these complex assemblages reveals a fundamental misunderstanding: the reductive assumption that these assemblages are simply about political arrangements, not the host of affective bonds and relations that characterize the UK's engagement with the continent in reality. Attempting to deal with Brexit in terms of Normal Politics does not deal with these concerns. In fact, we would argue that international relations more widely can best be understood through multiscalar assemblages of humans and non-humans, composed of affective and non-affective entanglements, and not through conventional assumptions about national identity or national interests. Both of these latter assumptions reveal the reliance on kinds thinking that characterizes Normal Politics.

Brexit, then, has proved an especially intransigent problem because of these three critical factors: populist politics which expresses matters of concern which are affectively powerful; the role of the non-human, whose movement cannot be separated from the humans whose travel they facilitate, and with whom they travel; and the dense stratification of relations at multiple scales that cannot be severed. These are intertwined, of course. However, their assembled nature is consistently obscured by Normal Politics, resting as it does on transcendent kinds thinking ('the will of the people', 'what the country wants', 'the needs of the economy') and a view of the non-human that reduces them to mere instruments obediently subordinate to human interests.

Moreover, we have claimed that Brexit is an assemblage converter, reanimating earlier assemblages and creating at least two new assemblages: Leavers and Remainers. Leavers and Remainers were produced by the referendum; they did not precede it. These are complex assemblages, which Normal Politics has converted into 'kinds'. Doing this allows some people to take their matters of concern and identify them with these different forms of kinds thinking. One could have voted Leave for any number of different reasons, but now one is a Leaver and critically not a Remainer, with interests in common with other Leavers (so the person living in poverty in Stoke forges a similarity with wealthy, Oxford-educated Jacob Rees-Mogg). None of this explains why Brexit is impossible to enact. This rests on our three factors identified earlier, and emphasizes why we need a greater understanding of the assemblages that constitute the politics of Brexit.

Climate Crisis

Our two examples so far have stressed several key features of an assemblage approach. Firstly, assemblage thinking emphasizes the crucial role of non-humans, which is in significant ways resistant to the analysis of Normal Politics. Secondly, Normal Politics privileges speaking politically ('look at the facts', 'speaking realistically', 'the national interest') over acting politically. Turning the ignition in a car is a political act, but one not spoken about politically. Presented solely as an issue of personal choice and freedom, car use is seen as without political consequence. Thirdly, with the example of Brexit we outlined how new ways of acting politically have emerged that are not responsive to the rhetorical appeals to evidence and 'the facts' characteristic of Normal Politics.

This dislocation in many polities between matters of fact and matters of concern means that Normal Politics finds it difficult to formulate a challenge to the current climate crisis. Indeed, it has struggled to even identify it as a political problem, as opposed to a 'natural' one. This is despite the long history of work identifying the enormous impact of human activity on the broader environment that has taken place since the Industrial Revolution. As early as 1962 Rachel Carson's *Silent Spring* identified the way in which chemical pesticides and other pollutants were becoming embedded in living organisms, including humans, well beyond their intended targets. Since then the evidence of the climate crises and environmental damage has been overwhelming. Beyond the 4.2 million killed annually by air pollution mentioned above, a further 4.8 million die by other forms of pollution (Steingraber 2018; IPBES 2019). These 9 million people represent 16% of deaths worldwide, 'three times more deaths than from AIDS, tuberculosis and malaria combined and 15 times more than from all wars and other forms of violence' (Landrigan et al. 2018, p. 562). Given the overwhelming evidence of the collapse of biodiversity, the melting of the polar icecaps, the transformation in climate and the fall in human and animal fertility it is remarkable that the climate crisis remains, despite the sterling efforts of the Extinction Rebellion and school children all over the world, a fringe issue in Normal Politics.

Normal Politics' denial of the non-human is critical here: it fails to recognize the powerful political agency of some actors—cars as we have seen, but also planes and plastics amongst others—which shape the policies of Normal Politics even as they are ignored by politicians. On the other hand, we have those elements of the world that cry out for political agency—oceans, glaciers, fish and countless insects—that are not given a voice within Normal Politics. Some of these non-humans intervene regardless of our views on their agency, but their voices remain ignored in the discourse of Normal Politics. The anthropocentrism which denies non-humans political agency, also underpins the widespread cultural phenomena of the emphasis on individual rights and choice, and the correlative denial of what we term 'the commons', the shared places, landscapes, ecologies and relations that make the world livable and upon which privatization depends (Tsing 2015, p. 270). As we saw with the car, driving is seen as a right that everyone has, not something to be restricted or controlled despite its widespread and well-evidenced damage. Positioning of the current climate crisis as anthropocentric, as the outcome of human intention, identifies many of the elements responsible for climate change, such as meat-eating, air travel, car driving, plastic use, solely as human choices. This means that although there is widespread acceptance of the reality and causes of our climate crisis, there is little in the way of a political response to them other than appeals to 'be greener'. Perhaps most intriguingly, it is the unrestricted consumption of resources made possible by wealth that has seen the fewest attempts at transformation. The behaviour of the wealthy middle classes as consumers is rarely seen as acting politically, because power and wealth enable the naturalization of choice; one of the capacities of being wealthy is to make your political actions a matter of choice not a matter of politics. Reducing the use of plastic bags, for example, has received extensive attention from government, whilst air travel is not only ignored, it is actively encouraged through budget fares and runway expansions (BBC 2018).

Climate crisis represents a clear example of something that has generated innumerable examples of speaking politically, but little recognition in Normal Politics of how it is caused by acting politically: turning on your car, flying in a plane, eating meat. These are political acts that drive climate change. Yet Normal Politics, lodged in the anthropocentric world of rights and choices, denies the political nature of these acts. Interestingly, although Normal Politics denies political voice to oceans and the climate, we are increasingly seeing that oceans and the climate are themselves capable of acting politically. Floods, droughts, tsunamis and hurricanes drive population movements that transform Normal Politics, regardless of whether Normal Politics sees these as political acts or not. The failure of Normal Politics to understand the nature of acting politically with regards to the climate crisis has been shown brilliantly by the US energy department's rebranding of fossil fuels as 'molecules of freedom' (O'Neil 2019). Here populist politics allows the non-human to speak politically and to have political agency, in a way that Normal Politics cannot conceptualize. The contempt shown towards this move by some politicians (for example, Jay Inslee, the Governor of Washington described the rebranding as 'a joke' on Twitter (O'Neil 2019)) shows precisely how little they understand the power of this radical translation; gas moves from being an inert chemical and apolitical to a major force in international relations, capable, according to the Energy Secretary Rick Perry, of liberating Europe (O'Neil 2019).

One objection that might be raised to this account is that environmental politics continually challenges car use, flying and consumption. Nonetheless, in our view they do so largely within the boundaries of Normal Politics. Urging people to alter their behaviour, make 'greener' choices, continues to place the burden of responding to climate change on the rights and responsibilities of the individual. Such a model, therefore, remains squarely within the anthropocentric assumptions of Normal Politics. This is not to say that such changes in personal choices and behaviour are to be dismissed, merely that they remain couched in terms which sustain, rather than undermine, the policies which they seek to challenge (Latour 2018).

The failure of climate crisis to emerge as a major political concern within Normal Politics raises more sharply than before the relevance of an assemblage politics for IR. Climate change has become a major and imminent threat to the future of the planet precisely because human concentration of wealth (a concentration that makes questions of inequality inseparable from the discussion of climate crisis and how it is to be managed—it cannot any longer be halted in any meaningful sense) has been made possible by making both humans and non-humans into resources for investment (see Tsing 2015; Moore 2015). Landscapes, atmospheres, technologies and other species have now to be recognized as active agents in human life, moving and changing in autonomous, often hidden, ways that shift the conditions of liveability for every species. There is no future for an IR, or sociology or archaeology for that matter, which has no space for the non-human, because without the non-human no serious challenge to climate crisis can be mounted. Without a response to climate crisis, there is no future for any of us.

Conclusion: An Assemblage Politics

In this chapter, we have ranged widely across topics from cars, to referenda, to the transformation of the global climate. Our central thesis has been that, if we are to address the international challenges we face today, we need to move away from what we have termed Normal Politics. This is the age of ecological emergency. Politics is about living and struggling on an already damaged planet, and the social sciences, including IR, must begin the work of plotting the processes now shaping the collective future of humans and non-humans. Normal Politics has emphasized the exteriority of nature to politics; in contrast we have argued for a vision of the world which embraces the non-human and recognizes that acting politically is a critical element in untold activities of everyday life. In place of the kinds thinking and anthropocentrism that Normal Politics is constructed upon, we propose instead an assemblage politics which celebrates the patchy, complex nature of the world and emphasizes the centrality and power of the non-human to politics at every scale, from the act of turning on a car ignition to negotiating international trade agreements. In place of the simplistic notion of scalability (that a referendum is simply the outcome of multiple votes added together; that car use is the outcome of multiple individual choices; that climate emergency can be stopped by policing individual behaviour) we have emphasized the complex way in which scales interact, folded in and through the non-human.

More broadly the assemblage politics we advocate challenges the manner in which IR approaches capitalism. It begins by placing front and centre the claim that we do not inhabit 'capitalism', a fully functioning system governing all aspects of life and with its own teleology, but rather a capitalism which has already badly, and in some cases irreversibly, damaged the world we inhabit (Tsing 2015). A contemporary politics should start by recognizing that, at this stage of capitalism's history, we are confronted by a world of ruins: climate crisis is rapidly threatening to overwhelm us, as it has already overwhelmed many other species; political institutions are compromised and susceptible to various forms of populist tyranny; social relations between humans are corroded by self-interest and individualism and those between humans and other animals are characterized by exploitation and depredation. This is the world in which politics now has to be made.

Such a politics would need to take into account other ontological principles that are entailed by the world of capitalist ruins. This is, first and foremost, a world of radical indeterminacy, in at least two senses. To begin with, a world of ruins is one in which patches of diverse interests can sometimes flourish, if often only briefly. Sometimes these interests are congruent with other, possibly larger, interests, but they are not reducible to them. The emergent nature of patches means that their origins are not predictable, and neither are their capacities, since these are also emergent from whatever relations a patch finds available to sustain it. The irreducible and partially autonomous capacities of patches make them difficult to manage within a larger overall logic of Normal Politics. There is a second sense in which the world of capitalist ruins is radically indeterminate. Patches, as emergent from historically specific contexts, are themselves fragile; their stability is a matter of endogenous and exogenous factors that may not be manageable (Tsing 2015). The failure to acknowledge the patch-like nature of the world is a major reason Normal Politics cannot address car use, Brexit or climate change.

In this assemblage approach, capitalism is not a unified, homogeneous entity, but, as Tsing puts it, 'a translation machine for producing capital from all kinds of livelihoods, human and not human' (Tsing 2015, p. 133). This form of capitalism is harder to challenge, since it no longer generates consistent and readily translatable experiences, such as those encountered by, say, workers in nineteenth-century manufacturing industries, or enslaved people on a plantation. No patch is 'representative' of other patches, let alone 'capitalism' as a whole. No group's struggles, therefore, taken alone will overturn capitalism or be a 'vanguard' for other groups. There is no longer an automatic urge to argue together. A different form of politics is required for a world of capitalist ruins; rather than the stories we are told by teleological narratives of Normal Politics ('the world is getting better', 'climate change does not exist', 'utopia will arrive'), or a politics based on a singular solidarity often offered in its place ('the workers', 'the oppressed', 'the nation', 'the true believers in the faith'), we need an assemblage approach to politics.

If we are to understand the history and dynamics of patches, we need to know what makes a patch a liveable space. All livelihoods are entanglements or, to use Tsing's (2015, p. 23) term, 'polyphonic assemblages', gatherings of ways of being. In her words, 'Assemblages are performances of livability…species-specific agilities are honed in the co-ordinations of assemblages…even humble creatures participate in making worlds…' (Tsing 2015, p. 157). This is a non-anthropocentric politics from the outset. It is not especially concerned with examining human relations with their favoured animal allies—dogs, farm animals and other denizens of anthropomorphic hierarchies as in most animal studies, for example. Tsing (2015, p. 158) again:

> Organisms don't have to show their human equivalence (as conscious agents, intentional communicators, or ethical subjects) to count. If we are interested in livability, impermanence, and emergence, we should be watching the action of landscape assemblages. Assemblages coalesce, change, and dissolve: this is the story.

Things become themselves in assemblages (another reason for rejecting a kinds thinking ontology), and in this sense they are always singular, indeterminate and have latent capacities that this form of politics can try to detect. This is a world in which human beings are never fully in control, but instead are parts of shifting moments of fugitive entanglements. Assemblage politics involves the search for these, the identification of their unrealized possibilities and the traces of not yet articulated common agendas.

The singularity of 'fugitive entanglements' also has political implications. In particular, singularity, in conjunction with emergence, means that efforts to scale up from any specific entanglement to a general instance (this strike is an expression of class struggle), or to scale down from a large entity to a singular instance (class position will determine class consciousness), have to be carried out with great care. Fugitive entanglements and patch politics by their nature resist scalability. This is a move from the notion of politics as social, and anthropocentric—Normal Politics—to a world of the geosocial, where human, world, animal, thing cannot be neatly purified into different camps, and the notion of politics applied to only one of them (Latour 1993; 2018).

Finally, it is our contention that a move to embrace the non-human, to recognize we already live in the midst of a ruined planet, and that our existence is necessarily patch-based and assemblage driven, is not, in fact, a message of pessimism. Instead, we would argue that this more ontologically accurate, and more ontologically humble, approach opens up the possibility for attending to the world in new ways, from learning from others in new ways, and of building new strategies for livelihood. The emphasis placed on the unique risk climate emergency poses to nation states reveals a particularly Eurocentric idea of both present and past: many people have already undergone transformations to their societies, including the deaths of huge swathes of their populations, mass displacement, economic deprivation and more. They went through this under the process we know today as colonialism. In the face of the climate crisis, one option might be for Western thinkers to turn to the knowledge and skills developed by Indigenous people to adapt and respond to the imposition of new political realities and new ways of living. This is not a call to mirror non-Western knowledge because it allows us to somehow live 'closer to nature' or any other form of primitivizing description of Indigenous people. Rather it is a call to listen and learn from experienced and sophisticated political players how, in the face of an overwhelming externally imposed crisis, they have developed strategies for *survivance* in Gerald Vizenor's (1999, 2008) term, the ongoing making of their worlds in skillful and productive ways, in a manner that combines survival and resistance. It is by using an assemblage approach, one that takes both other forms of knowledge and the non-human seriously, that IR can move beyond the limits of Normal Politics.

Acknowledgements We would like to thank the editors for their invitation to contribute to this volume, and for their patience during the writing process. The chapter has benefitted from the comments of, and discussions with, Alison Sealey, Barbara Misztal, Craig Cipolla, Lindsay Montgomery, Simon Dyson and Darryl Wilkinson, as well as the feedback from two anonymous reviewers.

References

Acuto, M., & Curtis, S. (2014). Assemblage thinking and international relations. In M. Acuto & S. Curtis (Eds.), *Reassembling international theory: Assemblage thinking and international relations* (pp. 1–15). Basingstoke: Palgrave Macmillan.

Ash Report (2017). *Smokefree: the first ten years. Tackling the smoking epidemic in England: The views of the public.* Action on Smoking and Health. Retrieved 14, June 2019, from http://ash.org.uk/wp-content/uploads/2017/06/170107-Smokefree-the-first-ten-years-FINAL.pdf.

Balkmar, D. (2012). *On men and cars: An ethnographic study of gendered, risky and dangerous relations* Linkoping: Linkoping University Press.

Barad, K. (2007). *Meeting the universe halfway: Quantum physics and the entanglement of matter and meaning Durham*. NC: Duke University Press.

BBC (2018). *Heathrow airport: MPs vote in favour of expansion*. BBC News. Retrieved 14, June 2019, from https://www.bbc.co.uk/news/uk-politics-44609898.

Bennet, J. (2010). *Vibrant matter: A political ecology of things*. Durham, NC: Duke University Press.

Bohm, S., Jones, C., Land, C., & Paterson, M. (2006). *Against automobility*. Oxford: Blackwell.

Braidotti, R. (2013). *The posthuman*. Cambridge: Polity Press.

Burke, A., Fishell, S., & Mitchell, A. (2016). Planet politics: A manifesto from the end of IR *Millenium. Journal of International Studies, 44*(3), 499–523.

Coole, D., & Frost, S. (2012). *New materialisms: Ontology, agency and politics*. Durham, NC: Duke University Press.

Cudworth, E., & Hobden, S. (2011). *Posthuman international relations: Complexity, ecologism and global politics*. London: Zed Books.

Cudworth, E., & Hobden, S. (2018). *The emancipatory project of posthumanism*. London: Routledge.

Cudworth, E., Hobden, S., & Kavalski, E. (Eds.). (2018). *Posthuman dialogues in international relations*. London: Routledge.

Chandler, D. (2013). The world of attachment? The post-humanist challenge to freedom and necessity. *Millennium: Journal of International Studies, 41*(3), 516–534.

DeLanda, M. (2016). *Assemblage theory*. Edinburgh: Edinburgh University Press.

Deleuze, G., & Guattari, F. (2004). *A thousand plateaus* (4th ed.). London: Continuum.

Douglas, M. J., Watkins, S. J., Gorman, D. R., & Higgins, M. (2011). Are cars the new tobacco? *Journal of Public Health, 33*(2), 160–169.

Eroukhmanoff, C., & Harker, M. (2017). *Reflections on the posthuman in international relations: the anthropocene, security and ecology*. Bristol: E-IR Publishing.

Friedman, S., & Laurison, D. (2019). *The class ceiling: Why it pays to be privileged*. Bristol: Policy Press.

Fuller, S., & Lipinska, V. (2014). *The proactionary imperative: A foundation for transhumanism*. Basingstoke: Palgrave.

Gabatiss, J. (2018, June 6). Each car in London costs NHS and society £8000 due to air pollution, report finds. *The Independent* Online no page.

Gardiner, B. (2019). *Choked: The age of air pollution and the fight for a cleaner future*. London: Granta.

Han, B.-C. (2017). *Psycho-Politics: Neoliberalism and new technologies of power*. London: Verso.

Haraway, D. (2016). *Staying with the trouble: Making kin in the chthulucene*. Durham, NC: Duke University Press.

Harman, G. (2014). *Bruno Latour: Reassembling the political*. London: Pluto Press.

Harman, G. (2018). *Object-oriented ontology: A new theory of everything*. London: Pelican.

Holt, D. B., & Thompson, C. J. (2004). Man-of -action heroes: the pursuit of heroic masculinity in everyday consumption. *Journal of Consumer Research, 31*(2), 425–440.

Horowitz, R. (Ed.). (2013). *Boys and their toys: Masculinity, class and technology in America*. New York: Routledge.

IPBES (2019). *Global assessment report on biodiversity and ecosystem services of the Intergovernmental Science- Policy Platform on Biodiversity and Ecosystem Services*. E. S. Brondizio, J. Settele, S. Díaz, & H. T. Ngo (Eds.). IPBES Secretariat, Bonn, Germany.

Landrigan, P. J., Fuller, R., Acosta, N. J. R., Adeyi, O., Arnold, R., Basu, N., & Abdoulaye, B. B., et al. (2018) The Lancet Commission on pollution and health. *The Lancet, 391*, 462–512.

Latour, B. (1993). *We have never been modern*. Cambridge: Harvard University Press.

Latour, B. (2004). Why has critique run out of steam? From matters of fact to matters of concern. *Critical Inquiry, 30*(2), 225–248.

Latour, B. (2018). *Down to earth: Politics in the new climatic regime*. Cambridge: Polity Press.

Lumsden, K. (2010). Gendered performances in a male-dominated subculture: "Girl Racers", car modification and the quest for masculinity. *Sociological Research Online*, *15*(3), 1–11.
McDonald, M. (2018). Climate change and security: Towards ecological security? *International Theory, 10*(2), 153–180.
Menasse, R. (2019). *The capital*. London: MacLehose Press.
Mitchell, A. (2014). Only human? A worldly approach to security. *Security Dialogue, 45*(1), 5–21.
Monbiot, G. (2019, March 7) Cars are killing us. Within 10 years, we must phase them out. *The Guardian* Online. Retrieved 13, July 2020, from https://www.theguardian.com/commentisfree/2019/mar/07/cars-killing-us-driving-environment-phase-out.
Moore, J. W. (2015). *Capitalism in the web of life*. London: Verso.
O'Neil, L. (2019, May 29). US energy department rebrands fossil fuels as 'molecules of freedom'. *The Guardian* Online. Retrieved 13, July 2020, from https://www.theguardian.com/business/2019/may/29/energy-department-molecules-freedom-fossil-fuel-rebranding.
Ong, A., & Collier, S. J. (2005). *Global assemblages: Technology, politics and ethics as anthropological problems*. Oxford: Blackwell.
Pereira, J. C. (2017). The limitations of IR theory regarding the environment: lessons from the anthropocene *Revista Brasileira de. Política Internacional, 60*(1), e018.
Redshaw, S. (2017). *In the company of cars: Driving as a social and cultural practice* London: CRC Press.
Salter, M. B. (Ed.). (2015). *Making things international 1: Circuits and motion*. Minneapolis: University of Minnesota Press.
Sassen, S. (2006). *Territory, authority, rights: From medieval to global assemblages*. Princeton, NJ: Princeton University Press.
Spelman, C., & Dromey, J. (2019, February 2). A Labour and Conservative MP explain why they are working together to ensure there must be a Brexit deal. *Birmingham Mail*. Retrieved 13, July 2020, from https://www.birminghammail.co.uk/news/news-opinion/labour-conservative-mp-explain-working-15861909.
Steingraber, S. (Ed.). (2018). *Introduction to 'Silent Spring' and other writings on the environment by Rachel Carson*. New York: Library of America.
Tsing, A. (2015). *The mushroom at the end of the world: On the possibility of life in capitalist ruins*. Princeton, NJ: Princeton University Press.
Vizenor, G. (1999). *Manifest manners: Narratives on postindian survivance*. Lincoln: University of Nebraska Press.
Vizenor, G. (2008). *Survivance: Narratives of native presence*. Lincoln: University of Nebraska Press.
Walby, S. (2007). Complexity theory, systems theory and multiple intersecting social inequalities. *Philosophy of the Social Sciences, 37*, 449–470.
WHO (2018). *Global status report on road safety 2018*. Retrieved 26, Dec 2019, from https://www.who.int/violence_injury_prevention/road_safety_status/2015/en/.
William, H. (1991). *Autogeddon*. New York: Arcade Publishing.

Across Species and Borders: Political Representation, Ecological Democracy and the Non-Human

Anthony Burke and Stefanie Fishel

Abstract Debates about democracy and ecology invariably lead to the question of the representation and membership of non-human animals, ecosystems and the biosphere in world politics. This chapter responds by interweaving two lines of inquiry. One considers the fundamental political theory that could generate an adequate account of how to give representation to the non-human in the polity. Key theoretical interventions will be considered alongside a posthumanist, new materialist account of the material agency of ecosystems. The second considers the dilemmas involved in designing ecological democratic institutions that could include the non-human in communicative systems of membership and accountability. We propose two new enabling structures for ecological inclusion and governance: 15 regional ecosystem assemblies to cover the Earth's major biomes, and an Earth System Council to coordinate integrated action, both of which include and channel representation from states, indigenous communities, and proxy guardians for the non-human. Such institutions require a deep commitment to the complexity and vitality of the biosphere, reflexivity and humility in proposing Earth system repair, and a constant awareness of the aporetic quality of political representation as such, in support of new forms of interspecies politics and governance that might work for and with the biosphere as a whole.

Introduction

At one point in his influential theory of international relations, *The Anarchical Society*, Hedley Bull seeks to refute advocates of a cosmopolitan and ecologically sustainable world order by saying that a "world society or community" is at best "an idea or myth that may one day become powerful." This is, in large part, because of

A. Burke
UNSW, Canberra, ACT, Australia
e-mail: a.burke@unsw.edu.au

S. Fishel (✉)
University of the Sunshine Coast, Sippy Downs, QLD, Australia
e-mail: sfishel@usc.edu.au

J. C. Pereira and A. Saramago (eds.), *Non-Human Nature in World Politics*, Frontiers in International Relations, https://doi.org/10.1007/978-3-030-49496-4_3

a lacuna in *representation*: "the great mass of political mankind does not have the means of interest articulation and aggregation, of political socialization and recruitment [outside] the mechanism of the society of sovereign states" (Bull 1977, p. 82; cf. Falkner 2017). While this is a fairly accurate statement of our current state-centric ontology, it founders on three contradictions. The first, which Bull concedes, is that interstate representation will fail because neither can the global common good be found through the "the views of states…assembled in international organizations" because "universal ideologies that are espoused by states are notoriously subservient to their special interests" and "agreements reached among states notoriously the product of bargaining and compromise rather than any consideration of the interest of mankind as a whole." The second and third he fails to see. These inhere in our Anthropocene situation and in the nature of political representation itself, which merely seeks to manage an ineluctable gulf between those who represent and those who are represented, between those who exercise power and those upon whom it is exercised, and between those who make law and those who are made by the law. This gulf is bridged by the *abstraction* into policy and law of the interests of those who are represented: by a speaking *in the name and the place of* the represented by parliaments, executives and bureaucracies; by *re-presentation* not translation (Latour 2005, p. 16).

In the Anthropocene—a situation in which it seems trivial and absurd to reduce the fate of the Earth's biodiversity and climate to another "interest" to be bargained—we are now forced not merely to find ways to understand and express the interests of humanity but of an entangled and complex *Earth* undergoing alarming system change. This requires finding ways to "represent," acknowledge, and include the non-human as a vital and living force in our politics in ways that cross borders, jurisdictions and communities. Numerous texts in Earth System Governance have incisively analyzed the many representational lacunas, democratic deficits and sustainability failures of states and international environmental regimes from a weakly anthropocentric perspective; focused on how exclusivist and elitist forms of environmental governance, by excluding key human voices from communities and civil society, have failed to protect ecosystems and human interests in environmental justice (Biermann 2014; Nicholson and Jinnah 2016; Stevenson 2012; Dryzek and Stevenson 2014). In this chapter, which takes a deep dive into political theory to examine ways beyond the constitutive exclusion of the non-human from the political, we seek to push beyond such anthropocentrism to ask how such institutions might include the force, vulnerability, and flourishing of the non-human more directly in their constitution and their work. Whereas writers such as Frank Biermann (2014, Chap. 3) have sought to understand the many (*non-state*) agents at work in governance, we argue for the recognition of the *non-human* as an actant across all scales of politics; this ought to be recognized with agency and representation within the polity.

We do this in the face of a politico-legal system that engineers a pernicious *in/visibility* for nature, making it available for appropriation and exploitation by the state and capital at the same time as its independent desires, lifeways and flourishing are erased. One key *constitutional* mechanism of in/visibility is the international customary law doctrine of Permanent Sovereignty Over

Natural Resources, which lies at the heart of international environmental law and provides states with "the inalienable right…to dispose of their natural wealth and resources in accordance with their national interests" subject to weak obligations to protect biodiversity and "ensure that activities within their jurisdiction or control do not cause damage to the environment of other states or in areas beyond the limits of national jurisdiction" (Burke 2019, p. 340; UNGA 1962; Atapattu 2015). In this politico-legal ontology, the non-human is mere property and matter; it has no independent existence outside the human use for it.

This chapter explores one potential line of movement beyond such a political in/visibility of the non-human by considering arguments from ecological democracy for the representation of ecosystems and animals in government and democratic politics, and adding to them our own. Building upon the model of "thing-systems" power we have developed elsewhere (Burke and Fishel 2019), we advance a posthumanist argument for political inclusion via recognition of the material presence and agency of ecosystems and non-human lives, and the resistant power of human/non-human assemblages. In this way, we not only cut through liberal demands that animals should possess the language to be considered as "political," we avoid using criteria for inclusion such as animal sentience or the possession of a "life project" (Nussbaum 2011, p. 159; cf. Schlosberg 2014, p. 79). Instead, we argue that the existence, vitality, and power of the biosphere and the Earth system *forces* a demand on the political. In a situation like the Anthropocene, when the human species claims the entirety of the Earth's land surface as a territory and has a planetary ecological effect that is simultaneously turning back on it, the living systems of biosphere demand recognition and representation in a polity that would extend below, above, and beyond the state. *Presence* requires representation.

In dialogue with a diverse group of thinkers in animal rights and ecological democracy, the chapter then builds a materialist argument for the political inclusion of the non-human in the polity. We first consider the ecological inadequacies of liberal democracy, and then critically examine normative arguments for an eco-centric transformation of governance practices and divisions of state sovereignty which nonetheless retain faith with the liberal state as a primary structure for ecological justice. Finally, we assert a post-Holocene ontology of inclusion, p. one that rests not merely upon an organism-focused emphasis on animal rights, but the prior existence, presence, and power of non-human lives and processes as they are manifested in vast and dynamic assemblages of beings across the Earth system. In common with these positions, we assert that listening and communication with the non-human is crucial, across the scale of all its processes, which raises questions of where ecological democracy would ground its ontology and how it can work across biomes and jurisdictions—across the boundaries between states and organizations, between borders and species—in ways that can honor the ever more dangerous entanglement of human and non-human lives.

We are not proposing a representation of the non-human that would amount to a minor structural innovation in the existing liberal–democratic state system, whose fundamental commitments and purposes would otherwise remain unchanged. Given the scale and intensity of the ecological crisis represented by the Anthropocene, liberal democracy must be refigured as ecological democracy and the underlying metaphysics of the human domination of nature profoundly set aside. This metaphysic is manifested in philosophy by a *hyper-separation* of human and nature that excludes the non-human as an agent, and in economics with a view of nature as inert matter easily transformed into "resources" or "ecosystem services" for human use (Plumwood 2002, p. 9; Burke 2019). Not only should the non-human have an equal—if not prior—status in the polity, it must be accompanied by a profound change in the nature and purpose of the political itself. A recognition of Anthropocene entanglement with the Earth forces us not merely to better "represent" the non-human in politics and governance, but to look beyond representation as such—insofar as it works in the liberal-democratic state—towards a politics of flow, sensing and affirmation that moves beyond divisions between inside and outside (Manning 2007, p. 134–61). A politics not of negativity or desperation—where one is either bargaining human extractivism against non-human flourishing or desperately attempting to forestall ecological disaster—but a joyful politics with and for the non-human in which new common worlds of interspecies flourishing are being created.

Just One Political Animal? Modes of Exclusion

In the Western tradition, the threshold of exclusion of the animal from politics has been based around three key qualities perceived to elevate humans over other animals: rationality and self-consciousness; the use of spoken and written language; and sentience and moral personhood. Such differences are baked into the foundations of politics with the post-Hobbesian view that only humans can emerge from the state of nature and enter into the original social contract. These loci of separation have a complex relation to the human's own animality—which, while acknowledged in ancient thought, is increasingly occluded after the Renaissance—but universally underpin an exclusion of the non-human from politics. In turn, this exclusion serves to rationalize nature's domination and appropriation by industrial capitalism.

The exclusion of the non-human from the political has deep roots in political thought dating back to classical times. The Greek and Roman Stoics asserted a strict division and hierarchy between the human and animal, on the basis that animals were incapable of moral judgement and reason, which "is taken to be a feature of human beings that sets them sharply apart from non-human animals and from their own animality" (Nussbaum 2006, p. 130). In his *Politics*, Aristotle famously stated that man (sic) is "by nature" a "political animal" because "man alone among the animals has speech." Unlike the Stoics, Aristotle acknowledges our biological continuity with other animals but asserts that "man is much more a political animal than any kind of bee or any herd animal" given that "speech serves to reveal the advantageous and the harmful, and hence also the just and the unjust….it is peculiar to man as compared to the other animals that he alone has a perception of good and bad and just and unjust and the other things of this sort; and community in these things is what makes a household and a city" (Aristotle 2013, p. 4). This capacity to desire justice through politics is later taken by Kant to deny animals personhood and independent value (Nussbaum 2006, p. 131).

Theorists have noted that Aristotle's political thought also contained the potential for the inclusion of animals in the work and practice of politics and a "possible recasting of this [human-animal] relation"; both humans and other animals conduct "work in common" and share "imagination, memory and voice" which ought to be the basis for an inclusive political community (Basnett 2016, p. 291, 295–6). In the *History of Animals*, Aristotle makes clear that he has chosen to "consider the parts of the human body" because man "must be the best known to us of all animals," and therefore we possess the best standards for us to understand other animals. He continues on to write that "Man" can be defined as tame, gregarious, viviparous, sanguineous, two-footed, and ambidextrous. Mankind is carefully compared with other animals: our spleens are shaped like the hog; our livers and kidneys like oxen (Aristotle 1883, p. 11).

It quickly becomes clear, however, that along with being useful as the best standard for comparison, Aristotle believes that humans hold special qualities above other animals. "Man is an animal," he writes, "but a perfect one" and also the only animal with a face: "the part immediately beneath the cranium is called the face in mankind alone, for we do not speak of the face of a fish or of an ox" (Aristotle 1883, p. 29, 12). Aristotle presents us with an early example of how, while there are promising aspects in how many Western philosophers have understood humans as part of the natural world, this hierarchical placement and defense of humans has also led to harm, both theoretically and actually. If only "man has a face," other animals—especially those who seem especially alien such as insects, reptiles, and fish—are easily selected out and treated as less than humans in importance, morally and otherwise. Furthermore, animal life as biodiversity (species among others) or as entangled species-being, cannot be imagined at all.

Another linked animal (and human) rights problem appears when relying upon hierarchical and essentialized understandings of humans *qua* non-humans. Many damaging institutions have been created and supported as "natural" or "essential" (mirroring the belief in the "natural" human body) by placing those that differ morphologically from white Europeans closer to non-human animals. This rationalization has been mobilized to support colonization, slavery, and other forms of oppression and terror. Additionally, biological taxonomies have long been used to argue for affinities between those humans enslave and non-human animals; the legal (non) status of animals is used to create rules and punish slaves. According to Colin Dayan, in *The Law is a White Dog*, the treatment of animals as chattel is a way to understand human cruelty both to animals *and* other humans: "[Animals] are used as examples when humans most need to categorize, to dominate, to justify slavery, genocide, and incarceration" (Dayan 2013, p. 116).

Caleb Basnett concludes that the potential in Aristotle for human and non-human animals to share a political community through their common work remains "a path not taken…the outline of some apparently unintended consequences of [his] thinking"—in part because Aristotle held that it is only in "our interactions with other human beings" that "some of us become just, and others unjust" based on a capacity for thought, reason and *logos* assumed to be solely human (Basnett 2016, p. 298). Therefore, while continuity between the human and the animal has been part of the historical narrative, it has been operationalized as a degraded continuity in both directions of the relationship. Rather than imagining a politics based on interdependence and recognition of animals and their potential rights (or our human responsibilities toward them), it inaugurated a politics that excised non-humans from political consideration due to their Otherness, their lack of predefined human qualities. The animal is thus a foil for upholding and defending a certain definition of humanity. This is easily transferred to human communities through social Darwinism, specious theories of racial differences like phrenology, and legal regimes that create hierarchies of legal personhood (de Fontenay 2012, Chap. 4).

Additionally, as argued by Carolyn Merchant, until the sixteenth century Europe was guided by organic metaphors wherein nature was understood through close-knit communities connected to the natural world intimately. The controlling image was "nature as nurture" and identified earth with motherhood: "a kindly beneficent female who provided for the needs of mankind in an ordered, planned universe." This identification of nature with the female sex was slowly overtaken by the Scientific Revolution as the dominant image morphed into "mechanism and the domination of nature." Of course, both of these images of nature had existed in philosophy and religion, but the introduction of new technologies drove the domination metaphor into social and political spheres (Merchant 1983, p. 2–3).

The image of nature as mother and nurturer carried with it certain cultural and normative constraints around human relationships with the natural world. The thinkers of the Scientific Revolution re-created and defended conceptions of nature that became based on misogyny and abuse, especially as they related to natural resources or ecosystem services that serve human consumption rather than care and restraint. This untied the normative constraint around respect for femininity and unleashed a justification for pillaging resources as one would control and use women's bodies. Therefore, we can see that femininity works in much the same way as race did in the above discussion. While ostensibly "nature as nurture" has positive effects, when societal and cultural understandings of femininity change so do our relations with nature.

Along with such structural notions of race and gender, another way in which humans define their relations with nature and animals is through sociality and the moral duties accorded to others within the *Polis*. This centers around theories of the social contract, from which Hobbes (like Rawls much later) excludes the non-human. Men, as explained by Aristotle, have "common employment," are social, and gregarious like bees, ants, wasps, and storks (Aristotle 1883, p. 4). Like Aristotle, Hobbes believes that animals and humans share sociability and are gregarious. Indeed, animals in complex societies can make a mutual agreement and "direct their action to a common end," but there is no evidence of the single will necessary for government. Additionally, for Hobbes, politics are the results of a contract that must include reason and language: politics is "thus a typical human product" and animals do not possess reason and only have voices, rather than speech, and therefore cannot rise to the political level (Paganini 2017, p. 190–94). Animals may have a weak form of contract, but this is not enough for them to emerge from the state of nature into the *polis*. Men also suffer from an equal love of liberty and dominion over others, complicating any forms of community that are not held in check by a "visible Power" "and tie them by fear of punishment to the performance of their Covenants" (Hobbes 1997, p. 93). Men are constantly in competition for "Honour and Dignity," whereas animals do not quarrel over envy and hatred. There is no war in the animal kingdom like in the human realm.

For Immanual Kant, this discord is the deciding factor in the difference between man and animal. For Kant, in contrast to Hobbes, conflict is a means for man to reach his full potential, rather than a flaw that must be controlled through human institutions and law. "Nature has planted in the species the seed of discord" and nature wills that man turn discord into Concord through generational and cultural growth, not individual change (Kant 1978, p. 238). In addition to being the only animal who can manipulate things, is clever in using others for his own purposes, and has a sense of morality focused outward toward living according to freedom, "the human species can work itself up to its destiny only through continuous progress within an endless sequence of many generations" (Kant 1978, p. 240–241). "Man," therefore, must live together with others in society to cultivate, civilize, and apply himself to moral purpose. Animals, on the other hand are excluded from moral personhood because of their inability to make themselves rational and govern their species as a whole. Put differently, non-human animals cannot hold a general will or imagine themselves as a group in the same manner as humans.

Contemporary contract theorists like John Rawls continue the tradition of excising animals from the political through their lack of moral personhood and language. Only moral persons are entitled to equal justice and rights; hence while animals can be owed duties of compassion and humanity because of their capacity for pleasure and pain, they are outside the scope of a political theory of justice. In fact, animals and nature show the very limits of a theory of justice. The social contract doctrine cannot be extended to them in a natural way. He bases this on the idea that justice is defined as fairness. While justice as fairness is not a complete contract theory, writes Rawls, an even wider conception of justness as rightness can still only include human relations. The scope of justice as fairness must be limited, and animals and nature, though important, must be placed outside. Our conduct with animals is also not regulated according to the principles of justice and equality. He asks: "On what grounds do we then distinguish between mankind and other living things and regard the constraints of justice as holding only in our relations to human persons?" (Rawls 1999, p. 15, 441).

The concept of equality can be distinguished on three levels, he replies. The first is as a public system of rules, or equality as regularity. Similar cases must be treated similarly. The second, and more difficult level, is that equality specified by the principles of justice require that "equal basic rights must be assigned to all persons." This is where he begins to explain the difficulty of including animals in his theory of justice. Even though animals are owed some protection their "status is not that of human beings" and one must consider "what sorts of beings are owed the guarantee of justice." The answer is that only moral persons have the right to equal justice. To be a moral being means that they are capable of having a conception of their good through a rational plan and our capable of having both a sense of justice and a desire to act upon principles of justice (Rawls 1999, p. 441–442). While Rawls' exclusion of the animal is not categorical, he shifts the solution to another branch of theory:

> A correct conception of our relations to animals and to nature would seem to depend upon a theory of the natural order and our place in it. One of the tasks of metaphysics is to work out a view of the world which is suited for this purpose; it should identify and systematize the truths decisive for these questions (448–449).

He concludes, however, by writing that if a "sound account of justice among persons" can be accounted for then "it cannot be too far wrong when these broader relationships are taken into account" (Rawls 1999, p. 449).

Finally, Aristotle, in the *Politics*, excludes animals on the basis that only humans possess speech. Thus, language is a universal threshold of exclusion through to the present day. Spoken and written languages have been the requirement for entry into politics and the polis. It is the capacity for language that sets humans apart from other animals; this is what allows them to have history, self-reflection, to create art and culture. Much later, Jürgen Habermas imports this classically Aristotelian exclusion into the very twentieth-century theory that many hoped could provide a basis for extending the space of deliberation and justice to the non-human world. Like Aristotle arguing that only humans can speak and reason about justice and form the polity, he suggests only "members of this community can place one another under moral obligations and expect one another to conform to norms in their behavior":

> Animals benefit for their own sake from the moral duties which we are held to respect in our dealings with sentient creatures. Nevertheless, they do not belong to the universe of members who address intersubjectively accepted rules and orders to one another. "Human dignity," as I would like to show, is in a strict moral and legal sense connected with this relational symmetry. It is not a property like intelligence or blue eyes, that one might "possess" by nature; it rather indicates the kind of "inviolability" which comes to have a significance only in interpersonal relations of mutual respect, in the egalitarian dealings among persons (Habermas 2003: loc. 677).

Worse, Habermas divides the emancipatory promise of critical theory and deliberative democracy from the emancipation of nature, which is in fact to be denied; he argues that the *telos* of human emancipation depends on the instrumental subjugation and utilization of nature through labor and technology. Hence, he rejects an earlier Marcusian vision of nature as a partner rather than an object of control (Eckersley 1992, Chap. 5). While both Robyn Eckersley and John Dryzek remain committed to building ecological democracy on the ethics of deliberation, Dryzek (1995, p. 20) is particularly blunt: we must "try to rescue communicative rationality from Habermas."

Ecological Citizens? Visions of Non-Human Inclusion

Green political thinkers working across ecological democracy and animal rights share a common conviction that liberal democracy fails to protect ecologies or to discharge the responsibilities that come with the human entanglement with the Earth and the utilization of its living forms. This failure forms a key justification for normative arguments to better understand, represent, incorporate, and support the interests of animals and ecosystems, along with the human interest in their survival and flourishing, in order to challenge the a priori exclusion of non-humans from the political.

Liberal democracy's ecological failures are, in part, failures of recognition and communication, i.e., of representation. John Dryzek argues that the fundamental modalities of government–society interaction in liberal democracy—preference aggregation, representation, lobbying, and partisan competition—fail to grapple with the "high degrees of complexity and uncertainty" and "collective action problems" inherent in ecological crises (Dryzek 1995, p. 15–16). In *The Politics of the Anthropocene,* Dryzek and Jonathan Pickering (2019, p. 23) describe liberal democracy as a "Holocene institution" that has developed a "pathological path dependency [which] decouples human institutions from the Earth system by embodying feedback mechanisms that systematically repress information about the condition of the Earth system, and systematically prioritize narrowly economic concerns." For Robyn Eckersley, such communication failures across species reveal a fundamental problem of justice: "liberal democracy is not accountable enough from the perspective of those suffering or concerned about present or future ecological harm" (Eckersley 2004: loc. 1154). One significant factor must no doubt be liberal democracy's historical anthropocentrism, which (in the words of Freya Matthews) "takes human interests as the measure of all value" and ensures that "only reasoning entities are accorded political standing" (Dryzek 1995, p. 19).

Both Eckersley and Dryzek build their argument out of a normative position drawn from critical theory and deliberative democracy. Using these starting points, Eckersley argues that ecological democracy will go beyond ecological sustainability as a norm to representing non-human "excluded others." She draws on the politico-normative principles of critical theory to insist upon "an other-regarding non-instrumental posture toward the nonhuman world": a "circle of moral considerability [that] is widened to the maximum to include all potentially affected others" and forecloses "the very possibility of arbitrarily displacing ecological costs onto innocent human and nonhuman others." She extends Habermas' view that "persons should not be bound by norms to which they have not given their freely informed consent" to "differently situated [nonhuman] others as ends in themselves." The principle Eckersley asserts is that "all differently situated others [human and nonhuman] ought to be free to unfold in their own distinctive ways and therefore should not be subjected to unjustified policies and decisions that impede such unfolding." This generates an *ecological ideal of political communication* that "seeks a kind of notional consent from these communities by means of representative thinking on their behalf, including imaginative role reversal, by those within the dialogue community" (Eckersley 2004: loc. 1465–1484). Dryzek (1995, p. 21–24) proposes a "non-anthropocentric democracy" based on an *ecologically* communicative rationality that treats "signals emanating from the natural world with the same respect we accord signals emanating from human subjects, and as requiring equally careful interpretation." This challenges models of liberal democracy based on hard boundaries between nations and "between the human and nonhuman world"; rather "it might be productive to start thinking about models of democracy in which the boundaries of communities are indeterminate." The aim is an "egalitarian interchange at the human/natural boundary; an interchange that involves progressively less in the way of human autism." Dryzek thus proposes ecological democracy as a "regulative ideal" that would aim for "effectiveness in communication that transcends the boundary of the human world."

We might interpret Dryzek's model of ecological democracy as an open one based on interspecies potentiality; an open "regulative ideal" that does not presuppose any one line of evolution in the polity. This contrasts with a model of the eco-centric polity that maintains firm categories and divisions but opens movement into them to the non-human: Donaldson and Kymlicka's *Zoopolis*. This model allocates non-human membership between (domestic and farm) animal "citizens," (urban and liminal) animal "denizens," and (wild) animal "sovereigns." It would generate a system of animal rights that challenges biopolitics but re-inscribes the state.

Firstly, Donaldson and Kymlicka argue that domesticated animals ought to "be seen as co-citizens…with the right to be represented through forms of dependent agency in our political decision-making." They assert that models of "dependent, assisted or interdependent agency" (drawn from disability rights theory) are highly suggestive for animal citizenship through political processes that "find ways of eliciting a person's sense of their subjective good, often through 'embodied' rather than verbal communication. In these new models, people with mental disabilities can enact their citizenship, but it requires other people…to help construct a 'script' of their conception of the good life, drawing on both verbal and nonverbal expressions of preference" (Donaldson and Kymlicka 2011, p. 59–60). This is a powerful move that pushes the rights of pets and laboratory and farm animals to the fore, challenges paternalistic models of care or exploitation, and demands multiple layers of representation: "in the legislative process, but [also] municipal land planning decisions, or the governance boards of various professions and public services (police, emergency services, medicine, law, urban planning, social services, etc.)." (Donaldson and Kymlicka 2011, p. 154)

Secondly, they describe denizenship as a situation in which the nuisance and interdependence of town and urban-dwelling species is managed by ethics of coexistence and harm minimization in cases where non-human populations need to be excluded or managed, supported by rights of residency and reciprocity. This category includes feral animals and introduced species in rivers and wilderness areas, which sometimes raise complex biopolitical dilemmas where they threaten biodiversity and ecosystem integrity. The ethics advanced by Donaldson and Kymlicka presses management practices to be less intrusive and harmful, along with a sharing of risks and the redesign of cities and transport systems to reduce damage to non-humans (Donaldson and Kymlicka 2011, p. 230–249).

Finally, they assert sovereignty for wild animals—those who avoid contact with humans and suffer from encroachments on their habitat—who should be considered "citizens of sovereign wild animal territory" to whom humans would have "nation-to-nation duties":

> (1) respect for territorial boundaries (e.g., an end to invasion and colonization); (2) limitation of spillover costs (e.g., transborder pollution or road deaths); (3) shared sovereignty of key international corridors (e.g., migratory routes); (4) respect for the basic rights of visitors; and (5) extension of assistance to refugees. In other words, our duty is to enable them to exist as wild animal communities, while limiting the negative costs of their inevitable contacts with us (Donaldson and Kymlicka 2011, p. 217).

In this model of the *Zoopolis,* we are obviously moving beyond calls to listen to and communicate with non-humans to a project that simultaneously seeks to transform the categories and exclusions of the nation-state, so that it will respect and safeguard their interests. However, the implications for the relation of international law to the non-human are not explored. On the one hand, the claim that wild animals should be treated as sovereigns—which implies strong rights to "communal flourishing that depends on securing lands and autonomy [and] specific traditional habitats" against "alien rule"—provides a strong normative principle against the state appropriation of nature enabled by the customary law principle of Permanent Sovereignty Over Natural Resources, which is the current and most powerful system that relates sovereignty to nature as domination and use (Donaldson and Kymlicka 2011, p. 174). On the other hand, the way in which state sovereignty anchors a complex international legal regime of human territoriality, property rights, and capitalist appropriation stands as a formidable barrier to non-human sovereignty. It is not clear how a *Zoopolis* would negotiate the system of sovereign states and an international environmental law that remains trapped between enabling exploitation and protection. We could, however, interpret it as an implicit challenge to the indivisible model of human sovereignty that is central to the social contract and the Montevideo convention, given their advocacy of forms of parallel, shared, and overlapping sovereignty, "sovereign communities of multi-species animal ecologies, rather than single species." They suggest transforming sovereignty in ways that "accommodate, rather than deny or suppress, the overlapping and mobile character of communities and territories. This will obviously require abandoning the idea that sovereignty must be univocal and absolute" (Donaldson and Kymlicka 2011, p. 190–91).

We are, however, left with questions. Would such overlapping sovereignties set aside state sovereignty over nature entirely, or create an uneasy system of coexistence that effectively bargains animal sovereignty against the human? The latter sovereignty reduces the nonhuman to a moral patient or an inert resource-in-waiting; the former would honor nonhuman agency, rights, and flourishing. Such "sovereignties" are radically different in kind.

Posthuman Representation: A Materialism of Shared Worlds

Existing visions of ecological democracy and citizenship have made admirable progress in rethinking the anthropocentric polity and asserting the normative weight of nonhuman agency, rights, and communication. At the same time, they are weakened by subtle fracture lines.

Donaldson and Kymlicka profoundly rethink the categories of inclusion and membership in the social contract but reinstate the sovereign state as a founding ontology and neglect the many ways it anchors an international system of ecological despoliation. Their hard allocation of rights, boundaries, and roles between "citizens," "denizens," and "sovereigns" falls short of the more holistic and transboundary model of political agency and inclusion that is needed now. The power of their notion of animal sovereignty is to assert the autonomy and inviolability of habitat as a key to animal flourishing, but these intuitions need to be extended beyond territorial divisions and rights of way—however nested, pooled, and flexible—to a genuinely holistic and planetary appreciation of the responsibilities and dangers that come with the human power to affect animal lives, habitats and ecosystems at multiple spatial and temporal scales: from the most local impacts on individuals and ecosystems to the biosphere and the Earth system, and from the most immediate events to a very long-term future measured in geologic time.

In an analogous way, Eckersley's green state offers a profound model of non-human agency and interspecies deliberation but dismisses cosmopolitan models of transnational ecological democracy and governance as a utopian "counsel of despair," banking instead on (no less utopian) hopes for an eco-centric normative transformation of the state. She corrals ecological democracy back into the state system rather than imagine how it might naturally flow outwards to transform and challenge the system's structuring ontologies. This hesitation is evident in the tension between, on the one hand, her challenge to States' presumption to *use* territory rather than protect it, and, on the other, her reluctance to directly challenge the international customary law doctrine of Permanent Sovereignty Over Natural Resources. While she asserts that "unit-level transformations provide the most likely way in which more general, system level transformations will take place," we would argue that they are powerfully co-constitutive—that international law, in fact, reifies and empowers state and extractivist ontologies of dominance over nature—and must be challenged and transformed at every scale in which they operate and appear (Eckersley 2004: loc. 2912).

Dryzek and Pickering avoid these problems by refusing to specify institutional blueprints, arguing instead for the cultivation of an "ecological reflexivity [that shows] foresight to anticipate and prevent catastrophes such as state shifts, as well as the ability to listen more effectively to ecological systems and rethink core values." Dryzek does, however, seem open to an ecological democracy that would flow across state ontologies and borders, saying that ecologically democratic institutions must work at the "appropriate scale" and "match the size and scope of problems," which suggests "numerous and cross-cutting *loci* of authority" (Dryzek and Pickering 2019, p. 48; Dryzek 1995, p. 24–26).

The scale and intensity of anthropogenic ecological transformations after the Holocene provokes complex questions: How much earthly and other-than-human reality can the state admit? Surely, it is time for the capitalist state's confidence that it can willfully construct and manipulate new realities—bordered polities, disciplined bodies, transmuted matter, and terraformed worlds—to be set aside as erroneous and dangerous? How long can it continue to assume that the entangled complexity, multi-variant agency, and planetary scale of the biosphere should organize itself according to the arbitrary boundaries and economic appropriations of human societies? Yet the primarily normative tenor of ecologically democratic theory can sometimes appear like an appeal to the castle guard for admission to the state, rather than a recognition that its walls are already being torn apart by the gales of earth system change, by the violent material processes such exclusions have set into the train.

Such a materialist, planetary perspective cuts through concerns that it is impossible to "know" nature's interests given the inability of non-humans to enter political processes by using human languages and organizational protocols. This view poses an epistemological problem that it may be difficult to ascertain or represent their interests faithfully or accurately. Eckersley challenges the conclusion that animals cannot be brought into political deliberation on normative grounds, but such exclusions are also a denial of ecological reality. Non-human animals possess complex systems of exchange, solidarity, and communication between and across species, some of which could be considered almost cultural when they take on sophisticated musical and artistic forms such as the songs of whales and the dances of birds. Animal and plant communities have not evolved to create sophisticated forms of what human societies would call politics, even if there is evidence of hierarchy, leadership, cooperation, and collective decision-making within some species. They have, however, evolved enormously complex and sophisticated forms of interspecies symbiosis and interdependence, and non-human communities engage in sophisticated forms of communication and messaging through the use of calls, sounds, gestures, eye contact, movements, and emissions of scents and chemicals. Such a recognition spurs Eduardo Kohn (2013: loc. 313–327) to argue that nonhuman life forms also engage in representation; that "signs also exist well beyond the human" and that "life is constitutively semiotic." The question then is how to make representation cross human and non-human worlds in ways that respond to their dangerous entanglement. Animals do not lack language or culture and have had little need for "politics" until humans began to have ever greater impacts on non-human lives and ecosystems that are now being played out on the vast and destructive scale of the Anthropocene. It is in this encounter between non-human and human that Clemens Driessen (2014, p. 91) argues there is already an "active animal politics": "practices [in which] animals in effect play a prominent role and are to some extent already taken seriously in ways that could be interpreted as political…as idiosyncratic beings engaged in a shared learning process [such as] nature conservation practices in which humans and animals learn to live together."

In pushing beyond the "hyperrelativist" arguments that it is impossible to know nature's interests, we can see ecological democrats reaching towards a materialist appreciation of the scale and agency not merely of particular sentient beings but non-human systems as such; it is for these we must change the polity, in order to honor and take account of them. Eckersley seeks to balance an appreciation of an "extra-discursive nature [that possesses] independent existence, agency and creativity" with an awareness that there remain dilemmas of knowledge and interpretation: "our constructed 'nature' is nothing other than our approximate, provisional attempted understanding the so-called real [i.e., extra-discursive] nature." She argues that this should "rule out obviously harmful interventions" and engage "care, caution and humility" when trying to "engineer what we might believe to be a flourishing nature" (Eckersley 2004: loc. 1544–1548). Dryzek extends a focus on nonverbal animal communication—body language, facial displays, pheromones—to "the ecological processes which transcend the boundaries of species, such as the creation, modification, or destruction of niches; or cycles involving oxygen, nitrogen, carbon, and water" (Dryzek 1995, p. 21). He and Pickering thus call for ecological democracy to create "entry points for nonhuman 'actants' into deliberative systems" by listening "to the screams of pain from the *Earth* system, from species extinctions, acidifying oceans, the death of waters that once were healthy, crashes in the number of insects in ecosystems" (Dryzek and Pickering 2019, p. 146–150).

Yet what responsible deliberative or governance system can now afford to ignore the planetary real—a real that organizes myriad living structures of symbiosis and interdependence, vast movements of ice, water, and gases, and powerful exchanges of energy, into a marvelous, dynamic, and chaotic totality? James Lovelock's "Gaia hypothesis" postulated that the Earth was a living whole whose evolutionary history was directly shaped by life, which affected the creation of water and ice, the Earth's anomalous composition of surface gases, its surface temperature, and the composition of its rocks (Lovelock 2000). Life has autopoetically engineered the Earth to make it hospitable for more life; this makes, in Lynn Margulis' (1998, p. 123) interpretation, the planet's surface "geophysiological…displaying the attributes of a living body composed of the aggregate of Earth's incessantly interactive life." In short, the Earth itself has a life-driven agency. Margulis insists on a strictly scientific vision of Gaia divorced from its mythic connotations, but climate change and the sixth extinction now exemplify a situation where the Anthropos is driving Gaia into the capricious and vengeful phase proper to her place in ancient Greek cosmogony, punishing her God-children and unleashing furies on the hubris of men (Latour 2017, p. 81–83).

What should this imply for representation? What must be "listened to" and included in the polity are organisms, lives, and processes across every scale, from the frog or marsupial endemic to a small ecosystem, to diminishing populations of insects and plankton and planetary changes in animal biomass, the oceans, and atmosphere. Such actants and thing-systems are not mere moral patients waiting for entry to the polity, and any politics that assumes it can ignore such systemic "screams of pain" will have to reckon with the forces and transformations that that refusal has unleashed: droughts, starvation, and super-typhoons; drowned farms and cities; conflict, mortality, and extinctions. We must not merely recognize that humans or obviously sentient and communicative animals can be understood to communicate, but non-sentient changes in matter, form, and energy as they interact with the biosphere and other components of the Earth system as well. In this light, the *in/visibility* of the other-than-human is a globally pathological situation. The demand that the non-human must either speak our political language or remain mute was always cast in the wrong direction; it is we who must learn the Earth's language and reimagine the polity in its idioms.

Visions of Transnational Ecological Democracy: Governing with the Non-Human

In this chapter, we have argued that the non-human deserves agency and representation in the (international) polity by virtue of its living vitality, its interconnection, and its sheer force and presence. If the ontic, biospheric real is accounted for in such a way, this changes the political question: from nature as an extension of our ourselves to one that asks how humans can construct an ethical project built on the understanding that humans are part of an entangled living whole where politics emerges from mutual experiences (Massumi 2014, p. 38–54). What changes would be wrought if humans reversed this understanding to one of integration and interaction rather than one of domination and control?

Our key argument is that politics more broadly, and political representation and agency more narrowly, must shift at the ontological level rather than just at the process level. To return to the provocateur for this chapter, Aristotle, to accept "man as a political animal" meant a concomitant creation of political community based upon this idea of man. As a "political animal," Aristotle believed that "man" had the capacity for thought, justice, and self-assertion, which led to the creation of political community, namely the Greek polis. The political animal and the political community are connected. As Eugene Garver asks, if Aristotle is to be politically relevant today, can the political community conceived differ from Aristotle's constitutional definition of that community? (Garver 2011, p. 115). A wider, more diffuse, and ecologically interconnected and entangled politics, or *demos*, will necessarily need to create new forms of community that exceed and differ from the state or the nation. In this spirit, Bruno Latour imagines a new kind of *demos*—an "object-oriented democracy," a "dingpolitik" rather than "realpolitik"—which assembles many kinds of people, roles, technologies, conflicts, and objects into a space of democracy (Latour 2005). Even as the Dingpolitik is concerned to point out how things and systems already constitute spaces of the *demos* beyond the lifeless parliaments of the liberal tradition, Latour stops well short of imagining a *demos* that would be fully ecological and include the non-human as both a participant and a constitutional source of origin. Given the scale and intensity of the planetary ecological crisis presented by the sixth extinction and the Anthropocene, a new vision of the *demos* must enable the full inclusion and representation not merely of non-human animals but ecosystems and the biosphere as a whole; a *democratic holism* that takes in the entire Earth. If we consider the complexity of the multifold social–ecological challenges that will arise as we move through the twenty-first century and into the twenty-second, and the gross inadequacies of our Holocene institutions, the nature of our politics must change.

With such a framing ethics in mind, below we briefly advance design proposals and principles that extend from the most local spaces of democratic governance to the global. Our analysis, and much of the moral–political philosophy it is informed by, strongly acknowledges the agency and power of non-human lives and systems; however, we still face the aporetic problem that animal communication has not evolved to take the form of political deliberation, meaning some form of proxy representation is required.

In general, the proxy representation of the non-human should be pursued through three means. Firstly, institutions must be *constitutionally* designed—or reformed—so that they have the primary purpose of protecting and restoring ecosystems, ensuring sustainability, and bringing forth the interests and needs of non-human lives into policy-making and deliberation (Voight). This should shape local governments, national constitutions, and state-level portfolio areas, especially those that affect the climate and ecology most profoundly: economic, agricultural, environmental, and military. Secondly, deliberative and governance institutions should be redesigned (or newly created) so as to allocate membership and seats to proxies for animals and ecosystems and ensure that affected human communities have stronger voices. While seeking elected proxies from within key biomes seems the most logical, Anthropocene ecological change shows that communities and ecosystems in multiple sites around the world will suffer effects; proxy and institutional design needs to respond to and work at such scales. Thirdly, polities and governance systems must be restructured to receive and interpret clear *signals* from the Earth; to better listen to and understand the nature and pace of ecological change and appreciate both the rights and needs of ecosystems and non-human lives, and the human communities that are entangled with them. This requires enormous investments in climate and ecosystem monitoring, science, and rapid information-sharing; a kind of global intelligence system for ecological security and reflexive governance (Dryzek and Pickering 2019, p. 35, 46; Burke and Fishel 2019, p. 104).

While these ideas should be pursued actively in existing national democratic spaces, we argue that global environmental governance and democracy could benefit from two innovations: a United Nations "Earth System Council," which would have two-thirds of its members drawn from a second new institution—the "Global Ecoregion Assembly"—of which we propose establishing 15 spread across the Earth's major biomes.

The Ecoregion Assembly would be a new transnational governance institution with a written constitution, anchored in the principles of Earth law and the rights of nature, that charges the body with responsibility for the protection of the ecosystems and biodiversity in its region, and for ensuring that ecologically sensitive activities (including greenhouse emissions) that occur in its region do not do wider damage to the Earth system. A transition from soft to hard law models of their operation can be considered—beginning with their functioning more like regional organizations of states, in which resolutions and statements would have normative guidance and force for member states or other actors, and moving to a model where Ecoregion Assemblies assume powers to legislate region-wide environmental laws and standards that must be policed and enforced. Regional member states would need to agree to yield or pool jurisdiction in environmental lawmaking, along the EU model, while ensuring their ability to make additional environmental laws or retain existing ones until the Ecoregion Assembly does so.

Ecoregion Assembly membership might be made up of the following: delegates of national governments with territory within the ecoregion; representatives of indigenous landowners from within the ecoregion; and a number of democratically elected representatives acting as proxies for human and non-human communities. Representatives from UNEP and the key international environmental treaty bodies should also be accredited as advisors. The number of members each assembly should have could vary by region but should be high enough to avoid risks of cooptation and corruption and to ensure responsive representation from local and regional communities, without becoming unwieldy. They should work by simple majority voting to avoid deadlock and spoiling. We propose 15 Global Ecoregions: the Arctic; Europe; the North Pacific; South Asia and Northeast Asia (including Myanmar, Malaysia, and the Mekong Basin); Central and North America; the Amazon Basin; South America; South Pacific (including South-Eastern Australia); Antarctica; Archipelagic Southeast Asia (including Northern Australia and PNG); the Indian Ocean (including West Australia); Southern Africa; Central Africa; North Africa and Middle-East; and the Mediterranean and Black Seas. These ecoregions cover areas that have broadly similar climate, a limited range of ecosystem/biome types, and a manageable number of associated national states. They also cover key oceans and ecosystems such as the poles and the Amazon. By being geographically contiguous, they can draw on regional concentrations of scientific expertise and, in most cases, successful patterns of diplomatic institutionalization and cooperation.

The United Nations Earth System Council would play a similar role in the protection of the global ecology as the Security Council plays in the protection of international peace and security (but would hopefully be far more representative, equitable, and effective). It would be permanently in session and, analogous to the way that Article 24 of the UN Charter confers on the Security Council "primary responsibility for the maintenance of international peace and security," the Earth System Council would have primary responsibility for the protection and security of the global environment and Earth system. The Council would have the power to make directions to states and its resolutions would be binding. It would also have the power to make recommendations to major treaty bodies such as the UNFCCC, CITES, and the Convention on Biodiversity. It would not have Chapter VII-like powers to impose sanctions or authorize the use of force, although in especially severe (and we would hope rare) cases, it could refer a situation to the Security Council for consideration. The Council's membership could be made up of 25 ambassadors: ten from UN member states and 15 from the Global Ecoregion Assemblies, which enables proxy non-humans to be represented at the highest levels of the UN system. Like the Ecoregion Assembly, the Council would have a constitutional purpose to represent and protect the Earth's ecology encoded in a new establishing article of the UN Charter.

These models assume (and drive) the evolution of international environmental law towards a more consistently ecological ontology, creating a politically complex interplay between a process that would better enforce and strengthen international environmental law, on the one hand, and, on the other, the perpetuation of resource extraction and pollution systems built upon its more ecologically problematic biases—such as the principle of permanent sovereignty over resources and continuing state freedoms to emit greenhouse gases and overfish. While we would argue such institutions will be very necessary by 2050 and beyond, we are aware that in the short-to-medium term they will be seen as radical and meet strong resistance from states. At the same time, we can also see state-based regional organizations like the South Pacific Forum—whose members have especially acute concerns about climate change and sustainability—beginning to act like ecoregional assemblies and advocate globally for new dispensations on climate action, forced migration, and sustainable development. This suggests an intriguing pathway: if states will not sponsor the creation of new ecologically democratic institutions such as the ecoregion assemblies, *communities and civil society can do so alongside the state system* as pilot experiments in new forms of ecological democracy that will protect the living complexity of planet Earth and the myriad human communities who depend on its survival. We doubt that, short of ecological catastrophe, states will initiate such change; it will take the emergence of a global community and civil society movement that is not merely asking for states to act but is resolved to reimagine and rebuild global governance and democracy as such.

References

Aristotle (2013). *Politics*. In Carnes Lord (Ed.) *Aristotle's politics*. 2nd Edition. Chicago and London: University of Chicago Press.

Aristotle (1883). *History of animals in ten books*, trans. Richard Creswell. London: George Bell and Sons.

Atapattu, S. (2015). The significance of international environmental law principles in reinforcing or dismantling the North-South Divide. In Alam Shawkat, Sumudu Atapattu, Carmen G. Gonzalez, and Jona Razzaque (Eds.) *International Environmental Law and the Global South* (pp. 74–108). New York: Cambridge University Press.

Basnett, C. J. (2016). Other political animals: aristotle and the limits of political community. *The European Legacy, 21*(3), 290–309.

Biermann, F. (2014). *Earth system governance: World politics in the Anthropocene*. Cambridge Mass. and London: The MIT Press.

Burke, A. (2019). Blue screen biosphere: The absent presence of biodiversity in international law. *International Political Sociology, 13*(3), 333–351.

Burke, A., & Fishel, S. (2019). Power, world politics and thing-systems in the anthropocene. In F. Biermann & E. Lovbrand (Eds.), *Anthropocene encounters: New directions in green political thinking* (pp. 87–108). Cambridge and New York: Cambridge University Press.

Bull, H. (1977). *The anarchical society: A study of order in world politics*. Basingstroke: Palgrave Macmillan.
Donaldson, H., & Will Kymlicka, W. (2011). *Zoopolis: A political theory of animal rights*. Oxford and New York: Oxford University Press.
Driessen, C. (2014). Animal Deliberation. In M. Wissenberg & D. Schlosberg (Eds.), *Political animals and animal politics* (pp. 90–104). Basingstoke: Palgrave Macmillan.
Dayan, C. (2013). *The law is a white dog: How legal rituals make and unmake persons*. Princeton: Princeton University Press.
de Fontenay, É. (2012). *Without offending humans: A critique of animal rights*. Minneapolis: The University of Minnesota Press.
Dryzek, J. S. (1995). Political and ecological communication. *Environmental Politics, 4*(4), 13–30.
Dryzek, J. S., & Pickering, J. (2019). *Politics in the anthropocene*. Oxford and New York: Oxford University Press.
Dryzek, J. S., & Stevenson, H. (2014). *Democratising global climate governance*. Cambridge and New York: Cambridge University Press.
Eckersley, R. (2004). *The green state*. Cambridge: The MIT Press.
Eckersley, R. (1992). *Environmentalism and political Theory*. New York: State University of New York Press.
Falkner, R. (2017). The anarchical society and climate change. In M. Carr, H. Suganami, & A. Humphreys (Eds.), *The anarchical society at 40: Contemporary challenges and prospects* (pp. 198–215). Oxford and New York: Oxford University Press.
Fishel, S. (2017). *The microbial state: Global thriving and the body politic*. Minneapolis and London: Minnesota University Press.
Garver, E. (2011). *Aristotle's politics: Living well and living together*. Chicago and London: University of Chicago Press.
Habermas, J. (2003). *The future of human nature*. Cambridge and Malden M.A: Polity Press. Kindle Edition.
Hobbes, T. (1997). *Leviathan*, Richard E. Flathman and David Johnston (Eds.). New York and London: W.W. Norton.
Hobe, S. (2015). Evolution of the principle on permanent sovereignty over natural resources: From soft law to a customary law principle? In M. Bungenberg & S. Hobe (Eds.), *Permanent sovereignty over natural resources* (pp. 1–13). Heidelberg, New York and London: Springer.
Kant, I. (1978). *Anthropology from a pragmatic point of view*. Carbondale and Edwardsville: Southern Illinois University Press.
Kohn, E. (2013). *How forests think: Towards an anthropology beyond the human*. Berkeley and London: University of California Press.
Latour, B. (2005). From realpolitik to dingpolitik—an introduction to making things public. In B. Latour & P. Weibel (Eds.), *Making things public-atmospheres of democracy* (pp. 515–539). Boston: MIT Press.
Latour, B. (2017). *Facing Gaia: Eight lectures on the new climatic regime*, trans. Catherine Porter. Cambridge and Malden MA: Polity.
Lovelock, J. (2000). *Gaia: A new look at life on earth*. Oxford and New York: Oxford University Press.
Manning, E. (2007). *Politics of touch: Sense, movement, sovereignty*. Minneapolis and London: University of Minnesota Press.
Margulis, L. (1998). *Symbiotic planet*. Amherst: Perseus.
Massumi, B. (2014). *What animals teach us about politics*. Durham: Duke University Press.
Merchant, C. (1983). *The death of nature: Women, ecology and the scientific revolution*. New York: Harper Collins.
Nicholson, S., & Jinnah, S. (Eds.). (2016). *New earth politics*. Cambridge Mass: The MIT Press.
Nussbaum, M. (2006). *Frontiers of justice: Disability, nationality, species membership*. Cambridge MA. and London: Harvard University Press.

Nussbaum, M. (2011). *Creating capabilities: The human development approach.* Cambridge MA: Harvard University Press.

Paganini, G. (2017). "Political animals in seventeenth century philosophy" in Stefanie Buchenau, Roberto Lo Presti (Eds.), *Human and Animal Cognition in Early Modern Philosophy and Medicine* (pp. 186–200). Pittsburgh: University of Pittsburgh Press.

Plumwood, V. (2002). *Environmental culture: The ecological crisis of reason.* London and New York: Routledge.

Rawls, R. (1999). *The law of peoples.* Cambridge Mass. and London: Harvard University Press.

Schlosberg, D. (2014). Ecological Justice for the Anthropocene. In M. Wissenberg & D. Schlosberg (Eds.), *Political animals and animal politics* (pp. 75–89). Basingstoke: Palgrave Macmillan.

Stevenson, H. (2012). *Institutionalizing unsustainability: The paradox of global climate governance.* Berkeley and London: University of California Press.

UNGA (1962). General Assembly resolution 1803 (XVII) of 14 December 1962, "Permanent sovereignty over natural resources".

Voigt, C. (Ed.) (2013). *Rule of law for nature: New dimensions and ideas in environmental law.* Cambridge and New York: Cambridge University Press.

A Quantum Anthropocene? International Relations Between Rupture and Entanglement

Cameron Harrington

Abstract The Anthropocene is marked as a paradigm shift in Earth history and politics. This has simultaneously led to two competing characterizations of the new epoch: as an age of "rupture" from the past and as a reflection of "entanglement" between the human and the non-human. In this chapter, I assess how the logics of rupture and entanglement create different, often competing understandings of the Anthropocene within the academic field of International Relations (IR). I conclude by challenging IR scholars to more closely engage with quantum social theory in light of the profound "spookiness" of the Anthropocene.

Introduction

> When we try to pick out anything by itself, we find it hitched to everything else in the universe. - John Muir (1911)

The Anthropocene, the so-called geological epoch defined by human action, has captured increasing amounts of attention from international relations (IR) scholars (Biermann and Lövbrand 2019; Chandler 2019; Dalby 2017; Fagan 2016; Hamilton 2016a; Hardt 2018; Harrington 2016; Hickmann et al. 2018; Mitchell 2014; Pereira 2017; Rothe 2017). Its scale(s), scope and conceptual complexity has left ample room for varying—often contrasting—ideas about what it is, when it began, who is responsible and how to respond to it. While it retains a marginal place in the IR canon overall, its emergence over the past decade has led to a series of critical interventions that have simultaneously decried the lack of IR engagement with the Anthropocene while also questioning whether IR, conventionally understood, is capable of faithfully engaging with its key conceptual insights (Burke et al. 2016; Harrington and Shearing 2017). This is because the Anthropocene has not simply been used as a convenient, catch-all term for environmental change, nor as a settled geological designation; it has also

C. Harrington (✉)
School of Government and International Affairs, Durham University, Durham, UK
e-mail: cameron.harrington@durham.ac.uk

J. C. Pereira and A. Saramago (eds.), *Non-Human Nature in World Politics*, Frontiers in International Relations, https://doi.org/10.1007/978-3-030-49496-4_4

been used as a conduit for a broader questioning of core metaphysical assumptions in the discipline ranging from human subjectivity and agency (Hamilton 2019) to the role of time in constructing worlds (Fagan 2019). Despite some scepticism,[1] the general consensus seems to be that the Anthropocene marks a radical paradigm shift in how we should view the world and our place (as humans) within it (Hamilton 2015). It offers a potentially radically different way of understanding geopolitics. If the Earth is defined by human action, and if humans themselves are constituted via nature itself, then "we need new practices, new ideas, stories, and myths" (Burke et al. 2016, p. 500). Holocene-bred theories that position the non-human world as a stable backdrop to dramas emanating from sovereign states and their international relations are no longer sustainable. Core shared fundamental assumptions about the state(s) of nature, agency, structure and the durability of sovereign decision-making are all troubled by Anthropocene discourses that emphasize complex spatio-temporal interdependence, indeterminant causes and the entangled nature of being between human and non-human entities.[2]

Perhaps unsurprisingly, much of the discussion surrounding the Anthropocene runs parallel with another leap in IR metatheorizing occurring concurrently, which attempts to bring insights from quantum physics into the social world (Der Derian 2016). Alexander Wendt's ground-breaking 2015 book, *Quantum Mind and Social Science: Unifying Physical and Social Ontology* asks readers to question the classical foundations at the heart of contemporary social theory and consider the effects of replacing them with insights derived from quantum reality. According to him, "the metaphysical assumptions of the classical worldview—materialism, determinism, locality and so on—[have been] deeply ingrained in the minds of social scientists" (Wendt 2015, p. 12). At least for mainstream, positivist IR, this is also true; it requires a classical metaphysics in order to function and make sense. International actors exist as distinct units that perform various localized actions based upon rational calculations of their known environment. The social structures that exist to order the world are reducible to the properties and interactions of individuals and their groupings. They can be studied by discrete observers, who can measure social life without interfering in it. Time and space exist as objective backdrops for action. Even critical and interpretivist approaches to IR, which challenge many of these claims, rely upon a view of the world that must be answerable to the standard laws of physics. Their rejection of naturalism is itself framed by a classical view (Wendt 2015, p. 13). Thus, the world of Holocene IR has been firmly classical in orientation. This is so, despite it being well over one hundred years since quantum discoveries first revolutionized physics. Wendt's book investigates the radical hypothesis that human consciousness is, in fact, a quantum phenomenon. The brain, in other words, is a quantum computer. By scaling subatomic quantum effects upward to the macroscopic

[1]Luke (2015, 2015).

[2]A number of important critiques suggest the Anthropocene concept depoliticizes the human in its desire to naturalize the species and envelop it within nature. Fagan (2019, p. 56) writes that "the idea of 'nature' (and therefore of 'human') itself is not innocent but entangled with discourses of race, extinction and geology emerging in the 18th century in which black life was treated as a form of animal life (Mirzoeff 2018); classing (some) humans as a 'part of nature' is nothing new."

level of the brain, Wendt is able to make on the one hand a rather parochial claim that dualism in the social sciences no longer holds true and combine it with a much more radical claim that social structures (even social reality itself) are emergent; they are not "actual realities existing somewhere above us in space, but potential realities constituted by non-local shared wave functions" (Wendt 2015, p. 33). This implies that "the relationship of agents and social structures is not a process of causal interaction over time, but a non-local, synchronic state from which both are emergent" (Wendt 2015, p. 260). The individualist accounts employed to build the world may be usefully augmented with models of entanglement or connection at a distance that lack classical causal mechanisms.

The nascent interest in applying quantum insights to IR has occurred at roughly the same time as the turn towards the Anthropocene. This is likely more than pure coincidence and could be read as indicative of some shared ideas and concerns regarding how to interpret a world that does not resemble the past or abide by seemingly settled theories of reality. Though the two approaches are not always commensurable and their insights are sometimes irrelevant to one another, their similarities may be usefully assessed. Thus, this chapter investigates how the use of the Anthropocene within IR presents two seemingly distinct discourses—separation and entanglement—to explain the relationship between humans and the world. The chapter argues that the distinction between the two discourses represents two unique ways of interpreting the world: one built upon a classical view of the world and another informed (if not explicitly guided) by insights gleaned from quantum worldviews. After all, if quantum theory holds at the subatomic level but eventually washes out at the macroscopic level (as explained via Bohr's correspondence principle), it would stand that the Anthropocene—perhaps the most "macro" of all "events"—should remain classically interpreted. This would no doubt be welcomed by some who are already sceptical that the pivot towards the Anthropocene in humanities and social sciences is reflective of "science envy," which dilutes and misinterprets the empirical findings as a way to repackage old ideas: a "paradigm dressed as epoch" (Baskin 2015). To add quantum social science to the mix may be a bridge too far. Yet, so much of what encompasses Anthropocene thought in IR—emphases on entanglement, complexity, non-linearity, emergence, vital materialism (Bennett 2010), etc.—bears some family resemblance to a quantum view of the world.

This chapter proceeds in two parts. It first examines the vision of the Anthropocene as a classical form of "rupture," whereby human actions have transgressed the limits of the Earth's carry capacity. In this reading, parts of the world are constituted by intrinsic properties that are observable and measurable. It measures the world, using various control variables, to conclude that the Anthropocene is defined by a radical separation from stable Holocene conditions. In this perspective, prominently amplified by earth system science, the steady state of the world is marked by radical upheaval and division. In fact, for the Anthropocene to be officially designated as a new unit of the International Chronostratigraphic Chart (the basis of the Geological Time Scale), sufficient evidence must be supplied to identify a clear break from previous eras. For the Anthropocene to exist, geologists need to find "a synchronous signal of the transformative influence of humans on key physical, chemical, and biological processes at the planetary scale" (Waters et al. 2018, p. 380). There must be a formal base of rupture—a boundary and point—that separates the Anthropocene from the Holocene. The hunt for the so-called "Golden Spike," the geological marker that can formally characterize the deviation from the Holocene, positions the Anthropocene as a radical coming-apart from the past; a "no-analogue" state (Steffen et al. 2004). The logical next step for many in international relations is to construct new forms of governance and scientific co-production that can tackle the consequences of Earth's system shifts. From this are born research programmes like Earth System Governance (Biermann 2007) and Future Earth (van der Hel 2016). This discourse of Anthropocene IR derives from classical, neo-Newtonian notions of the world.

Second, the chapter examines another discourse of Anthropocene IR, one which is premised on non-classical forms of metaphysics to locate and observe a diverse set of worldly realities. Out of this come forms of "entanglement," as a way to characterize the collapse of the nature/culture divide and as a reflection of deep enmeshment between humans and other beings and processes in the Earth system. This perspective has achieved a certain resonance by offering a relational ontology built from the collapse of historical subject-object binaries. The article engages with recent insights from the so-called "quantum turn" in social theory, placing it alongside other concepts in the ongoing Anthropocene literature like complexity, which focuses on creative potentiality and possibility as a way to better understand the profound "spookiness" of the Human Age. These approaches disrupt classical notions of locality, discreteness, space and time. Here, the Anthropocene offers a particularly non-classical lens through which to further engage with worldly reality.

Understanding the differences (and occasional intertwinement) between the conceptual framings of the Anthropocene within IR underscores an ongoing, often-obscured subplot in discussions about the discipline's response to global environmental upheaval: how the seemingly benign nature of contestation and critique may belie deeper metaphysical allegiances. This chapter does not answer its own question whether the Anthropocene is quantum or classical or some combination of both, nor does it debate the merits of any particular approach like Earth System Governance. Instead, it offers a provocation to think creatively about the different realities in which human and non-human worlds interact in the midst of unfolding ecological disasters.

The Anthropocene as Rupture

One of the principal ways to interpret the Anthropocene is as a great separation. Officially, it is the transition from one geological epoch (the Holocene) to another; one that is now fully human-made. In popular collective imagination, it represents a shift away from a harmonious and historically benign relationship with the Earth. In its stead, humans are now viewed as transformative actors that have the capacity to reshape the planet. Here, the Anthropocene ideal frames a type of estrangement from the natural harmony that once existed between humans and the world. In this sense it represents a radical break from the past.

In order for the Anthropocene to be officially designated it requires geologists to locate evidence of a fundamental *rupture*. Convincing evidence must be produced showing rapid and profound changes to the Earth's geology (Zalasiewicz et al. 2017), significant enough to designate the epoch as "functionally and stratigraphically distinct" (Waters et al. 2016). Geologists, therefore, work to tear apart the Earth's layers and show in various ways the seemingly epochal effects of industrial activity. In both senses—in popular imagination and in scientific formalization—the declaration of the Human Age takes us beyond the edge of the once-known world and into something fundamentally different. The Anthropocene depends on the loss of form—on the rupture between an old world and a new one.

This sentiment is expressed in a few different ways. One common tendency is to proclaim in stark terms the magnitude and intensity of changes across the planet. The great acceleration of Earth system indicators suggests that a fundamental shift in the makeup of the planet is ongoing (Steffen et al. 2015). The amount of carbon released into the atmosphere is unprecedented over the last 66 million years (Zeebe et al. 2016) leading to a rapid warming trend experienced across the world. In 2018, surface temperatures were the fourth warmest since modern recordkeeping began in 1880 (NASA 2019). That same year global ocean temperatures were 0.66 °C (1.19 °F) above the twentieth-century average, making it the 4th warmest year for ocean temperatures on record. Taken together, the past 5 years (2014–2018) have been the warmest years on record (NOAA 2019). Emerging out of and alongside these climatic changes is a host of harms—including but not limited to the further loss of sea ice, sea-level rise, food and water insecurity, and increases in the number and intensity of severe weather events like heat waves, storms, droughts and floods (Herring et al. 2019). Beyond the precipitous warming of the oceans and of surface temperatures, human actions have led to sharp declines in biodiversity, atmospheric increases of nitrous oxide, and acidification of the oceans. While notoriously imprecise, it is estimated conservatively that overall extinction rates are between – 100 and 1,000 times higher than natural background rates of extinction and future rates are likely to be 10,000 times higher (De Vos et al. 2015). Given the extent of species loss across the plant and animal kingdom, we may be in the midst of Earth's sixth mass extinction in its 3.7-billion-year history—and the first to be caused by a single species (Kolbert 2014). This "biological annihilation" (Ceballos et al. 2017) portends a radical disruption in the functioning of the planetary biosphere.

These proclamations arise from a conceptual vision of the Earth as a holistic, dynamic whole, system (Maslin and Lewis 2015b). This conception and objectification of an Earth System is produced, disseminated and reinscribed through the language and technologies of Earth System Science (ESS), together with the global institutions created to coordinate its work and build corresponding, enactable governance policies. In contrast to previous scientific conceptions that compartmentalized the natural world for study, ESS views humans as part of a large, integrated system, "in a constant state of movement driven by interconnected cycles and forces, from the planet's core to the atmosphere and out to the moon, and powered by the flow of energy from the sun" (Hamilton 2016b, p. 2). By focusing on the system as a whole rather than focusing on disparate aspects like sediment layers, ESS is able to proclaim radical, paradigm-shifting events like the Anthropocene and attach human causes to it. The upscaling of systems thinking to the climate and the biosphere (Steffen et al. 2004) allows for the envelopment of the Earth into an integrated assemblage and is in many ways a necessary technique for the Anthropocene concept to emerge. The ESS research paradigm, which first emerged in the 1980s is meant to link knowledge from the physical and biological sciences and demonstrate the growing impact of human activities upon the Earth system. The two scientific bodies perhaps most responsible for its emergence and spread were the International Geosphere-Biosphere Program (IGBP), founded in 1983, and the International Human Dimensions Programme (IHDP), founded in 1990. At the outset, IGBP was set up to understand and comprehend human impacts on a global scale, and originally excluded social scientists from discussions. Its focus was assessing the "natural background of change" on which humans were then imposed (Uhrqvist and Lövbrand 2014). However, it was not long after commencing activities that the IGBP began building institutional bridges with social scientists. The connections between the biophysical drivers of Earth change (particularly land use change) and human actions were too stark to ignore and precipitated interdisciplinary work. Thus, through the 1990s, natural scientists and social scientists from the IGBP and the IHDP collaborated in earnest to bring the communities together to assess and evaluate the status and trends of ecosystems through time (Mooney, Duraiappah, and Larigauderie 2013). As Steffen et al. make clear, "It is no accident that the proposal for the Anthropocene Epoch arose out of the Earth System science community, in particular out of the synthesis project of the IGBP" (2016, 334). Its aim to produce a holistic vision of the Earth and humanity as a single, interconnected organism with shared responsibilities is also a quest for worldly synthesis, a classically informed science of modernity. In 2012, a new global change research programme, Future Earth, was created by merging together IGBP, IHDP and DIVERTAS: a programme on biodiversity science. Future Earth has grown into an influential programme that brings together scientists and other societal actors to "co-produce" knowledge about the challenge of global sustainability (van der Hel 2016).

The scientific unity between social, physical and biological sciences that defines ESS, together with its attachment to systems thinking, has led to the use of influential metaphors like tipping points and planetary boundaries to illustrate the human dimensions of the Earth system. Tipping points first emerged in the 1970s and 1980s through dynamic systems theory to describe nonlinear, abrupt system behaviour (Russill and Nyssa 2009). The concept as it relates to climate change became popularized in the early years of the twenty-first century, specifically via a 2005 speech by former NASA scientist James Hansen to the American Geophysical Union (AGU). Hansen declared that,

> IPCC uses a burning embers diagram to quantify reasons for concern about climate change. The impression created is that 2- or 3-degrees Celsius warming, relative to the present is probably dangerous. The burning embers are usually interpreted with a probabilistic approach, which has certain merits. However, I suggest that the burning embers are a fuzzy concept

> that discourages action, action that is needed urgently, because we are on the precipice of climate system tipping points beyond which there is no redemption (Hansen 2005, p. 7–8).

Hansen's warning about tipping points derived in no small measure from the success of Malcolm Gladwell's book *The Tipping Point*, published in 2000. In typical Gladwellian fashion, the popular philosopher took the original scientific concept and used it to examine the way certain aspects of life suddenly shift from obscurity to ubiquity (Gladwell 2000). For him, the tipping point was the moment at which an idea or behaviour catches on and spreads to new situations. Climate change researchers saw an appealing resonance for a variety of reasons, not least because they sought to generate popular and political responses to increasing environmental threats.

Not long after Hansen's original call, the editors of *Nature* acknowledged the contagion-like spread of tipping points literature, and noted its utility as a both a description of radical ecological shifts and as a way to attract attention to the social dimensions of climate change:

> It is possible to make people change their minds and behaviours, and for those changes to spread like a contagion. 'Look at the world around you,' Gladwell argues. 'It may seem like an immovable, implacable place. It is not. With the slightest push – in just the right place – it can be tipped (Nature, p. 785; also quoted in Russil and Nyssa, p. 338).

The scientific study of tipping elements in the Earth's climate system examines the critical threshold, whereby a slight disturbance could have qualitatively great effects and alter the state or development of a system. It is the point at which human activity pushes components of the Earth system past critical states into qualitatively different modes of operation (Lenton, Held, Kriegler, Hall, Lucht, et al. 2008). After the threshold is crossed, it will be difficult or impossible to return it to the former conditions which have been previously relied upon (Bosselmann 2010). At the local and regional level tipping points are the abrupt, irreversible shifts in the state of biodiversity and ecosystems (Secretariat of the Convention on Biodiversity 2010). It is exceedingly difficult to predict with any accuracy just how close ecosystems are to a specific tipping point and how much additional pressure might cause a shift. Likewise, it is possible for an ecosystem to have tipped into another state without being known. A number of prominent theoretical and popular examples of tipping points for subsystems of the Earth system include the potential collapse of the Atlantic thermohaline circulation (THC), cryosphere, marine ecosystems, dieback of the Amazon rainforest and decay of the Greenland ice sheet (Lenton et al. 2008; Brook et al. 2013; Pereira and Viola 2018). Tipping point literature was successful in amplifying the notion of thresholds which, once transgressed, could cause a radical ecological rupture. It focused on how threshold effects cause critical transitions, which in turn lead to state shifts that abruptly override trends and produce unanticipated biotic effects (Barnosky et al. 2012).

Corresponding with the spread of tipping points has been the emergence of the "planetary boundaries" framework (PB), which attempts to define "a safe operating space for humanity" (Rockström et al. 2009a, b). One should be careful not to fully conflate tipping points and planetary boundaries because PB does not imply a global synchronous threshold or tipping point (Steffen et al. 2015). It instead offers a wider zone of safety, "upstream" of the tipping point. This is intended to allow society time to react to early warning signs that it may be approaching a threshold.

After its emergence in a 2009 article by Johan Rockström and colleagues, the PB approach has rapidly spread across a variety of scientific and policy domains and now comprises a significant portion of ESS and its governance offshoot of the Earth System Governance. It also embodies a particular corner of Anthropocene imagination—one that characterizes it as a dramatic rupture between planetary stability and radical upheaval without the prospect of control. PB is designed for the global level and is intended to demonstrate both quantitatively and qualitatively the evidence of human disruptive influence on all planetary systems. Rockström et al. (2009a, b) specify nine important boundaries for Earth system processes: climate change; biodiversity loss; the nitrogen cycle; the phosphorus cycle; stratospheric ozone depletion; ocean acidification; global freshwater use; land use change; atmospheric aerosol loading; and chemical pollution. It is a perfect marriage of political and scientific aims. It captures the notion of the whole earth, including its subsystems, and advocates for the creation of stable institutions to guarantee a safe transition into the Anthropocene. Rockström et al. are offering a vision of the world that is rapidly overshooting a number of boundary conditions which, if crossed, will lead to major disruptions in parts of the system and which will transition to a separate state (Biermann 2012).

As Biermann (2012) and Galaz et al. (2012) make clear, planetary boundaries are not the first scientific attempt to define biophysical boundaries for human development. Earlier attempts include seeing human actions as embedded in Earth's life-support system (Odum 1988), and determining global biophysical constraints for the expansion of the economic subsystem (e.g. Ciriacy-Wantrup 1952; Daly 1977). Some of the most influential environmental policy documents, including the Club of Rome's report Limits to Meadows and Club of Rome (1972), and later the German Advisory Council's Global Change's initiative, attempted to define "Tolerable Windows" for human action on the Earth. The rapid diffusion of the concept in the wake of tipping points literature is not surprising for a number of reasons. First, the concept captures multiple global-scale environmental stresses within one integrated framework at the same time as it offers preliminary quantified goals to support decision-making. PB can thus be implemented alongside the Sustainable Development Goals (SDGs) as an indicator for a thriving society, even if it lacks deeper social contexts and singularizes the planetary experience. Additionally, PB stresses the urgency of political action through its emphasis on the risks associated with transgressing critical Earth system thresholds or tipping points. It is widely acknowledged that setting the boundaries via control variables is an intensely political process,[3] as is the call for human populations to live within the boundaries. Oran Young and Falk Schmidt argue that despite the inherent politics to the concept, it is an effective way "to come to grips

[3]Take, for instance, the debate whether to quantify climate change via average temperature increases above pre-industrialized levels or in terms of concentrations of CO_2 (parts per million) measurable in the atmosphere. Whether one chooses to focus on 2 degrees Celsius or 350 ppm is not simply a scientific decision but reflective of attitudes towards risk and various social tradeoffs (Young and Schmidt 2019).

with the challenges of the Anthropocene" (Young and Schmidt 2019, p. 420). They suggest that a political interpretation means focusing on the trade-offs and incentives of living within the boundaries and emphasizing how the outcomes of a healthy earth system are non-rival and non-excludable.

Though tipping points and PB differ in substantial ways, they both represent similar understandings of the Earth system, where deviations from the steady state of the Holocene give way to a new version of nature: disjointed and separate from the past. This new form of nature through a variety of techniques, metaphors and measurements becomes "pre-reflectively projected, quantified and conceived as a calculable and carbonic human construction..." (Hamilton 2018, p. 33). In its original framing, the Anthropocene was meant to capture the recent split in Earth history, created from the impact of human activity on the Earth system. For some, the lesson of the Anthropocene is that our new epoch contains no precursor; it is a *rupture* that requires new understandings of the human relationship to the Earth and the dangers that flow from it (Hamilton and Grinevald 2015; Nicholson and Jinnah 2016; Parr 2018). The Anthropocene is not the continuation, nor even simply the intensification of human impacts upon the Earth. According to Clive Hamilton (2016b, p. 100),

> we are witnessing not continuous change but rupture – a rapid transition to a new geological epoch, or perhaps an era, that is *permanent*... The Earth has now crossed a point of no return; its great cycles have changed; the chemical compositions of air and ocean have been altered in ways that cannot be undone except on a millennial timescale. In short, the Earth System is now operating in a different mode and nothing humans can do now, even ending the burning of fossil fuels in short order, can turn the geological clock back to the Holocene.

For Hamilton, this realization is nothing short of a gestalt switch, which must lead to a paradigm shift in knowledge, ethics and action. To use the Anthropocene in any other way is to distort the meaning and significance of the Anthropocene as it was originally framed by Earth System Science.

All told, these particular interpretations make explicit the classical impulse to observe, map, measure, monitor and order the planet into a system according to divisible flows, material supplies and connected processes. It "replicates a Western secular cosmology that works by explicitly measuring the distance *between* an 'objective' nature and humanity" (Hamilton 2017a, p. 591). Goals like tipping points and planetary boundaries employ rational calculations of how the Earth system functions to set the limits of a sustainable and good life. This may, in fact, be a useful way to manage complexity and risk in the Anthropocene but it is reflective of a particular merger of classical metaphysics and Western science. One of its dangers is to further empower what Uhrqvist and Lövbrand label as the northern "expertarchy" and an associated eco-managerialism (Uhrqvist and Lövbrand 2014) in a process that eradicates the political economy of the Earth system. The boundaries and limits of planetary processes and acceptable political behaviour become mapped onto existing biopolitical administrations of life. Everything from personal action to the parameters of multilateral agreements can be subsumed into a hierarchical order of integration. It is, therefore, no distant leap to the increasing proposals to geoengineer Earth system processes across knowable global scales. These calls are enacted and deemed acceptable when rapid ecological shifts are detected across a world rendered relatively simple, knowable and predictable machine with interlocking constituent parts.

The Anthropocene as Entanglement

Mark Maslin and Simon Lewis write that there can be multiple definitions of the new epoch, versions that are not beholden only to the strictures of geological record-keeping. For them, there can be a historian's Anthropocene, a political scientist's Anthropocene, a geographer's Anthropocene (2015b), etc. They make clear that the choice of selecting a date for the Anthropocene is not simply a matter of scientifically discovering a golden spike but is also a political and philosophical choice. Accepting the so-called "1610 Orbis hypothesis," when the "new world" and the "old world" collided would imply the world-making power of colonialism, global trade and coal. Likewise, to select 1964 and the nuclear fallout event horizon marker would imply that the legacies of elite-driven cold war rivalries built amidst the nuclear age are to live forever. As the authors put it, "The event or date chosen as the inception of the Anthropocene will affect the stories people construct about the ongoing development of human societies" (2015b, 178). Even amongst the 34-scientist Anthropocene Working Group, which is tasked with compiling the evidence to formally declare the epoch, there is a minority belief in a time-transgressive Anthropocene. According to this logic, it is better to think of Anthropocenes with multiple beginnings rather than an epoch derived from a single moment of origin, whether it be nuclear fallout or genocide (Subramanian 2019).

Though the Anthropocene is indeed a geological phenomenon, it cannot be held securely in the stratigraphic record, separate from the wider world(s) from which it emerges. Instead (and I imagine all Anthropocene scholars would accept this), it points to a broader understanding of how complex interconnectedness functions across a dynamic Earth system through space, time, being and matter. Accordingly, it becomes possible to consider the connections between the Anthropocene and a quantum worldview which highlights the dynamics of entanglement. Though the two views are often complementary (e.g. complexity theory and quantum social theory share much of the same language), studies that focus on Anthropocene entanglement point less to the ecological emergencies born from planetary shifts and instead focus on the destabilization of classical, Western forms of metaphysics and reality which order our understanding of IR (Hamilton 2017b). While the former is generally held to be authoritative and valuable, the latter conceptions of the Anthropocene seem to be rendered as speculative theorizing (at best) or cultural literary exercises (at worst).[4] This section will primarily focus on entanglement, but other related quantum processes and concepts also bear a family resemblance to Anthropocene theory. A longer piece could conceivably expand on the corresponding mechanics of

[4]Clive Hamilton (2016b, p. 97) has referred to the feminist theorist Donna Haraway's renaming of the Anthropocene as Capitalocene, Plantationocene, or Chthulucene as "terminological incontinence".

non-locality, or other quantum aspects like emergent properties, quantum probability, complementarity, indeterminacy, strict causality and even pluriversal cosmology as a way to link the Anthropocene with quantum theory. It is clear that both the Anthropocene and a quantum worldview can offer an alternative view of the universe and consider new, strange metaphysical possibilities. In other words, a non-classical form of rupture.

The concept of quantum entanglement was initially developed by Schrödinger (and later John Bell) in response to a 1935 paper by Albert Einstein, Boris Podolsky and Nathan Rosen. Schrödinger showed how when two quantum particles interact their states will depend on each other instantly, across any distance, which challenged the basic notions of realism and localism. Realism here refers to the idea that physical systems possess inherent properties that are independent of the observer and the methods they use to measure them. Localism holds that local events cannot be affected by actions in space—like separated regions (Gröbachler et al. 2007). For Einstein and other classically trained physicists, the closer objects are to one another, the easier it is to influence one another. In contrast quantum entanglement argues that non-classical correlations can exist between two spatially separated quantum systems—particles can influence each other across massive distances and across time. In basic terms, reality is intrinsically non-local. In a quantum world, there exists a kind of vital force that connects and forms physical reality. In recent decades, a variety of studies have confirmed Schrödinger's view that nature behaves in quantum ways. "In particular, the laws of physics have been found to be inherently non-local: The results of a measurement at one position in space can dictate the possible outcome of a measurement performed at a different position" (Karimi and Boyd 2015, p. 1172). To maintain realism in the midst of these results would necessitate the introduction of "spooky" actions that defy locality (Gröbachler et al. 2007). Instead we must consider the notion that individual particles, when entangled, bind together into an indivisible whole. Entangled systems are not divisible, and thus they disrupt simplistic concepts of distance and space as containers of action. They do not "communicate" in the classical sense. They do not exchange information faster than the speed of light. Communication is instantaneous and infinite. Speed, time and causality fade away in entangled quantum systems. All of this contrasts with classical systems, which are always divisible. Any collective properties held in the classical system arise from constituent components that themselves have certain properties (Vedral 2011).

The logical extensions from this are challenging because they contradict some fundamental assumptions about our grasp of reality. Once one reaches the quantum level, it becomes clear that our taken-for-granted beliefs in classical reality—atomism, materialism, determinism—can be supplemented by quantum insights. Atomism, or metaphysical individualism, holds that the world is composed of individual units with innate, privately held properties, which are nonrelational. Quantum indeterminism means that there are no individually determinate entities to be discovered. Meaning and matter (i.e. words and objects) are no longer cleanly separable. Where does this leave us? According to Karen Barad, the absence of strict determinism does not lead to free will or a collapse into discourse (however defined).

Instead, we should see reality not as rolling, unravelling stasis (i.e. waiting to be ruptured), but a

> dynamism that operates at an entirely different level of "existence," where "existence" is not simply a manifold of being that evolves in space and time, but an iterative becoming of spacetimemattering. Space, time, and matter are intra-actively produced in the ongoing differential articulation of the world. Time is not a succession of evenly spaced intervals available as a referent for all bodies and space is not a collection of preexisting points set out as a container for matter to inhabit (Barad 2007, p. 234).

This is all intriguing exposition, but does it accurately portray the reach of quantum physics? What if quantum entanglement "washes out" past the subatomic level, as many physicists still believe? As Vedral (2011, p. 41) explains,

> Most demonstrations of entanglement involve at most a handful of particles. Larger batches are harder to isolate from their surroundings. The particles in them are likelier to become entangled with stray particles, obscuring their original interconnections. In accordance with the language of decoherence, too much information leaks out to the environment, causing the system to behave classically.

Yet, ongoing work has advanced quantum theory into new realms, including biology, psychology and international relations. Studies that investigate macroscopic quantumness are now commonplace (Fröwis, Sekatski, Dür, Gisin, and Sangouard 2018). As Smith (2016) explains, Heisenberg's famous uncertainty principle has been detected in a drum visible to the naked eye and quantum entanglement has been observed in a pair of spatially separated diamonds two millimeters in size, at room temperature. Other recent findings suggest that "birds exploit non-local connections with the earth's magnetic field to help them navigate, that plants exploit quantum effects in photosynthesis, that fruit flies' sense of smell relies on the ability to detect quantum vibrations in smelly molecules, and that quantum processes might even facilitate social learning by protozoa" (Wendt 2015, p. 134).

While acknowledging that quantum macroscale implications remain at this point largely speculative,[5] a number of scholars, including a handful of international relations scholars, have begun to extrapolate quantum physics into social and political theory (Wendt 2015; Smith 2016; O'Brien 2016; El Khoury 2015; Der Derian 2011; Zanotti 2017; Aarfi and Kessler 2018). Entanglement in this context deviates from its original, strict laboratory definitions and may be better seen as analogous rather than *truly* indicative of combined quantum effects. Nonetheless, entanglement offers a valuable alternative paradigm in which to make sense of the utter strangeness of the Anthropocene. Such a reading offers a potentially reflexive account of the complexities and challenges that emerge from the Anthropocene, while also offering alternatives to classical interpretations of the human age as rupture.

---

[5] As a recent paper describes, "The common intuition is that, while nature might allow for quantum effects on macroscopic scale, it makes them practically impossible to observe. This is due to technical limitations that forbid one to perfectly isolate a system from its environment and to perform measurements with unlimited precision. This leads to an effective quantum-to-classical transition, which can be ideally derived from the quantum laws themselves" (Fröwis, Sekatski, Dür, Gisin, and Sangouard 2018).

It is well-accepted that a variety of Earth systems make up an integrated global ecosphere. To theorize and conceive of the Anthropocene as a quantum phenomenon means it has to be seen as more than a networked connection spanning the world. Quantum entanglement refigures "one-ness" and "two-ness" and implies that "duality, unity, multiplicity, being are undone. 'Between' will never be the same. One is too few, two is too many" (Barad 2010, p. 251). It entails a non-classical relation to multiple possibilities. Thus, if seen in this way, as a form of "intra-action" the Anthropocene evacuates strict notions of causality in favour of iterative processes. Entangled systems are in fact "super-correlations" that go well beyond any kind of correlations created on the basis of classical physics (Barad 2007). They force us to question the idea that any entity—human or non-human, sentient or non-sentient, bio-physical or physical, macro or nano—can be truly distinct. Entanglement instead suggests that "systems must always be conceptualized within collective terms" (Holm and Taffel 2017, p. ix).

The Anthropocene can provide us with a striking illustration of this condition, making apparent the ways in which human societies are entangled with the processes of something we once called "nature." While nature has never been an entity that existed outside of humans and society state-system shifts for climate change, ocean acidification, stratospheric ozone depletion and land-system changes highlight how deeply human action can, and has, reshaped our biophysical realm, and similarly, how biophysical systems have shaped human life. A recognition of entanglement in this instance compels us to reconsider the diverse ways in which subjects and objects co-exist and are co-constituted intersubjectively. Instead of a world comprised of distinct entities (e.g. states, individuals, non-humans) interacting in a rational pursuit of their needs, entanglements inform us of the indistinction of actors in the first place and perhaps even their existence in multiple spaces simultaneously. It is not simply that humans are a part of nature. The social and the natural have become indistinguishable. One can find this way of thinking even in Bruno Latour's words, "[t]o be a subject is not to act autonomously in front of an objective background, but to share agency with other subjects that have also lost their autonomy" (2014, 5, italics in original).

Classical forms of IR derive from seemingly timeless realities of matter and life: physical objects are bounded (distinct), measurable and calculable. For an organism to be "alive" and coherent, it requires some shielding (boundaries) from its outside environment, otherwise its wave functions inevitably collapse. However, objects are also autopoietic, which means they require an open system in order for the exchange of energy to withstand thermodynamic decay (Wendt 2015, p. 134). From these assumptions, it becomes possible to construct an IR that is thoroughly predictable. Emilian Kavalski writes (2012, p. 138) that the discipline's neo-Newtonian foundations have left international life as "a closed system, changing in a gradual manner and following linear trajectories, which can be elicited through discrete assessments of dependent and independent variables. What IR intends to produce in this way is a nearly mechanistic model of international politics that is perceived to be as rigorous and robust as the one of the natural sciences." Kavalski draws from John G. Ruggie and Richard Ned Lebow to exhort IR scholars to open up to new ontologies that can engage with the *reality* of international life, in all its complexity. This means understanding the vast, and heterogeneous realities of global affairs, and not simply repackage old, failed, theories with new questions.

Fundamentally, the Anthropocene opens IR scholars up to a metaphysics of entanglement and non-linear causality. It teaches us that what exists comes to be through the process of intra-action rather than the atomistic ontology of mainstream IR that represents social phenomena as quantifiable and predictable (Kurki 2008). The characteristics of the Anthropocene—with garbage patches twice the size of France floating in the Pacific Ocean; with microplastics (which contain potentially toxic chemicals) flowing from synthetic fibres we wear on our body out into the ocean and back into us through seafood and tap water; with radionuclide fallout from mid-century nuclear weapons test burying itself in Arctic ice cores for eternity—leave us fundamentally at odds with classical ideas about time, space and agency. The Anthropocene's quantum potential for IR is clear. Laura Zanotti encapsulates this sentiment when she writes, "we do not act as monads endowed with *potentia*, but as entangled phenomena that are part of what exists. We do not pre-ordain material dynamics as a result of aspirations dictated by reason. Instead, we participate in processes of materialization of which we are a part but do not control" (2017, 372).

Conceptualizing Anthropocene IR as entanglement instead of rupture leads to specific political and ethical implications. While it creates openings or cuts for alternative social orders, it also requires us to realize that our "abstract idealistic aspirations are entangled in complex clusters of causation and processes of materialization that we as humans may contribute to but do not determine" (Zanotti 2017, p. 372.) Where this leads ethically remains open because a quantum Anthropocene is filled with an infinite set of possibilities and, "a force of justice is available with-in every moment, every place, every bit of matter" (Barad 2019, p. 544). But the world is not inherently just. Responding with diverse forms of worldly care (Harrington 2017; Harrington and Shearing 2017) or with radical hospitality (Barad 2019) are but a few responses that could open up possibilities for ethical forms of relational practice.

One should also be aware of the use of entanglement as a metaphor within classical approaches. One may find numerous examples within the ESS and ESG literature of the entanglement metaphor being used without challenging the underlying assumptions of atomistic determinism. Scott Hamilton warns, "Declaring classical sciences and renderings of nature to be "entangled," therefore, does not actually make them so. It actually masks the certainty of a classical Newtonian causality still working beneath the Anthropocene's discursive surface" (2017b, p. 590).

Conclusion

Ralph Waldo Emerson labelled dead metaphors in language as "fossil poetry." The deadest word was "once a brilliant picture" (Emerson 1844). One can find all sorts of fossil poetry in IR: from anarchy to the tragedy of the commons. But what happens when we, as IR scholars, encounter, alternative paradigms that challenge our fundamental worldviews? How do we make sense of an entangled world that is defined by its rupture? Is it possible or desirable to repurpose old realities for new epochs?

This chapter offered a preliminary assessment of how IR scholars might begin to incorporate the "patchiness" of the Anthropocene which envelops us and contemplate the heterogeneity of space and time (Tsing 2015). It used two compelling visions of the Anthropocene—as rupture and as entanglement—to ask whether they can offer a pathway towards new metaphysical insights that see a world that "sits alongside other ones" (Seth 2013, p. 150).[6]

It began by tracing one influential and popular discourse of Anthropocene thought: the Anthropocene as "rupture." This view emanates from different directions. Perhaps most influentially, it is reinscribed by the International Commission on Stratigraphy, the body tasked with officially designating its arrival at the Anthropocene. It also flows from dominant institutions of Earth system science like the IGBP and the IHDP. All rely upon a view of a holistic Earth comprised of constituent, intertwined parts that are rapidly undergoing a damaging transformation as a result of human action. Through metaphors, metrics like "tipping points" and "planetary boundaries,"[7] these international science programmes have constituted global change as a problem that needs intervention and rectification (Uhrqvist and Lövbrand 2014). In the words of Clive Hamilton (2016b, p. 2), "The idea of the Anthropocene was conceived by Earth System scientists to capture the *very recent rupture* in Earth history arising from the impact of human activity on the Earth System as a whole." This labels the new Human Age as a paradigm shift, whereby the world undergoes a great separation and humanity overshoots the boundaries of Holocene-bred safe operating spaces. Its translation fits nicely into most forms of international relations, as evidenced by the continued success of the Earth System Governance programme.

The chapter then engaged with the recent turn to quantum social theory and contemplated whether Anthropocene thought can offer a type of respite from the fatigue from repetition in IR. Visions of complex and emergent entanglements between humans, culture and the non-human world across space and time, allows for a certain level of creativity within the discipline. It stands so far outside the realm of coherent IR thought that it has allowed for new types of speculative theorizing. The divisions which for so long have held sway (subject/object, inside/outside, agent-structure, etc.) are problematized, if not fully challenged. It is this disruption of dualism that has prompted some to turn to quantum theory as a radically new paradigm for us to view and understand the deep relations being uncovered in the midst of ecological calamity.

---

[6]This sentiment from Seth and echoed by Rojas (2016) is expressed in relation to contesting colonial logics embedded in IR. Neither piece directly deals with quantum social theory and I do not wish to reassamble their critique in the service of my own quantum interest here. That said, there may be fruitful conversations to be had between those engaged in decolonial (or postcolonial) projects that destabilize the modern world and quantum social theorists.

[7]Schmidt (2019) has recently labelled the planetary boundaries framework as a *grundnorm* (a norm basic to all others) for international programmes of environmental law and governance.

An extended comment on whether the two perspectives are in any way compatible or commensurate is something better left for another day. It may be that it is not necessary to connect the two perspectives at all and yet still arrive at similar conclusions. Both accept as given the radical implications of the Anthropocene. Earth System Science and quantum approaches both focus on complexity and the placing of humans within webs of practice. Indeed, it is possible to see elements of quantum ontology within the discourses of classically informed Anthropocene IR. Take, for instance, how quantum physics challenges the possibility of a subject-object distinction (Zanotti 2019). Such a proclamation is commonplace within most Anthropocene IR literature. Yet a few, if any, of these interventions actually engage with the "quantumness" of such a claim. Instead, they tend to connect the blurring of the subject and object with evidence of human impacts on the non-human world rather than the philosophical ramifications of ontological indeterminacy. Finally, it is true that both classical and quantum interpretations have the potential to efface important worldly differences in their desire for operational boundaries on the one hand or in their promotion of planetary entanglement on the other. Further scholarly interventions may usefully add to this conversation.[8]

On and on we can go, declaring the advent of momentous, unprecedented trends. Some have proclaimed that the Earth has entered "uncharted territory" (World Meteorological Organization 2017). Others have suggested that the "end(s) of the world" or the "end of nature" is at hand (Barad 2019; Harrington 2016; Morton 2013; Danowski and Viveiros de Castro 2017). In both readings, the world is transformed from something known, anticipated and respected. What comes after is obscured, unfamiliar and dangerous. Given what we now seem to know about the Anthropocene, with its multi-scalar systems made up of individuals, communities, ecosystems and technologies that flow across boundaries, there is a pressing need for interdisciplinary, empirical and theoretical work that encounters entanglement and the Anthropocene. This is so despite the fact that many remain sceptical that either the Anthropocene or quantum entanglement are appropriate for IR scholars to discuss or whether they offer anything profoundly new. Nonetheless, the Anthropocene holds multitudes and contradictions within it that do not produce easy or comfortable answers. But it offers us a pathway for encountering new cosmopolitical ideas that are simultaneously new yet immeasurably ancient—for perceiving our entangled existence as kin (indeterminant but coupled) on a damaged planet without obscuring the worldly differences of culture, gender, class, space, species and time (Fishel et al. 2017).

[8]Drawing upon Norbert Elias, Quilley and Loyal's (2005) article "Eliasian Sociology as a 'Central Theory' for the Human Sciences" offers a useful framework for thinking through potential 'bridges' between fields of investigation.

References

Aarfi, B., & Kessler, O. (2018). Forum introduction: Social theory going quantum-theoretic? Questions, alternatives, and challenges. *Millennium: Journal of International Studies, 47*(1), 67–73.

Barad, K. (2007). *Meeting the Universe Halfway: Quantum physics and the entanglement of meaning and matter*. Durham: Duke University Press.

Barad, K. (2010). Quantum entanglements and Hauntological relations of inheritance: Dis/continuities, SpaceTime Enfoldings, and justice-to-come. *Derrida Today, 3*(2), 240–268.

Barad, K. (2019). After the end of the world: Entangled nuclear colonialisms, matters of force, and the material force of justice. *Theory and Event, 22*(3), 524–550.

Barnosky, A. D., Hadly, E. B., Bascompte, J., Berlow, E. L., Brown, J. H., et al. (2012). Approaching a state shift in Earth's biosphere. *Nature, 486,* 52–58.

Baskin, J. (2015). Paradigm dressed as epoch: The ideology of the anthropocene. *Environmental Values, 24*(1), 9–29.

Bennett, J. (2010). *Vibrant matter: A political ecology of things*. Durham: Duke University Press.

Biermann, F., et al. (2012). Navigating the anthropocene: Improving earth system governance. *Science, 335,* 1306–1307.

Biermann, F. (2007). 'Earth system governance' as a crosscutting theme of global change research. *Global Environmental Change, 17*(3–4), 326–337.

Biermann, F. (2012). Planetary boundaries and earth system governance: Exploring the links. *Ecological Economics, 81,* 4–9.

Biermann, F., & Lövbrand, E. (Eds.). (2019). *Anthropocene encounters: New directions in green political thinking*. Cambridge: Cambridge University Press.

Bosselmann, L. (2010). *Earth Governance. Trusteeship of the Global Commons*. Cheltenham, UK: Edward Elgar.

Brook, B. W., Ellis, E. C., Perring, M. P., Mackay, A. W., & Blomqvist, L. (2013). Does the terrestrial biosphere have planetary tipping points?". *Trends in Ecology & Evolution, 28*(7), 396–401.

Burke, A., Fishel, S., Mitchell, A., Dalby, S., & Levine, D. (2016). Planet politics: A manifesto from the end of IR. *Millennium: Journal of International Studies, 44*(3), 499–523.

Ceballos, G., Ehrlich, P. R., & Dirzo, R. (2017). Biological annihilation via the ongoing sixth mass extinction signaled by vertebrate population losses and declines. *Proceedings of the National Academy of Sciences (PNAS), 114*(30), E6089–E6096.

Chandler, D. (2019). *The Transvaluation of Critique in the Anthropocene, 33*(1), 26–44.

Ciriacy-Wantrup, S. V. (1952). *Resource conservation: Economics and policies*. Berkeley: University of California Press,

Dalby, S. (2017). Anthropocene formations: Environmental security, geopolitics and disaster. *Theory, Culture and Society, 34*(2–3), 233–252.

Daly, H. E., (1977). *Steady-state economics: Science and society*. W. H. Freeman and Co, New York.

Danowski, D., & Viveiros de Castro, E. (2017). *The ends of the world*. Cambridge: Polity.

Der Derian, J. (2011). Quantum diplomacy, German-US relations and the psychogeography of Berlin. *Hague Journal of Diplomacy, 6,* 373–392.

Derien, J. (2016). Project Q Vision. *Project Q*. Retrieved 25, Nov 2019, from http://projectqsydney.com/q-vision/.

De Vos, J. M., Joppa, L. N., & Gittleman, J. L., et al. (2015). Estimating the normal background rate of species extinction. *Conservation Biology, 29*(2), 452–62.

El Khoury, A. (2015). *Globalization development and social justice: A propositional political approach*. London: Routledge.

Emerson, R. W. (1844). The poet. *Poetry Foundation*. Retrieved 25, Nov 2019, from https://www.poetryfoundation.org/articles/69389/from-the-poet.

Eroukhmanoff, C., & Harker, M. (Eds.), *Reflections on the Posthuman in International Relations: The Anthropocene, Security, Ecology*. E-International Relations.

Fagan, M. (2016). Security in the anthropocene: Environment, ecology, escape. *European Journal of International Relations, 23*(2), 292–314.
Fagan, M. (2019). On the dangers of an Anthropocene epoch: Geological time, political time, and post-human politics. *Political Geography, 70,* 55–63.
Fishel, S., Burke, A., Mitchell, A., Dalby, S., & Levine, D. (2017). Defending planet politics. *Millennium: Journal of International Studies, 46*(2), 209–219.
Fröwis, F., Sekatski, P., Dür, W., Gisin, N., & Sangouard, N. (2018). Macroscopic quantum states: Measures, fragility and implementations. *Reviews of Modern Physics, 90*(2), 025004.
Galaz, V., Biermann, F., Folke, C., Nilsson, M., & Olsson, P. (2012). Global environmental governance and planetary boundaries: An introduction. *Ecological Economics, 81,* 1–3.
Gladwell, M. (2000). *The tipping point: How little things can make a big difference.* Little Brown.
Gröbachler, S., Paterek, T., Kaltenbaek, R., Brukner, Č., Żukowski, M. A., et al. (2007). An experimental test of non-local realism. *Nature, 446,* 871–875.
Hamilton, C. (2015). Getting the Anthropocene so wrong. *The Anthropocene Review, 2*(2), 102–107.
Hamilton, C. (2016a). The anthropocene as rupture. *The Anthropocene Review, 3*(2), 93–106.
Hamilton, C. (2017a). *Defiant earth: The fate of humans in the Anthropocene.* Cambridge, UK: Polity.
Hamilton, C., & Grinevald, J. (2015). Was the anthropocene anticipated? *The Anthropocene Review, 2*(1), 59–72.
Hamilton, S. (2016b). The measure of all things? The Anthropocene as a global biopolitics of carbon. *European Journal of International Relations, 24*(1), 33–57.
Hamilton, S. (2017b). Securing ourselves from ourselves? The paradox of "entanglement" in the Anthropocene. *Crime, Law and Social Change, 68*(5), 579–595.
Hamilton, S. (2018). The measure of all things? The anthropocene as a global politics of carbon. *European Journal of International Relations, 24*(3), 33–57.
Hamilton, S. (2019). I am uncertain, but We are not: A new subjectivity of the Anthropocene. *Review of International Studies, 45*(4), 607–626.
Hansen, J. (2005). Is there still time to avoid "dangerous anthropogenic interference" with global climate? A tribute to Charles David Keeling. *American Geophysical Union.* Retrieved 25, Nov 2019, from http://www.columbia.edu/~jeh1/2005/Keeling_20051206.pdf.
Hardt, J. (2018). *Environmental security in the Anthropocene: Assessing theory and practice.* London: Routledge.
Harrington, C., & Shearing, S. (2017). *Security in the anthropocene: Reflections on safety and care.* Bielefeld: Transcript Verlag.
Harrington, C. (2017). Posthuman security and care in the anthropocene. In C. Eroukhmanoff & M. Harker (Eds.), *Reflections on the posthuman international relations: The Anthropocene, security and ecology* (pp. 73–86). Bristol: E-International Relations.
Harrington, C. (2016). The ends of the world: international relations and the Anthropocene. *Millennium: Journal of International Studies, 44*(3), 478–498.
Herring, S. C., Christidis, N., Hoell, A., Hoerling, M. P., & Scott, P. A. (2019). Explaining extreme events from 2017 from a climate perspective. *American Meteorological Society,* 100(1): Special Supplement.
Hickmann, T., Partzsch, L., Pattberg, P., & Weiland, S. (Eds.). (2018). *The anthropocene debate and political science.* Milton Park: Routledge.
Holm, N., & Taffel, S. (Eds.). (2011). *Ecological entanglements in the anthropocene.* Lanham: Lexington Books.
Kavalski, E. (2012). Waking IR Up from its 'Deep Newtonian Slumber.'*Millennium: Journal of International Studies, 41*(1), 137–150.
Kolbert, E. (2014). *The sixth extinction: An unnatural history.* New York: Henry Holt & Company.
Karimi, E., & Boyd, R. W. (2015). Classical entanglement? *Science, 350*(6265), 1172–1173. https://doi.org/10.1126/science.aad7174.
Kurki, M. (2008). *Causation in international relations: Reclaiming causal analysis.* Cambridge: Cambridge University Press.

Lenton, T. M., Held, H., Kriegler, E., Hall, J. W., Lucht, W., et al. (2008). Tipping Elements in the Earth's climate system". *Proceedings of the National Academy of Sciences, 105*(6), 1786–1793.
Lepori, M. (2015). There Is No Anthropocene: Climate change, species-talk, and political economy. *Telos, 172,* 103–124.
Luke, T. W. (2015). Introduction: Political critiques of the anthropocene. *Telos, 172,* 3–14.
Maslin, M., & Lewis, S. (2015a). Anthropocene: Earth System, geological, philosophical and political paradigm shifts'. *The Anthropocene Review, 2*(2), 108–116.
Maslin, M., & Lewis, S. (2015b). Defining the anthropocene. *Nature, 519,* 171–180.
Meadows, D. H., & Club of Rome. (1972). *The limits to growth: A report for the Club of Rome's project on the predicament of mankind*. New York: Universe Books.
Mirzoeff, N. (2018). It's not the anthropocene, it's the white supremacy scene. Or, the geological color line. In R. Grusin (Ed.), *After extinction*. Minneapolis: University of Minnesota Press.
Mitchell, A. (2014). Only Human? A worldly approach to security. *Security Dialogue, 45*(1), 5–21.
Mooney, H. A., Duraiappah, A., & Larigauderie, A. (2013). Evolution of natural and social science interactions in global change research programs. *Proceedings of the National Academy of Sciences, 110*(1), 3665–3672.
Morton, T. (2013). *Hyperobjects: Philosophy and ecology after the end of the world*. Minneapolis: University of Minnesota Press.
Muir, J. (1911). *My first summer in the sierra*. Boston and New York: Houghton Mifflin Company.
National Aeronautics and Space Administration (NASA) (2019). 2018 fourth warmest year in continued warming trend, according to NASA, NOAA. Retrieved 29, April 2019, from https://climate.nasa.gov/news/2841/2018-fourth-warmest-year-in-continued-warming-trend-according-to-nasa-noaa/.
Nature (2006). Editor's summary. *Nature, 441*, 785.
National Oceanic and Atmospheric Administration (NOAA) (2019). Global Climate Report—Annual 2018. Retrieved 25, Nov 2019, from https://www.ncdc.noaa.gov/sotc/global/201813.
Nicholson, S., & Sikina, J. (Eds.). (2016). *New earth politics: Essays from the anthropocene*. Cambridge, MA: MIT Press.
O'Brien, K. (2016). Climate change and social transformations: is it time for a quantum leap?. *WIREs Climate Change, 7*(5), 618–626.
Odum, H. T. (1988). Self-organization, transformity, and information. *Science, 242*(4882), 1132–1139.
Parr, A. (2018). *Birth of a new earth: The radical politics of environmentalism*. New York, NY: Columbia University Press.
Pereira, J. C. (2017). The limitations of IR theory regarding the environment: Lessons from the Anthropocene". *Revista Brasileira de Política Internacional, 60*(1), e018.
Pereira, J. C., & Viola, E. (2018). Catastrophic climate change and forest tipping points: Blind spots in international politics and policy". *Global Policy, 9*(4), 513–524.
Quilley, S., & Loyal, S. (2005). Eliasian sociology as a 'Central Theory' for the human sciences. *Current Sociology, 53*(5), 807–828.
Rockström, J., Steffen, W., Noone, K., Persson, Å., Chapin, F. S., III, et al. (2009a). A safe operating space for humanity. *Nature, 461,* 472–475.
Rockström, J,. Steffen, W., Noone, K., Persson, Å., Chapin III, F. S., et al. (2009b). Planetary boundaries: Exploring the safe operating space for humanity. *Ecology and Society, 14*(2), 32. Retrieved 25, Nov 2019, from http://www.ecologyandsociety.org/vol14/iss2/art32/.
Rojas, C. (2016). Contesting the colonial logics of the international: Toward a relational politics for the pluriverse. *International Political Sociology, 10*(4), 369–382.
Rothe, D. (2017). Global security in a Posthuman Age? IR and the Anthropocene challenge. In C. Eroukhmanoff, M. Harker (Eds.), *Reflections on the Posthuman in International Relations: The Anthropocene, Security, Ecology* (pp. 87–101). E-International Relations.
Russill, C., & Nyssa, Z. (2009). The tipping point trend in climate change communication. *Global Environmental Change, 19,* 336–344.

Schmidt, J. J., (2019). The moral geography of the Earth system. *Transactions of the Institute for British Geographers, 44*(4), 721–734.
Seth, S. (2013). Once was blind but now can see: Modernity and the social sciences. *International Political Sociology, 7*(2), 136–151.
Smith, T. S. J. (2016). What ever happened to quantum geography? Toward a new qualified naturalism. *Geoforum, 71*, 5–8.
Steffen, W., Sanderson, R. A., Tyson, P. D., Jäger, J., Matson, P. A., et al. (2004). *Global change and the earth system: A planet under pressure.* Berlin: Springer.
Steffen, W., Richardson, K., Rockström, J., Cornell, S. E., Frazier, I., et al. (2015). Planetary boundaries: Guiding human development on a changing planet. *Science, 347,* 6223.
Steffen, W., Leinfelder, R., Zalasiewicz, J., Waters, C. N., Williams, M., et al. (2016). Stratigraphic and Earth System approaches to defining the Anthropocene. *Earth's Future, 4,* 324–345.
Subramanian, M. (2019). Anthropocene now: Influential panel votes to recognize Earth's new epoch. *Nature*. 21 May 2019. https://doi.org/10.1038/d41586-019-01641-5.
Tsing, A. L. (2015). *The mushroom at the end of the world: On the possibility of life in capitalist ruins.* Princeton: Princeton University Press.
Uhrqvist, O., & Lövbrand, E. (2014). Rendering global change problematic: the constitutive effects of Earth System research in the IGBP and the IHDP. *Environmental Politics, 23*(2), 339–356.
van der Hel, S. (2016). New science for global sustainability? The institutionalisation of knowledge co-production in Future Earth. *Environmental Science & Policy, 61,* 165–172.
Vedral, V. (2011). Living in a quantum world. *Scientific American, 304*(6), 38–43.
Waters, C. N., Zalasiewicz, J., Summerhayes, C., Fairchild, J.J., Rose, N. L., et al. (2018). Global Boundary Stratotype Section and Point (GSSP) for the Anthropocene T Series: Where and how to look for potential candidates. *Earth Science Reviews, 178,* 379–429.
Waters, C. N., Zalasiewicz, J., Summerhayes, C., Barnosky, A. D., Porier, C, et al. (2016). The Anthropocene is functionally and stratigraphically distinct from the Holocene. *Science, 351*(6269), aad2622.
Wendt, A. (2015). *Quantum mind and social science: Unifying physical and social ontology.* Cambridge: Cambridge University Press.
Zalasiewicz, J., Waters, C. N., Summerhayes, C. P., Wolfe, A. P., Barnosky, A. D., et al. (2017). The Working Group on the Anthropocene: Summary of evidence and interim recommendations. *Anthropocene, 19,* 55–60.
Zanotti, L. (2017). Reorienting IR: Ontological entanglement, agency, and ethics. *International Studies Review, 19*(3), 362–380.
Zanotti, L. (2019). *Ontological entanglements, agency and ethics in international relations: exploring the crossroads.* Abingdon: Routledge.
Zeebe, R. E, Ridgwell, A., & Zachos, J. C. (2016). Anthropogenic carbon release rate unprecedented during the past 66 million years. *Nature Geoscience, 9,* 325–329.

Ecologies of Globalization: Mountain Governance and Multinatural Planetary Politics

Rafi Youatt

Abstract Mountain governance has emerged as new node of planetary management of non-human nature. Referencing the common ecological fates suffered by mountains around the world because of climate change, it aims to fuse ecological, political, and cultural perspectives together in a new political object. One of its central organizing categories is the sacred landscape, which is neither a purely natural object nor a socially valuable one, but rather, an exceptional sacred one to be protected, conserved, and/or developed. This chapter enquires into the production of global mountains by secular governance regimes. Against a number of accounts of environmental governance, it first argues that mountains are multiple objects, not singular natural ones. It then suggests that sacredness is a particularly problematic way of rendering these multiple mountains as singular, in ways that undermine the very ends of mountain governance. Instead, in the service of finding a more politically plural, but also more multinatural, planetary politics, the chapter proposes an analytic and political focus on routes and routing, which better show the multiple mountains that are in play in these regions, drawing on the transboundary region surrounding Mt. Kailash in the Himalayas as a case study.

Introduction

Over the past few decades, mountains have emerged as a new planetary object of management and governance. While mountains have been subject to forms of governance for many centuries, this valence references the common global fate shared by mountain ecosystems around the world due to climate change, and emphasizes a shared culture among mountain peoples around the world that is seen as both threatened and a potential salve for environmental woes. Just as the "end of nature" has been declared, we are seeing a resurgence of geographical categories, like mountains, to frame the spaces of governance as global objects of concern. Mountains have thus become one in a series of environmental "globals," including global climate, global

R. Youatt (✉)
Department of Politics, New School for Social Research, 79 5th Ave, New York City, NY, USA
e-mail: youattr@newschool.edu

J. C. Pereira and A. Saramago (eds.), *Non-Human Nature in World Politics*,
Frontiers in International Relations, https://doi.org/10.1007/978-3-030-49496-4_5

oceans, and global biodiversity, which characterizes the particular ways that they are subject to forms of liberal governance—the integrated, multi-scalar management and negotiation of problems, by states, international institutions, and a range of civil society actors.

Central to this form of governance are ideas of sacred mountains and sacred landscapes. These claim to simultaneously account for the religious relationships of local people to the mountains, create exceptional conditions for political interventions on behalf of secular sustainability, and reinforce cultural pluralism as a guiding ethos for political relations. While in many ways an improvement on the devoutly secular and modernizing developmentalist projects of previous eras, this chapter argues that the pairing of the sacred and the sustainable in contemporary mountain governance nonetheless functions with its own blind spots, with both negative political and ecological consequences. In particular, it shows how the sacred landscapes and cultural values concepts tend either to over-assume a singular, transparent global nature that certain cultures are said to hold sacred, or to fall into a troubling account of cultural pluralism without nature. Instead, the chapter argues that we need to grasp the multi-faceted mountain as inassimilable fully into liberal governance thought, and it points to an imaginary of multiple mountains as a different grounding for contemporary ecological politics.

Drawing on the work of geographers, the first part of the chapter grounds this argument in an account of how mountains have been understood in environmental governance. It suggests, in particular, that they have first been pulled up to the conceptual level of global objects and then reterritorialized in regional and local mountainous spaces, infused with liberal pluralism, secular and sacred binaries, and scientific management of both culture and nature. Whereas this practice of governance tends to work with an understanding of singular mountains overlaid by multiple cultural understandings (including sacredness), the second part of the chapter suggests some important reasons why mountains might be better understood as plural objects. In thinking about multiple mountains that are partly incommensurable with one another, the implication is that global mountain governance might be more attentive to the outsides and limits of its conceptual apparatus, and to conceiving of socioecological relations on mountains that are not easily reducible to sacredness in contrast to secular governance. Mountains in this sense pose a more fundamental challenge to the politics of climate change, conservation, and development than their exceptional but marginalized sacred status in secular governance would suggest.

To illustrate the specific workings of mountain governance regimes under conditions of plural mountains, the third part of the chapter looks at a particular mountain region—the mountainous tri-corner boundary region in the Himalayas between China, India, and Nepal—where such understandings are operative.[1] It first considers the centrality of the language, politics, and practices of Mt. Kailash as "sacred" and some intersections with sustainability. Though not universally used in Kailash

[1] Kangrinboqe is the name used by Tibetan Buddhists, while Tise is the name of the mountain used by local residents. Kailash, or Kailas, is the name in Sanskrit. For purposes of readability, I use Kailash here, arguably the best known of the three names in English speaking world.

mountain governance, sacredness is frequently invoked as the justification for protection, intervention, and ecological management of mountains, while also a way of speaking for the relationships of local communities to the mountains. I suggest that sacred status is paradoxically sometimes used to dissuade us of the plural being of mountains, rather than to reaffirm it. The fourth and final section then argues instead for the utility of "routing" as part of an alternative grammar for understanding the politics of plural mountains. Routes, in a physical and abstract sense, exist throughout mountains, and are perpetuated by many different sorts of forces, including mountains, institutions, local culture and economy, global climate, religious relations, and development projects, which can together be understood as involved in routing.

Mountains in Environmental Governance

In post-1960s environmental politics and until fairly recently, mountains were generally slotted into other environmental and political spaces, such as climate change, conservation, or resource extraction, and therefore approached as something that existed on the side of the larger imperatives of security politics, global economy, and environmental governance, driven by the exigencies of nation states as a political form. In international relations, this meant a focus on the ways that international institutions and agreements could help smooth out and make agreements on issues of collective environmental concern possible among states (Young 1989; Barkin and Shambaugh 1999; Mitchell 2003). While mountain regions from the Caucasus to the Andes and the Alps certainly involved a number of geopolitical issues, and to a certain extent also involve questions about interstate cooperation, the perspectival heights of this approach have been increasingly challenged by work emphasizing the importance of multi-scalarity under globalizing forces (Brenner 2004; Swyngedouw 2004). That is, it is in the particular ways that scales connect to and transform one another—international, local, and regional—that we need to understand in order to make sense of contemporary global transformations. Methodologically, one way that this has been approached has been via a growing commitment to researching the "everyday" and "practices" (Soloman and Steele 2017) and by focusing on specific development and conservation projects where the interactions of these multiple scales can be observed (Agrawal 2005; Goldman 2005; Buscher 2013). Such an approach, sometimes inflected by ethnographic sensibilities, already suggests a way to think differently about political scales of the international—not as a level of political activity above and beyond domestic politics, which in turn encapsulates a number of smaller local scales, but rather, as a constitutive layer of places.

In policy-oriented academic work, a second framing comes from the language of global environmental governance. Here, multiple stakeholders flatten their differences into differences of values or culture, and agreed upon rules of order provide a possible scaffolding for deliberation and regulation, in order to iron over political and social fractures and inequalities (Haas et al. 1993; Keohane 2001; Esty and Ivanova

2002; Biermann 2014).[2] More specifically, this approach works within an ontological space where the sacred refers to religious values that are expressible as interests; the sustainable is scientifically knowable, including a final translation of indigenous knowledge into scientific form; and contested landscapes are understood as shared spaces upon which conflicting values supervene. As such, mountains come to be framed in liberal governance as a singular natural space containing different values (Cadena 2010). Mountains have often been translated here into the language of state and global governance, but in ways that mean that there does not always appear to be an outside—all values are included. This appearance, as suggested below, is misleading, as the conditions of entry actually require bracketing off significant aspects of lived worlds that matter for environmental issues.

But what, precisely, is it that makes mountains global? Mountains have unique features as a geo-governance formation compared to their other planetary geo-cousins. They are different sorts of global objects than the global climate, which has an emergent global systemic quality and is arguably a new kind of "planetary real" (Burke et al. 2016, p. 501). Global climate works as a whole, or as a system, with emergent effects that push downward, while global mountains largely do not have this cohesive, effect-creating quality. Global mountains are also different kinds of objects from global biodiversity, which is a pattern of species richness and distribution seen at a global level, but is also a scalable concept that can refer to scales as small as a handful of dirt (Mitchell 2016).

Mountains are global in a different way, and in particular, in a way that is emblematic of a different aspect of globalizing processes. As geographer Eric Swyngedouw (Swyngedouw 2004) has noted, globalization, both economic and political, is not just the organization of processes at a global scale. Rather, he suggests it should also be understood as a process of reterritorialization that involves intensifying regionalization and even localization through the language of the global. By first pushing sociomaterial processes upward to the language of the global, Swyngedouw suggests that existing local ties are broken; but because globalization cannot easily sustain itself purely at the level of the global, it touches back down, settling into new visualizations of the regional, the transboundary, and the local. It is in this way that mountains are being forged as global. Among the key effects that globalization of this kind has in mountain governance are a working political vision of larger regional areas as "mountainous" (often via a re-emphasis of pre-existing geographical categories), and a vision through the idea of "mountain people" within a particular region around a presumed set of commonalities (Debarbieux and Rudaz 2015, p. 139, 140). While the natural boundaries of the mountain and scientific knowledge about the space are a guiding parameter, the emphasis here is on the meaning-making practices of local communities who may not share deep cultural ties. For example, Gilles Rudaz (2009) examines the way that the Caucasus Mountain Alliance links eight mountain

[2]Those more critical of global governance for its apparent abdication of politics in favor of management and its generally uncritical stance on power relations include (Paterson 2000; Gill 2003; Duvall and Barnett 2005); more specific critiques of global environmental governance as green governmentality include (Luke 1999; Goldman 2005).

villages in four countries (Azerbaijan, Georgia, Russia, and Armenia) together under the rubric of development. Here, as Rudaz notes, mountain governance is now explicitly set up not by cultural ties, but rather by "the fact of belonging to a community of problems…, related to belonging to a 'mountain' environment or communities" (Rudaz 2009, p. 29). This politics of rescaling, then, is the key to understanding the global mountain.

But if this account of the way mountains are globalized is right—that they are first scaled upward to the global, and then reterritorialized in somewhat new regional and mountain community form, what is the content of that reterritorialization? Alongside geopolitical and transnational considerations, the language of the "sacred" is a key node here. Although mountain governance projects deal with a wide range of issues, both social and ecological, one particularly important element has been a pronounced turn toward using the sacred to underpin their special status, and their intersections with the sustainable. Initiatives aim to preserve and catalogue the world's *sacred mountains* (Bernbaum 1997), collect the world's *sacred landscapes* (UNESCO), and name global *sacred natural sites* (Verschuren et al. 2010). Working across many religions, sacredness here means a set of values that are religious or spiritual in nature, and that do not inhere in the landscape (except for target populations), so much as supervene on it. For example, writing about sacred mountains around the world, the mountain writer, climber, and scholar Edwin Bernbaum (Bernbaum 2010, p. 33, 34) writes: "Because of the way they stand out from the surrounding landscape and soar up toward the sky, mountains have a natural tendency to attract attention and inspire visions of another, higher realm of existence…. In that aura of mystery and power, people of diverse cultural backgrounds sense the presence of something of ultimate value that gives meaning and vitality to their lives. As a consequence, many cultures and traditions regard mountains as sacred places of special significance…" This approach follows the religious gaze to the mountain peaks themselves.

In the case of the tri-corner boundary region discussed later in this chapter, Mt. Kailash is the peak at which religious pilgrims, sacred art, and culture train their gaze. Having arrived at Mt. Kailash, one Hindu pilgrim "saw the shape of Om in every pebble around Lake Mansarovar."[3] The circumambulation of Mt. Kailash itself is particularly important, with pilgrims moving in opposite directions (clockwise for Hindu, Jain, and Buddhist; counter-clockwise for Bonpo) (Wollmer 2014). Finally, the mountain peak—in the singular—tends to dominate mountaineering perspectives, with their focus on peaks, and to the extent that such approaches have informed conservation and development politics (e.g., Everest), even a broader view of the mountain in the region still has a peak-focus. For all these reasons, the mountain itself is one important site of study, raising questions about the rising impact of tourism and development on both people and nature, and asking questions about what multi-religious pilgrimage looks like in practice.

Yet focusing solely on the mountain as a sacred site—a site that is almost over-invested with conflicting meanings, to the point that mountain seems obscured as much as illuminated—risks hiding the broader sets of interconnections that make the

[3]Personal interview.

mountain possible.[4] *Sacred landscapes* is a framing device that aims to understand the interconnections in the broader area as a kind of shared zone of religious values, one tending toward a new mode of sustainable development. While sacred mountains are part of a sacred landscape, and act as a linchpin for the greater area, sacred landscapes de-emphasize the mountain itself, in favor of identifying a series of nodal points within a broader region, or socioecological ("biocultural," in development worlds) processes that are shaped by sacred (meaning religious) processes. But like the sacred mountain, sacred landscapes produce a set of inclinations that are worth unpacking.

Perhaps a more accurate way to put it is that the sacred landscape is the product of secular governance's reframing of elements it cannot directly capture, with two consequences. The first is that it sets up an indirect process of partial reincorporation on the grounds of liberal pluralism. The sacred landscapes concept does not necessarily relegate religious activities or individuals to a mundane, if nominally celebrated, sphere of life, but it does reframe and reincorporate them on liberal secularist terms, as a problem of competing *values* on a common plane.[5] The second is that it creates an historical version of the secular itself. As Asad (2003, p. 25) notes, "the secular is neither singular in origin nor stable in its historical identity, although it works through a series of particular oppositions." Rather, it shifts over time through affiliations and particular grammars. In the case of mountain governance, its particular grammar includes sustainability via livelihoods, biocultural diversity, and sacred landscapes, overlaid with a global, geographically oriented socioecological management perspective.

Where, then, does this critique of sacred landscapes leave the mountain itself? The next section turns to the ontological questions underlying the global and planetary politics of mountains, which are key to the imaginary of mountain governance, and to its alternatives.

What are Mountains?

Mountains are supposed to be obvious things. Like rocks, gravity, and seminar tables, they tend to be invoked as paradigmatic exemplars of brute facts. They are perennial and unchanging, at least relative to human time scales, and their size, visibility, and solidity suggest a place where human lives, values, and meanings are at best secondary stories evoked by mountains. Yet in a number of ways that are interesting and politically significant, mountains are also changing and unstable sorts of entities,

[4] A focus on Mt. Everest, for example, hides the complexities of tourist economies, Sherpa culture, and the fabric of social life that exist beyond the mountain (Sherpa 2014).

[5] In the context of deep political divides in India and Nepal about the proper role of secularism in *state* politics (if any), the role of sacred landscapes in multi-layered governance narratives (which includes parts of those states) is a particularly complex one. As Sen (2005) has argued in Indian context, a different version of secularism is where the state is in *equal relationship* to multiple religions (i.e., Muslim law has legal standing), rather than a strict separation of church and state.

historically and politically contingent as a class of objects (emerging as what Latour (2004b) calls "matters of concern"—things that form central points of contestation in knowledge controversies, and are perhaps even best understood as multiple objects rather than singular ones).

Before getting to the ways their materiality weighs on social life, we should start with the idea of mountains. Mountains are an idea with a perspective—that of flatlanders, particularly city dwellers, looking at high elevations from a distance and fixed point (Debarbieux and Rudaz 2015, p. 4).[6] The mountain is what is up, based on a vantage point from below, and thus dependent on the location of the observer. Though natural scientists have rooted around for decades in search of an objective definition of mountains based on absolute elevation cutoffs, steepness, and vegetation zones, they do not seem to have succeeded in agreeing and generally seem to have given up trying. Much like the concept of species in biology, mountains are operationally functional, even as they are definitionally problematic. Meanwhile, public and vernacular usages regularly refer to nearby rises of only a few hundred feet as mountains. Mount Beacon, near New York City, is a perfect example, with a total height of 1600 ft; Mount Davidson in San Francisco is a local highpoint, at all of 980 ft.

As a number of recent accounts of mountains have suggested, including Debarbieux and Rudaz's powerful recent (2015) book *The Mountain*, the fact that mountains exist independently of human thought and phenomenal experience is itself an historical idea, rooted in the emergence of eighteenth-century Enlightenment natural sciences and politics.[7] Through the sciences of geology and geography, and through the slow political projects of empire and colonialism, they show that mountains "became a category of comparable and commensurable physical objects characterized by a set of attributes, all purportedly objective and hence independent of the particular points of view from which the inhabitants of one place or another might see and describe their surroundings" (Debarbieux and Rudaz 2015, p. 15). As natural objects, mountains were brought into new governance regimes within Europe, including state-led resource management schemes; scientific study of alpine habitats with unique flora and fauna (and sometimes people); and new modes of imagining the nation with mountains at its heart.

The invention of *altitude* as an objective measure also had significant consequences, leading to wide attention to the highest summits from scientists and mountaineers. In the context of colonial projects of territoriality, mountain sciences were modes of standardization that not just permitted but constituted colonial expansion. At the same time, they also involved a shift in the way that mountains were understood in Europe and America. On this telling, mountains were previously popularly conceived in relational terms, based on contingencies of use or cultural perceptions, before European natural historians and geographers enacted a radical break with these pre-scientific views by ushering in the mountain as objective entity. In the process, mountains lost a variety of local human meanings—economic use,

[6]See also Ruggie (1993) on single point perspective and territoriality.

[7]See also MacFarlane (2003).

sacred personhood, locally relative quality—and became defined instead by their "proper" characteristics (Debarbieux and Rudaz 2015, p. 15, 16). Just as the modern state started to move from a political project oriented around sovereign control and territoriality, and toward governmentality and biopolitical forms of control, European mountains were recast as objects of natural science, acting as "laboratories" of modernity and crucibles of "experimentation" for modernist projects, both natural and social (Debarbieux and Rudaz 2015, p. 9, 46).

While powerful in opening up a better account of the competing field of meanings around mountains, this account involves a number of difficulties. It first involves a problematic subscription to a kind of global history that emanates slowly outward from Europe. It locates the idea of the mountain fully in the decisive break of Euro-American modernity, and then continues to follow its globalization in regrettable but nonetheless evident fashion through colonialism and empire, before concluding by examining contemporary forms of global mountain governance. While such a progression is significant as one history of the mountain in Enlightenment thought and politics, it risks creating a history of mountains as a global object in which large swathes of humanity appear, in Wolf's (1982) memorable phrase, as "people without history."[8] While Europeans were exploring the world's mountains, history and politics were happening in those places, with their own modes of governance and figurations of mountains. Sometimes those came into contact with Euro-American ideas of mountains, particularly in colonial periods, yet it would be a mistake to assume a "pebble in the pond," outward concentric circles model of idea diffusion.

In this context, mountains cannot just be brute reality around which humans create shifting representations, constructions, or rescalings. Rather, their "brute reality"—if it even exists—is instead inaccessible in any definitive way. Chasing what a mountain is, "in and of itself," misses the mark, in that it denies the inherent relationalities between humans and mountains that could form the starting point of inquiry, and in that it denies the possibility that this relations might be made of something other than cultural valuation of an object.

Such a telling also makes it difficult to understand a deeper political challenge that thinking about, with, or for mountains can pose. If we understand mountains as figures that bring both material and semiotic aspects together,[9] then we can start to see that there are literally multiple mountains. These multiples may or may not be *commensurable*—that is, they may not exist side by side and be compared against

[8]The critiques mentioned here are, lamentably, not new. Starting at least with Edward Said (1979), waves of postcolonial scholarship have made such points, and written alternative histories, including (Chakrabarty 2000; Ling 2002; Muppidi 2004).

[9]Figures are like "constructions" but they do not presume an outside, independently existing thing called mountain that is then constructed. The figure itself is the material-semiotic terrain (Haraway 1988; Law 2007). While we might well want to slip into working on the *idea* of mountains, in ways that are partly decoupled from figures, such work does not imply a single thing called mountain in the background of multiple ideas of a mountain.

one another on a single plane.[10] Such an evaluation depends on a reading of there being multiple cultures operating on a single nature—a mononaturalism and a multiculturalism, in Latour's (2004a) terms. For example, a mononaturalist view would see the existence of Mt. Kailash from a Nepali Hindu perspective, a Tibetan Buddhist perspective, and a syncretic Hindu-Bonpo perspective, all against the backdrop of a single physical mountain. But the precise commensurability or incommensurability of these multiple ontologies—which we might stylize as those of secular global governance and those organized along multiple religious lines—is not only a theoretical question, but also an empirical matter of social research, even if it is true that not all ontologies are precisely and transparently knowable. To hold open the possibility of incommensurability of these worlds, we might at least approach the agency mountains in mountain governance as a question to be answered, rather than a starting point of analysis (Mitchell 2002, p. 53)—what happens when we engage the agency of mountains-as-persons in the social and political lives of pilgrims and residents of the area, in ways that secular governance cannot entirely admit to?[11] In particular, how would it matter for questions about development, heritage, and sustainability in an important area of the Himalayas?

Here I therefore suggest a slightly different version of mountain as agent, one that does not itself take up alternative human/non-human relations through forms of religious personhood, which I leave to those more deeply versed in specific particular regions, religions, and cultures. Instead, this chapter proposes that we should find ways to *understand the agency of mountains as figures that emerges in relation with global mountain governance thought*. In what ways might finding these kinds of agencies thus contravene the working idea that mountains are part of a natural landscape with sacred values attached to them? Might taking the agency of mountains more seriously lead not just to alternative accounts of personhood, but also to rethinking the secularist view of mountains as natural objects? The ontology of mountain governance, too, is partly incommensurable with others, and it too has room for equivocations of its own, but it does not yet entirely accept that it has an outside. Liberal pluralism and singular mountains need not necessarily be either starting points or closed systems. Through this, we might also complicate usefully the stark lines that are sometimes drawn between the political ontologies of secular and indigenous cosmologies.

---

[10]On the problems of commensurability, see Povinelli (2001), Asad (2003), Cadena (2010), Blaser (2012). Some of these problems as they relate to IR are developed in Blaney and Tickner (2017), Youatt (2017).

[11]See Chakrabarty (2000), which prods us to think about the agency of spirits, divinities, and gods in peasant rebellions in India, in a more literal sense.

The Greater Kailash Transboundary Region: A Case Study

One area of particular interest for such projects is the greater transboundary region around Mount Kailash.[12] The mountain is best known outside the region as an emerging site of pilgrimage for at least four major religions (Hindu, Buddhist, Bonpo, and Jain), attracting pilgrims from multiple countries to circumambulate, but never climb it. However, the greater Kailash transboundary region is arguably an indeterminate area until it is fixed politically, and the three countries, major intergovernmental organizations, and local peoples all view its boundaries differently, if indeed they see it as a coherent region at all. While Mt. Kailash lies culturally at the core, the greater Kailash transboundary region has pilgrimage routes that extend like tendrils deep into each of the three countries, fanning out and dissolving into multiple starting points. Moreover, the relationship between the physical Mt. Kailash and multiple ideas of Mt. Kailash in different religious traditions, as they have evolved over time, is also contested (McKay 2015). Finally, as a space of global mountain governance, this area involves politically difficult questions about religious tourism and heritage; emerging experiments in conservation, development, and conservation-led development; questions about the effects of climate change on people and nature (Sherpa 2012; Wapner 2014); and the longer political histories of the relations of mountainous spaces to state capitals and colonialism.

Before turning to these issues, a brief overview of the region is helpful. The region is politically contested by nation states, though it is also one of the few sites of explicit border cooperation between and India and China. From an international perspective, it lies at the three-way intersection of India, Nepal, and China. While Mt. Kailash itself is in the Tibetan Autonomous Region of China (TAR), the pilgrimage routes extend into India and Nepal, both by land, and, increasingly, by air. This sacred geography intersects uneasily with the geopolitical geography. At the center of this geopolitical triangle is the contested tri-corner boundary of Lipulekh Pass, which permits north-south passage from the Tibetan Plateau to the high hills of the Indian state of Uttarakhand. This region is also the headwaters of the Kali River, which, since the Sugauli Treaty of 1862, has marked the ambiguous India–Nepal border. The highest headwaters of any river tend to dissolve into a fan of small springs and rivulets, and creeks, and the border near Lipulekh Pass is no exception, leaving it open to be contested—waves of British surveyors since the Sugauli treaty marked the head of the Kali River in different places. While Nepal insists the border is marked by the headwaters of the river, India claims that the ridgeline to the east of Kalapani is the boundary. The Indian security forces (ITBP) have controlled the territory since the 1962 Sino-Indian War, and the pilgrimage route to Mt. Kailash through Lipulekh Pass goes through this area, with a temple to Kali and a pool in front of it sometimes considered to be the headwaters of the Kali River. In a recent controversy, a Maoist student political group from Nepal marched to plant the Nepalese flag further west, claiming that the boundary was improperly identified;

---

[12] I have also learned greatly from Mukta Tsing Lama, Tsewang Lama, and Shekhar Pathak. See Lama (2002), Pathak (2019).

the Indian government responded by tightening security at all the border crossings between India and Nepal. On the ground, this tightening meant that the Nepali day laborers on whom the Rung communities on the Indian side rely for cheap labor were unable to cross easily.

While the region can be understood as the periphery of the three nation states, transnational affiliations in place are a second set of connections that also define this region. The region is composed of a patchwork of peoples who are both ethnically distinct and deeply interrelated as trading partners and laborers. Syncretic religion is one key marker of the area, with local deities intermixing with Hindu gods in the Indian and Nepalese side, with Buddhist elements creeping in the closer one gets to the Chinese border. Transnational affiliation also describes the long-standing trading that has gone on this area, in all directions. The closure of the India–China border after the 1962 Sino-Indian War has diminished the trade between Rung and Tibetan peoples, but trade continues (more circuitously) via Nepal. Economic migration also continues—not just regionally, but also in ever larger national and global circuits. Peoples also move from one side of the Nepal border to the Indian side with regularity, often for seasonal work, including the provision of services for Kailash pilgrims and the picking of naturally growing cash crops like *yarsagumba*, a caterpillar–fungal hybrid valued for its aphrodisiac qualities in China.[13] In short, this view of the region sees it linked not by geopolitical rivalry, but by a denser web of interaction among peoples with lengthy historical ties. In the absence of an explicitly claimed collective identity, it begs the question of why and where a line might be drawn around this particular region. And, the transnational linkages here remain constituted by the geopolitical borders, which, rather than receding in importance, remain intermittently powerful presences.

In this vein, one particular way this region has been framed in mountain governance is as the Kailash Sacred Landscape, an area spanning approximately 18,500 miles, and covering a high-altitude area of the Himalayas and the peak of Mt. Kailash itself (over 21,000 ft.). Designated by the International Centre for Integrated Mountain Development (ICIMOD) as a conservation and development landscape, it is also under consideration as a possible UNESCO World Heritage site. This policy framing of the region deploys a number of important grammars of sacredness, sustainability, and conservation-led development. As one report on the KSL puts it, in the inimitable language of development, "sacredness values attached within the landscape elements have been, from time immemorial, the driving force for the existence of some of the most natural sites, and time tested traditions of management and sustainable harvest of natural resources."[14] This version of the sacred landscape, then, is as a complex product of secular development and conservation projects, producing a certain set of inclinations, both about its object of knowledge and within societies that see themselves as distant from such landscapes.[15]

---

[13]On yarsagumba in Tibet, see (Yeh and Lama 2013), Stewart (2014); also Negi (2006).

[14]India KSL Feasibility Report, preface, p. 1.

[15]On secularism as historical formation rather than general political principle of government, see Asad (2003).

Routes

To better grapple with the multiplicity of mountains and the problematics of sacredness, this section suggests the idea of *routing* as a part of larger grammar of power that emphasizes mobility as its central empirical space. It is explicitly agnostic about the "kinds" of things that can engage in routing—anything can route, whether material, virtual, divine, political, as long as the networks of things involved in producing it can be delineated. In this frame, multiple mountains are starting points for investigation of routing, rather than an obstacle to be formatted and singularized. Routes are central to mountains in general, given the way that the geography tends to direct movement along shared pathways. Given the historical centrality of routes to this area of the Himalayas specifically—such as trade routes (Harris 2013), pilgrimage routes, road building efforts, and snow leopard routes, among many—an analytic of routing is also particularly apt.

A route is, in its most common usage, the way that someone or something gets from a starting point to an endpoint. It is generally a regularized pathway, as routes are traveled repeatedly, though it may be flexibilized, as in onboard navigation tools. Some routes have existed for centuries, while others can be established quickly and provisionally. A related sense of route is as a method or process leading to a particular result—the many routes to health, or salvation, or the route to a balanced federal budget. This meaning of route suggests a road with side paths that should not be taken, though individuals may always deviate from them.

Route is also a verb, which means "to send or direct along a specified course." When one is routed, one is steered or sent. Routers for the Internet draw on this sense of the word, selecting among the best paths within a network to move packets of data along. Routing can take place without a central authority, as in the development of wildlife routes over mountain passes, or it may have a directing agent, such as a traffic warden rerouting traffic, or it may be highly directed, as in the establishment of pilgrimage routes by the state.

In both senses, routes mean enabling movement or mobility, but also limiting, directing, or setting of endpoints. Rather than understanding mountainous areas as a region, then, which appears only in the governance frame, and not as a shared identity or nascent nation state, its political geography might be better mapped as routes and mobilities of people, things, capital, and species across space.[16] Following Sheller (2011, p. 1, 2), we might look at this spatial mobility not on its own, but also by asking about the "means for movement"—which infrastructures, systems, vehicles, linkages allow mobility, and for whom—and the "circulation of information, images, and capital." It is not just movement, but also the impediments to movement, forms of dwelling and place-making, and immobilization that are crucial for investigation. In sum, routes are a form of mobility regime that can enable or disable movement, redirect it, or channel it.

[16]Mountain governance and other global governance projects have moved in this direction themselves, emphasizing routes in heritage preservation, such as the Silk Road Project, which helps foster certain kinds of mobilities—tourists and capital, in particular—in the name of past mobility.

Unlike a focus on geopolitics in the region, routes do not take the nation state as the starting point of analysis, nor do they assume that routes are inherently nationalized, though state power and authority is certainly important to routes, and route-making has been central to making states as well. At the same time, against a strong regional transnationalism that might seek to emphasize cultural connections between groups in the region as its endpoint, a focus on routes does not assume the normative priority of these associations, as they jostle with other forms of politics, economy, and identity that are national and global. Routing is thus a multi-scalar activity. Routes are localized, but the forces of routing can come from afar—an important question is to ask how are these distances bridged, and how power works from a distance.[17] Finally, routes and routing are particularly open for thinking about the sources of agency, human and non-human, in creating socioecological worlds, and in themselves creating the things that pass through them. As Elizabeth Povinelli (2011) puts it:

> …things do not simply move. Routes *figure* space—they create worlds—and are figured by figurated space, by the worlds through which they move. They are the condition of the previous circulatory matrixes and become part of the matrix that decides which other things can pass through and be made sense of within this figured space. And routes configure things—they shape them into 3D manifestations. What "things" are—what counts as an entity—should be understood broadly. Whether container ships, kin or stranger socialities, psychic expectations, affective intensities, linguistic forms: all form, conform, and deform existing cultures of circulation. Social institutions, or 'demanding environments,' emerge around these material and affective curvatures, effectively controlling the further fabrication of things and their movements.

Routes, then, are not just pathways for pilgrims, tourism dollars, local livelihoods, though they are that; the process of routing also actively shapes those very things.

To take one example, one pilgrimage route, used primarily by a smaller group of state-sponsored Hindu pilgrims from India but also doubling as route for local Rung people, animals, and goods, runs from Darchula (India) to Lipulekh Pass. It goes largely along the west (Indian) side of the Kali River, before crossing into the Tibetan Autonomous Region (TAR) of China and going up to Mt. Kailash. The route, which runs across the closed India–China border, was recently reopened, after negotiations between India and China. While this route is sometimes portrayed as an ancient path that Hindu pilgrims have traveled for hundreds of years, the routing forces here are more varied and more contemporary, and the mountain they reference is not a singular one.

Fused with ethno-religious nationalism, sacred markers along the route are not only a means of marking an ancient route, but also for asserting a Hinduized state

[17] Scale considerations are particularly important—though routing is spatial, it is not localized. Thus, multiple scales come to bear in routing of Hindu pilgrims from India on their way to Mt. Kailash—global tourism, ethnonational religion, heritage politics, and transnational labor circuits. This "flatter" ontology means taking larger "scales" as heterogeneous networks and associations of things that are often sustained over time (Marston and Woodward 2005; Latour 2007). While this has its imperfections as a method, it allows a greater questioning of the operation of multiple scales in a research site, and the generative interaction of human and non-human things.

authority over territory. For example, the painted signs and spray-painted rocks along the pilgrimage route along the Kali River to Lipulekh Pass, written in both Hindi and English, exhort Hindu pilgrims onward with spiritual messages—they are painted onto the rocks and maintained by India's Public Works Department. The Kali temple at Kalapani likewise lays claim to the headwaters of a sacred Hindu river.

Pilgrims move through this transboundary space only through significant state and interstate efforts to both make their mobility possible and to regulate it. The newly reopened Lipulekh Pass pilgrimage route involves a lottery system in India, the accompaniment of pilgrims along the route through Uttarakhand by Indian authorities, and the transfer of pilgrims to Chinese authorities at the border, who accompany them to their circumambulation around Mt. Kailash. For other pilgrims, the transboundary quality of the route comes increasingly through skipping over large parts of the terrestrial elements of the pilgrimage, taking increasingly faster hops through air travel to Kathmandu or Lhasa, and then going by helicopter and jeep, through new pilgrimage routes. At the same time, many pilgrims are unable to engage in transboundary pilgrimage at all—those unable to secure permits to cross borders and those who do not win the lottery or have sufficient funding. What was once a once-in-a-lifetime (usually end of lifetime) journey that would take two months can now take as little as a few days for those with adequate resources and inclination.

Other routing forces are at work along this path, too. The Kali River (which also marks the India–Nepal border) has carved the routes that the pilgrimage routes now follow, at a geologic time scale. Though the Kali River itself does not lead to Mt. Kailash, the mountain is understood as the source of the major rivers of the region—Indus, Sutlej, Brahmaputra, and Karnali—both in an ecological sense and a cosmological one, with the river Ganges falling onto the head of the Hindu god Shiva, who is said to reside at Kailash, and forming braids.

It is interesting to note that these rivers mostly appear to be considered as sacred as the Ganges by the local populations, which is not true in the "pan-Indian" context—as an ICIMOD KSL report notes (121):

> "Like mountains, the rivers in Pithoragarh region also have a special place in the life of local population. The rivers of the landscape have been playing significant role in shaping the cultural life of the landscape. Unlike river Ganga, these rivers did not receive much reverence in pan Indian context, however they are considered as sacred as Ganga by people living in their valleys. For them each tributary is synonym of Ganga and carries equal sacred notion."

Understanding Kailash to be the center of a sacred landscape is a flatlanders geography—a vision that sees rivers in an upstream direction, leading to its divine source. But in other ways, for those living in the region, the downstream view of the Kali River is a more potent one, including its capacity to route and re-route.[18] In June 2013, landslides and flashfloods involving moving walls of water and soil killed nearly 6000 people in the state of Uttarakhand who were unable to evacuate, both local residents and visiting pilgrims, and including a number of people harvesting the highly prized medicinal caterpillar–fungus *yarsagumba.* This led both to calls

[18] Acciavatti (2015).

for early warning systems and a questioning of development projects, including local dams, that appear to have contributed to the flooding. The river itself, too, routes by forming significant obstacles to mobility at the high elevations, even as it enables transportation mobility further downstream. Because bridges are few and far between, the river is sometimes crossed by people and by goods, via rigged ropes systems. Bridges at the border take on added power as well, given the efficacy of the river in keeping people separate.

Finally, responses to climate change are one significant challenge that mountain governance aims to tackle, where it encounters the power of hydrological routing and rerouting: the melting of glaciers that have sustained large populations in areas below means that the status of these mountains as the "water towers of Asia" will change and the change in habitat conditions will increase the likelihood of massive landslides, the range of snow leopards, and the availability of new ranges for livestock and medicinal plants.

Conclusion

To conclude, this chapter has suggested that mountain governance, as an emerging node of planetary governance, has moved further toward the making of transnational and regional objects, of which the Kailash case is one example. This "rescaling" has involved a two-step dance. First, it has involved the positing of a global class of objects of concern called mountains, in the decades since the 1992 Rio Earth Summit, partly in an effort to go around the state through governance arrangements. Second, it has meant a series of reterritorializing moves, in which mountains (and increasingly, mountain ranges) have emerged as the adequate level of specificity for governance activities. Unlike the global climate, which continues to exist in more or less singular terms, global mountains do not hold such a coherence, and have resorted back to more regional geo-forms, such as ranges and peaks, and a vocabulary of sacred exceptionalism. Understanding the politics of particular mountains like Mt. Kailash, then, means taking account of this rescaling move—both within global governance, and in the effects that such rescalings have in the circulations of "everyday mountain governance," where multiple mountains appear more powerfully. By contrast, by focusing on routing, and the kinds of forces involved in it, we can rethink what routes in mountains are. They are not just material pathways, but places with multiple entities and forces involved in steering and even shaping the very things that move through them.

What are the advantages to this understanding of mountains as multiple? One benefit is that it arguably moves us closer to a better empirical understanding of both the politics and ecology of mountains spaces, and of planetary questions more broadly. Rather than reading downward from a planetary scale that is itself already infused with particular normative commitments and empirical frames, it builds up a different picture of the planetary that is rooted in a combination of partly overlapping worlds and partly unknowable worlds. In each of these worlds, there are ways to understand relationships with nature differently, but these cannot be stipulated in a single language, nor made fully commensurable in a single frame. They must be worked out, through forms of political communication and strategic equivocation, and through acceptance of the elements and forces that move some but not others.

Second, the implication of this analysis is not that mountain governance ought to take more factors into account. Indeed, as Povinelli has suggested, it is first and foremost a question of understanding how governance conceives the inconceivable, and to continue to insist that there is an irreducible outside to global mountain governance, no matter what its inclusionary logics are. The multiplicity of mountains does not suggest a specifically different approach to doing mountain governance—rather, given the existence of this irreducible outside, it is not a question that should be asked with the aim of resolving it, but instead about coming to understand its limits.

Finally, this seems to me a particularly promising place for a planetary politics, in spite of the discomfort it provokes for its step away from a singular nature, on one hand, and in spite of its skepticism about the possibility of using the "sacred" as a grounds for political change, even as a matter of strategic essentialism. Planetary politics desperately needs alliances and intersections, not fragmentation into mutually inaccessible boxes nor a totalizing politics, but such alliances and intersections cannot be known in a single plane. Perhaps then can we start to understand Aldo Leopold's (1949) famous question—how to "think like a mountain"—differently, and begin to think and learn with the mountains, as many as there may be, and work to reverse the brutal degradation, flattening, and reformulation of natures and cultures in the Anthropocene.

Acknowledgements The author is thankful for support from the India–China Institute at the New School, and the Graduate Institute for Design, Ethnography, and Social Thought. Michael Dobson, Sarah DeMartino, and Iver "J" Johnson provided invaluable research assistance.

References

Acciavatti, A. (2015). *Ganges water machine: Designing New India's ancient river.* Applied Research and Design Publishing.

Agrawal, A. (2005). *Environmentality: Technologies of government and the making of subjects.* Raleigh: Duke University Press.

Asad, T. (2003). *Formations of the secular: Christianity, Islam, modernity.* Stanford: Stanford University Press.

Barkin, J. S., & Shambaugh, G. (1999). *Anarchy and the environment: The international relations of common pool resources.* Albany: SUNY Press.

Bernbaum, E. (1997). *Sacred mountains of the world.* Berkeley: University of California Press.

Bernbaum, E. (2010). Sacred mountains and global changes: Impacts and responses. In B. Verschuuren, R. Wild, J. A. McNeely, et al. (Eds.), *Sacred natural sites: Conserving nature and culture* (pp. 107–123). Washington, DC: Earthscan.

Biermann, F. (2014). The anthropocene: A governance perspective. *The Anthropocene Review, 1*(1), 57–61.

Blaney, D., & Tickner, A. (2017). Worlding, ontological politics and the possibility of a decolonial IR. *Millennium, 45*(3), 293–311.

Blaser, M. (2012). Ontology and indigeneity: On the political ontology of heterogenous assemblages. *Cultural Geographies,* 1–10.

Brenner, N. (2004). *New state spaces: Urban governance and the rescaling of statehood.* Oxford: Oxford University Press.

Burke, A., Fishel, S., Mitchell, A., et al. (2016). Planet politics: A manifesto from the end of IR. *Millennium, 44*(3), 499–523.

Buscher, B. (2013). *Transforming the frontier: Peace parks and the politics of neoliberal conservation in Southern Africa.* Durham: Duke University Press.

Cadena, M. (2010). Indigenous cosmopolitics in the Andes: Conceptual reflections beyond 'Politics'. *Cultural Anthropology, 25*(2), 334–370.

Chakrabarty, D. (2000). *Provincializing Europe: Postcolonial thought and historical difference.* Princeton: Princeton University Press.

Crews, C. (2019) Social movements and earthbound people: Towards a new politics of the Earth in the Anthropocene. Dissertation, New School for Social Research.

Debarbieux, B., & Rudaz, G. (2015). *The mountain: A political history from the enlightenment to the present.* Chicago: University of Chicago Press.

Duvall, R., & Barnett, M. N. (2005). Power and global governance. In R. Duvall & M. N. Barnett (Eds.), *Power and global governance* (pp. 1–32). New York: Cambridge University Press.

Esty, D., & Ivanova, M. (2002). Revitalizing global environmental governance: A function-driven approach. In D. Esty & M. Ivanova (Eds.), *Global environmental governance: Options and opportunities* (pp. 181–204). New Haven, CT: Yale School of Forestry & Environmental Studies.

Gill, S. (2003). *Power and resistance in the new world order.* New York: Palgrave Macmillan.

Goldman, M. (2005). *Imperial nature: The World Bank and struggles for social justice in the age of globalization.* New Haven, CT: Yale University Press.

Haas, P. M., Keohane, R. O., & Levy, M. A. (1993). *Institutions for the earth: Sources of effective international environmental protection.* Cambridge, MA: MIT Press.

Haraway, D. (1988). Situated knowledges: The science question in feminism and the privilege of partial perspective. *Feminist Studies, 14*(3), 575–599.

Harris, T. (2013). *Geographical diversions: Tibetan trade, global transactions.* Athens, GA: University of Georgia Press.

Keohane, R. (2001). Governance in a partially globalized world. *American Political Science Review, 95,* 1–13.

Lama, T. (2002). *Kailash Mandala.* Humla Conservation and Development Association.

Latour, B. (2004a). *Politics of nature: How to bring the sciences into democracy.* Cambridge: Harvard University Press.

Latour, B. (2004b). Why has critique run out of steam? From matters of fact to matters of concern. *Critical Inquiry, 30*(Winter), 1–17.

Latour, B. (2007). *Reassembling the social: An introduction to actor-network theory.* Oxford: Oxford University Press.

Law, J. (2007). *ANT and material semiotics.* Retrieved July 30, 2019, from http://www.heterogeneities.net/publications/Law2007ANTandMaterialSemiotics.pdf.

Leopold, A. (1949). *A sand County Almanac.* Oxford: Oxford University Press.

Ling, L. H. M. (2002). *Postcolonial international relations: Conquest and desire between Asia and the West.* Basingstoke: Palgrave Macmillan.

Luke, T. (1999). Environmentality as green governmentality. In E. Darier (Ed.), *Discourses of the environment* (pp. 121–151). Malden, MA: Blackwell.

MacFarlane, R. (2003). *Mountains of the mind.* New York: Pantheon.

Marston, J., & Woodward, S. A. (2005). Human geography without scale. *Transactions of the Institute of British Geographers, 30,* 416–432.

McKay, A. (2015). *Kailas histories: Renunciate traditions and the construction of Hiamlayan sacred geography.* Boston: Brill.

Mitchell, T. (2002). Rule of experts: Egypt, techno-politics, modernity. Berkeley: University of California Press.
Mitchell, R. B. (2003). International environmental agreements: A survey of their features, formation, and effects. *Annual Review of Environmental Resources, 28,* 429–461.
Mitchell, A. (2016). Beyond biodiversity and species: Problematizing extinction. *Theory, Culture, and Society, 33*(5), 23–42.
Muppidi, H. (2004). *The politics of the global.* Minneapolis: University of Minnesota Press.
Negi, C. (2006). Yar tsa gumba (Cordyceps sinensis): A call for its sustainable exploitation. *International Journal of Sustainable Development and World Ecology, 13,* 1–8.
Paterson, M. (2000). *Understanding global environmental politics: Domination, accumulation, resistance.* New York: St. Martin's Press.
Pathak, S. (2019). Kailas-Manasarovar sacred landscape: Understanding a multicultural trans-boundary region. *Economic and Political Weekly, 54*(10), 31–40.
Povinelli, E. (2001). Radical worlds: The anthropology of incommensurability and inconceivability. *Annual Review of Anthropology, 30,* 319–334.
Povinelli, E. (2011) Routes/Worlds. *E-Flux,* 27. Retrieved July 30, 2019, from http://www.e-flux.com/journal/27/67991/routes-worlds/.
Rudaz, G. (2009). Territorial redefinition and the governance of mountain regions. *Journal of Alpine Research, 97*(2), 27–37.
Ruggie, J. (1993). Territoriality and beyond: Problematizing modernity in international relations. *International Organization, 47*(1), 139–174.
Said, E. (1979) Orientalism. New York: Vintage Books.
Sen, A. (2005). Secularism and its discontents. In *The argumentative Indian. Writings on Indian history, culture and identity.* London: Penguin Books.
Sheller, M. (2011). Mobility. Sociopedia.ISA. Retreived May 29, 2020 from: http://www.sagepub.net/isa/resources/pdf/Mobility.pdf
Sherpa, P. (2012). *Sherpa perceptions of climate change and institutional responses in the Everest region of Nepal.* Dissertation, Washington State University.
Sherpa, P. (2014). Climate Change, perceptions and heterogeneity in Pharak, Mount Everest region of Nepal. *Human Organization, 73*(2), 153–161.
Soloman, T., & Steele, B. J. (2017). Micro-moves in international relations theory. *European Journal of International Relations, 23*(2), 267–291.
Stewart, M. (2014). Caterpillar governance in developing Yunnan. In E. Yeh & C. Coggins (Eds.), *Mapping Shangri-La: Contested landscapes in the Sino-Tibetan Borderlands* (pp. 175–198). Seattle: University of Washington Press.
Swyngedouw, E. (2004). Globalisation or 'glocalisation'? Networks, territories, and rescalings. *Cambridge Review of International Affairs, 17*(1), 25–48.
Verschuren, B., Wild, R., McNeely, J. A., et al. (2010). *Sacred natural sites: Conserving nature and culture.* Washington DC: Earthscan.
Wapner, P. (2014). Climate suffering. *Global Environmental Politics, 14*(2), 1–6.
Wolf, E. (1982). Europe and the people without history. Berkeley: University of California Press.
Wollmer, W. (2014). *The inner and outer paths of Mt. Kailash: Pilgrimage to the abode of Chakrasamvara*: Vajra Books.
Yeh, E. T., & Lama, K. T. (2013). Following the caterpillar fungus: Nature, commodity chains, and the place of Tibet in China's uneven geographies. *Social and Cultural Geography, 14*(3), 318–340.
Youatt, R. (2017). Personhood and the rights of nature: The new subjects of contemporary Earth politics. International Political Sociology. 11(1): 39–54.
Young, O. (1989). The politics of international regime formation: Managing natural resources and the environment. *International Organization, 43*(3), 349–375.

Conflicting Temporalities and the Ecomodernist Vision of Rewilding

Rasmus Karlsson

Abstract Ecomodernism offers a progressive and humanist vision of the Anthropocene, one in which publicly funded innovation has made possible both universal prosperity and planetary-scale rewilding. However, given the present primitive state of technology, ecomodernism is surely guilty of fabulism as its realisation would depend on technologies that may not be available for many decades. Despite this, ecomodernists argue that there is an overriding moral imperative to accelerate the transition to a fully integrated high-energy planet even if this accentuates the short-term need for solar radiation management. The aim of this chapter is to review the debate between ecomodernists and traditional environmentalists in relation to these conflicting temporalities. It is suggested that science may be of surprisingly little help in settling the underlying macro-political disputes.

Introduction

Considering how far we have come as a species over the last two or three hundred years, it would be very sad if the journey would end here. Or if the rest of the journey would be only for the few, leaving behind a planet that has been made largely unrecognisable and uninhabitable. It is thus not surprising that the natural impulse is one of scaling back (Crist 2018), to think that there is a way of unmaking modernity and harmonising society with non-human nature. According to ecomodernism (Asafu-Adjaye et al. 2015), that impulse, however, is fundamentally misguided. As a progressive and humanist vision of the future, ecomodernism instead suggests that only by doubling down on an Enlightenment trajectory of democratically guided scientific progress will it be possible to meet global human needs while restoring an ecologically vibrant planet (Symons 2019).

Essentially, the ecomodernist vision is about reaffirming pluralism, about envisioning an open-ended democratic future where some can be punk rockers and others stockbrokers, where people can live different lives and where the human condition

R. Karlsson (✉)
Department of Political Science, Umeå University, Umeå, Sweden
e-mail: rasmus.karlsson@umu.se

J. C. Pereira and A. Saramago (eds.), *Non-Human Nature in World Politics*,
Frontiers in International Relations, https://doi.org/10.1007/978-3-030-49496-4_6

can remain as contradictory and confusing as ever. Whereas traditional environmentalism seeks to impose a singular ethic of sustainability (Crist 2020), ecomodernism is cautioned by historical attempts to ensure ethical homogeneity and social purity. Rather than to fundamentally remake or rewire humanity, the ecomodernist vision seeks to reconcile the broad emancipatory forces of the last centuries with the natural environment through publicly funded breakthrough innovation. Yet, given the current primitive state of technology, ecomodernism is here surely guilty of fabulism and its optimism about the future may appear "not just self-delusional but outright dangerous" (Arias-Maldonado 2018, p. 139).

To make any sense, ecomodernist optimism has to be looked upon from the vantage point of the deep future, when technology exists to freely manipulate matter at the molecular level (Umbrello and Baum 2018), rendering both the production and transportation of material goods unnecessary and when abundant energy has not only overcome practically all forms of environmental determinism but also made possible climate restoration (Dorr 2016). Fantastic or even unrealistic as such a future may seem, ecomodernism is premised on the view that the *relative* distance from where humanity is today to that future is far shorter than the distance to anything remotely resembling "sustainability" as conceived by traditional environmentalism.[1] In this, ecomodernism is in fact founded on a grim reading of current environmental trends which fully acknowledges the risk of catastrophic climate change (Pereira and Viola 2018). If long-term sustainability could be achieved by, say, a twenty or thirty per cent reduction in economic activity, then "degrowth" would be the obvious and preferable response (Kallis et al. 2018). However, in a world of almost eight billion people, the level of depopulation and economic contraction needed is so great that it would be politically and socially unfathomable, especially in the light of competing national interests and widely shared material aspirations as the world desperately seeks to bounce back from the SARS-CoV-2 epidemic.

As such, rather than seeing a more integrated world as a threat to sustainability (Myers and Kent 2003) and erecting new barriers to migration (Cafaro and Staples 2009), ecomodernism takes the opposite view, that only in an open world of equalised opportunities will it be possible to summon the political momentum necessary for effective environmental action and avoid the kind of costly distractions over migration policy that has nearly derailed liberal democracy itself in recent years. Captured in the metaphor of a "high-energy planet" (Karlsson 2018), ecomodernism seeks to fully eschew Malthusian reductionism and its polarising demand-side framing of environmental problems and replace it with a unifying commitment to a supply-side energy revolution (Shellenberger et al. 2008). Yet, it is clear from the onset that there is a considerable temporal distance between the present and the hypothetical fruits

[1] While obviously diverse, "traditional environmentalism" is here used to denote the widely shared belief that human societies must harmonise with nature in order to avoid economic and ecological collapse. Specifically, Thomas Princen writes that "if there were a single philosophical position in environmental thought, adhered by all who are concerned about environmental destruction, it is that at the root of that destruction is human's separation from nature" (Princen 2010, p. 82). Ecomodernism differs from traditional environmentalism in that it rejects this analysis and instead advocates further "trophic detachment" (see Quilley 2011).

of that revolution. In the meantime, any attempt to take global development seriously would lead to further emissions overshoot and accentuate the need for controversial forms of climate engineering.

The aim of this contribution is to problematise these conflicting temporalities and, in particular, their implications for non-human nature. However, prior to that, the terms of the debate between ecomodernists and traditional environmentalists are reviewed with an emphasis on the controversy surrounding nuclear energy. On this basis, it is then argued that science is of surprisingly little help when it comes to settling these macro-political disputes or even for determining what should count as empirical evidence of failure and success, respectively. The chapter concludes by suggesting that, as the environmental crisis worsens, the stakes in these debates are likely to rise further yet, with both nuclear energy and proposals for climate engineering taking centre stage.

The Terms of the Debate

With regard to non-human nature, ecomodernism[2] represents a philosophical shift from the *management* of nature to the *liberation* of nature (Symons and Karlsson 2018). In the very long run, ecomodernism envisions a comprehensive decoupling of the economy from the environment, either through atomically precise manufacturing or large-scale space colonisation. In the meantime, ecomodernists propose a gradual substitution of natural products with synthetic ones and the use of agricultural intensification in order to spare marginal lands. In essence, while traditional environmentalism seeks to re-embed society in nature, ecomodernism advocates *greater separation* as a pathway to planetary-scale rewilding.[3] Among its critics, this idea has been misunderstood as some kind of opposition to family farms or wilderness hikes whereas it is rather about minimising large-scale interventions in nature, such as those caused by mining, forestry or energy production. Similarly, other critics have suggested that ecomodernists "celebrate the end of nature" (Fremaux and Barry 2019, p. 10) when a more open-minded reading of ecomodernist texts would make it clear that what is driving the ecomodernist project in the first place is precisely a moral concern for non-human nature and a desire to end human domination.

Beyond such elementary misunderstandings, ecomodernists diverge from traditional environmentalists in how they understand the broader cosmic setting of the human enterprise. Early pioneers of what is today known as ecomodernism, such as

---

[2]For the purpose of this chapter, I have used *The Ecomodernist Manifesto* from 2015 as the primary source of what ecomodernism entails (Asafu-Adjaye et al. 2015). The manifesto was an effort to forge a consensus position among 18 leading ecomodernist thinkers. This means that I do not specifically engage with post-ecological perspectives but rather build on the seminal work of Martin Lewis (1992) with regards to the politics of non-human nature.

[3]For ecomodernists, rewilding is a future-oriented process by which land is returned to nature, and the degree of human intervention and management is reduced. For a broader conceptual discussion, see Gammon (2018).

the 2019 recipient of the "Breakthrough Paradigm Award"[4] Stewart Brand, expressed a keen interest in space colonisation (Brand 1977; McCray 2013), a prospect that fundamentally alters the ethical context in which the ecological crisis is to be understood (Gott 1993, p. 319). Rather than accepting the subtractionist understanding of traditional environmental ethics in which the best would be for humanity not to exist, the possibility of humanity becoming a multi-planetary species and thereby securing not only its own long-term survival but that of life more generally, profoundly recasts the terms of the debate. Similarly, when taking an astrobiological perspective (Frank and Sullivan 2014), today's changes in atmospheric chemistry become less of a final verdict on the moral failures of humanity and more of fairly predictable stumbling blocks on the road towards becoming a technologically mature civilisation.[5]

Introducing such alternative end-points also makes it far more difficult to determine what is actually promoting versus impeding sustainability in the present. For instance, "luxury" emissions from aviation may turn out to be essential for driving aerospace innovation while demand for "unnecessary" consumer electronics may lay the foundation for the next revolution in electronics. All this brings back the argument over the *relative* distance towards a long-term sustainable socio-ecological regime, especially in the context of world politics with its deeply fractured epistemologies. Whereas a traditional vision of sustainability would always remain threatened by re-emerging unsustainable practices and its regime of ecological austerity would have to be policed indefinitely, the ecomodernist vision is about finding long-term solutions that are inherently compatible with the "broken timber of humanity", to speak with Immanuel Kant. Instead of rationing the consumption of individual natural resources at the local level, ecomodernism sees the wise use of resources as a stepping-stone to a global post-scarcity future in which resource extraction has either become unnecessary due to atomically precise manufacturing or can take place outside the biosphere. While probably not entirely historically accurate (York 2017), ecomodernists often use the example of whale oil to illustrate this point about the transformative role of technology and how, instead of politically determining what would be a sustainable rate of whaling or trying to "correctly" price the ecological services provided by whales as suggested by neoliberal conservationism (Büscher et al. 2014), technological innovation ultimately rendered whale oil obsolete as a fuel. According to ecomodernists, the example of whale oil fits into a larger narrative which suggests that humans primarily spare nature for which they have little practical use (Blomqvist et al. 2015) and, as such, the path to ecological flourishing does not go through better valuing nature in monetary terms but instead through technological innovation that makes nature *less* economically valuable. Ecomodernists also note that, as societies become more affluent, post-materialist values tend to become more pronounced, and the ability to set aside land for conservation purposes increases.

[4]The Breakthrough Paradigm Award is an annual prize offered by the Oakland-based ecomodernist think tank *The Breakthrough Institute*.

[5]For a generalised discussion on the possible implications of technological transformation and astronomical trajectories, see Baum et al. (2019).

Ecomodernist criticism of traditional environmentalism often comes down to debates over scale. For instance, while residential photovoltaics may be sufficient for powering household appliances or charging personal electronics and thus give an optical illusion of decarbonisation, that says very little about the energy needed to build the house in the first place or the energy embodied in the different consumer products. This is why the concept of "Energy Return on Investment" or EROI has become crucial for ecomodernist thinking as it engages with the question of what technologies could facilitate a rapid global decarbonisation (Karlsson and Symons 2015). Reviewing historical emission trends and to what extent alternative energy sources have actually succeeded in displacing fossil fuels (Cao et al. 2016; Loftus et al. 2015; York 2012), it is worth noting that sustained reductions that would be proportionally commensurate with the targets of the Paris Agreement have so far only come from two sources, either from severe economic contraction as in the wake of the SARS-CoV-2 epidemic (Le Quéré et al. 2020) or the deployment of nuclear energy. Calculations suggest that if the rest of the world would be able to replicate the build-out of nuclear energy in Sweden and other leading countries in the 1970s, fossil-fuel electricity worldwide could be displaced within a couple of decades (Qvist and Brook 2015). In contrast, and despite a much publicised effort, the German "Energiewende" based on renewable energy has not led to any significant decarbonisation as emissions have stayed roughly flat since its inception in 2011, prompting the German government to admit that it will miss its 2020 climate target. The intermittent nature of renewable energy sources means that they depend on fossil fuels for load-balancing while their low EROI (which becomes even lower in case batteries are used for load-balancing) makes them incapable of doing the heavy lifting needed to decarbonise the global economy as a whole (Moriarty and Honnery 2016). Often discussions over these issues tend to get stuck over particular technical details or price estimations that are taken out of context. For instance, it is not uncommon for authors who advocate for a post-capitalist future to dismiss nuclear energy on the basis that it is "too expensive" (Nisbet 2014), which, after all, is a rather odd concern since the EROI of an all-renewable energy systems would be "well below the range of the thresholds identified in the literature as necessary to sustain high levels of development in current industrial and complex societies" (Capellán-Pérez et al. 2019, p. 18). Important as these controversies may be, they ultimately reflect different orientations towards the Enlightenment and debates that go back all the way to the making of modernity (McMahon 2001).

For ecomodernists, access to clean and abundant energy is essential to eradicate poverty through transformative social change and comprehensive modernisation (Bazilian and Pielke 2013). With regard to nature, a high-energy future is seen as equally crucial for protecting biodiversity and avoiding the kind of energy sprawl associated with renewable energy sources (Trainor et al. 2016). More specifically, ecomodernists point to how mass desalination of seawater for agriculture can spare natural aquifers and how direct air capture of carbon dioxide can restore atmospheric concentrations to pre-industrial levels (Keith et al. 2006). On their end, traditional environmentalists believe that the planet cannot sustain such energy levels (Moriarty and Honnery 2009) and that there is an urgent ethical need for an "energy descent" in order to "unlearn abundance" (Alexander and Gleeson 2019). Again, these are not so much technical debates as they are conflicts between different worldviews and ways of understanding the human condition.

Whereas traditional environmentalists often believe that the future will be marked by crisis and energy scarcity (Friedrichs 2013) and that community resilience can be strengthened by relocalising the economy and adopting "low-tech" options (Alexander and Yacoumis 2018), ecomodernists fear that such an outlook may quickly become self-fulfilling. Instead of "reshoring" and retreating from the global, ecomodernists see international trade as instrumental for overcoming local scarcity and ensuring food security in times of extreme weather events (Karlsson 2017). In this, ecomodernists point to the need for a progressive and coherent theory of globalisation, one that can equalise opportunity and, to quote the mission statement of the Breakthrough Institute, "accelerate the transition to a future where all the world's inhabitants can enjoy secure, free, prosperous and fulfilling lives". More than anything, it is this commitment to global equality of opportunity that sets ecomodernism apart from existing theories of sustainable development but also mainstream political debate which often rests on the silent assumption that universal prosperity is simply ecologically impossible. Historically, the construction of modern welfare states has depended on the creation of high-productive off-farm jobs and rapidly rising energy consumption. Ecomodernists see no compelling reason why the developing world should not be allowed to follow a similar trajectory and that, rather than making climate mitigation contingent on the poor never "catching up", ecomodernists argue that global economic convergence should be the natural starting point for all climate mitigation strategies.

One of the most tired tropes of ecological economics is that "infinite growth is not possible on a finite planet". Obviously, in an open universe, that truism becomes slightly less convincing (Bostrom 2003). The problem, however, is how to make the transition from where humanity is today to a prosperous, post-scarcity future in which technology has advanced sufficiently to overcome environmental determinism. Beyond the fictional accounts of *StarTrek*, few attempts have been made to articulate what such a future would even look like.[6] In a post-utopian political landscape deprived of political imagination, visions of a bright multi-planetary future are readily dismissed as hopelessly naïve or hubristic. Part of this dismissal has to do with declining trust in our capacity for collective action. Whereas the same lack of purpose and direction has clearly not befallen the enemies of the Enlightenment (Lilla 2016), the political Left has long abandoned the Promethean imagination that used to be one of its defining characteristics (Meyer 2016).

How can the temporal gulf that separates a worsening ecological crisis in the present from a future of ecological restoration and rewilding be bridged? Ecomodernists generally agree that the state is the only actor with both the political legitimacy and institutional capacity to facilitate the necessary acceleration of science and technology (Symons 2019). However, in the present, there is no constituency

---

[6]For a recent such attempt which, if nothing else, has received considerable traction in social media, see Bastani (2019).

that demands such innovation efforts. Instead, much of the climate debate remains trapped around personal emissions reductions and the relative morality of this or that lifestyle. Similarly, as witnessed most recently at COP24 in Katowice, there is no serious discussion about the energy needs of a universally developed world (Arto et al. 2016) or to what extent the actions undertaken by the most ambitious countries are actually possible for others to replicate (or if their perceived success depends on the offshoring of energy-intensive industries to developing countries). In fact, even accounts that specifically engage with the offshoring of emissions through trade tend to remain focused on how to reduce *consumption* in rich countries rather than how to make *production* less polluting everywhere. For lack of a better description, the inability to take global scalability issues seriously represents a form of "climate nationalism" that ecomodernists believe distorts the standards by which progress towards the goal of climate stabilisation is measured (Karlsson and Symons 2015). One flashpoint in this debate has to do with the use of energy forestry which, although perhaps useful for reducing emissions in rich and sparsely populated countries in the Global North, would have far-reaching consequences for biodiversity and food production if used in the tropics (Pimentel et al. 2009). A similar argument can be made surrounding the use of large-scale hydropower which depends on the existence of a finite number of fast-flowing and dammable rivers. As such, even if countries like Iceland or Costa Rica may appear to be "climate leaders" with nearly 100% of their electricity supply coming from low-carbon sources, that leadership is dependent on unique geographical features that other countries do not possess. This helps to explain why ecomodernists so wholeheartedly have come to see breakthrough innovation as the only way that the bourgeoning energy demand of the developing world can be reconciled with the need for effective climate action (Green 2017). However, such innovation efforts will not happen as long as it is commonly believed that existing renewable energy sources are sufficient to fight climate change or if the richest and most motivated countries were to succeed in partially decarbonising their economies using non-scalable technologies. It is thus not surprising that one of the most heated exchanges between ecomodernists and traditional environmentalists has concerned precisely the feasibility of a 100% renewable energy future (Clack et al. 2017) and the question of what technological pathways that are most suitable for facilitating a deep decarbonisation (Jenkins et al. 2018).

Love Your Monsters

Writing in *Breakthrough Journal*, the French sociologist Bruno Latour argued that, rather than running away in panic from the world that it has created, humanity needs to learn to "love its monsters":

> But our sin is not that we created technologies but that we failed to love and care for them. It is as if we decided that we were unable to follow through with the education of our children … The dream of emancipation has not turned into a nightmare. It was simply too limited: it excluded nonhumans. It did not care about unexpected consequences; it was unable to

> follow through with its responsibilities; it entertained a wholly unrealistic notion of what science and technology had to offer (Latour 2011, p. 22, 28).

No monster is more emblematic of modernity than nuclear energy. Despite that civilian nuclear power has historically saved close to two million lives and prevented 60 gigatonnes of carbon emissions (Kharecha and Hansen 2013), traditional environmentalists typically see it as an undemocratic and ungovernable "mega-technology" (Barry 2016) which only serves to reinforce the hubristic separation from non-human nature that they believe is the root cause of the ecological crisis (Princen 2010, p. 82). In contrast, ecomodernists point to the experience of Sweden in particular to show that not only is a high-nuclear pathway eminently compatible with democracy but also capable of producing unparalleled ecological benefits (Goldstein and Qvist 2019; Pierrehumbert 2016). Turning the separation argument around, ecomodernists see the relative "disembeddedness" of nuclear energy as one of its principal merits as its high density allows for the production of massive amounts of clean and dispatchable electricity without the need for the kind of large-scale interventions in nature associated with biomass and renewable energy sources.[7] For these reasons, nuclear energy has become increasingly central to the ecomodernist agenda as a way of ending human domination and making room for rewilding, especially in a future with greater global equality.

While there may be some truth to the suggestion that if nuclear energy was discovered today, it would immediately be recognised as *the* solution to climate change, existing nuclear technologies come with the long shadow of the Cold War and are held to standards that are very different from those of fossil or renewable energy sources. While next generation nuclear technologies capable of safely recycling existing nuclear waste and eliminating the need for further uranium mining may assuage some of those concerns (Brook et al. 2015), nuclear energy still has a profound *cultural* problem as embracing it also means making peace with the masculinity, artificiality and hyper-rationalism associated with the Enlightenment. It is important to recognise how secondary "hard" scientific data is to such cultural concerns. Just as the suburban university professor who takes her Mini Cooper to the Farmers' Market and Instagrams organic produce is seen as a trailblazer of sustainability while the effective altruist in his high-rise eating Soylent is seen as hopelessly estranged from nature, the perceived *unnaturalness* of nuclear energy puts it at odds with prevailing cultural logics and risk perceptions. Similarly, while few people would probably give much thought about biking, many people have intense anxieties about flying even as all "scientific" risk assessments show that flying is infinitely safer.

The democratic use of nuclear energy depends on the existence of a high-trust society and strong professional ethics among those set to regulate it. As such, it is not particularly surprising that small-scale renewable energy sources, even if dependent on the mining of rare earth minerals and a large component of natural gas in order to provide energy when the wind does not blow or the sun does not shine, are seen as a

[7] The surface footprint of nuclear power is around 2 km^2/GWe compared to 50 km^2/GWe for photovoltaics, 300 km^2/GWe for wind turbines and 2500 km^2/GWe for biomass (Berger et al. 2017, p. 65). Note that the footprint of wind and solar is even higher if intermittency is accounted for.

safer bet, especially as overall trust in public institutions is declining. However, in the same way, that meaningful social resilience does not come from encouraging individuals to become "preppers" but from strengthening our collective institutions in order to prevent collapse in the first place, ecomodernists see a high-energy future as a way of going all in with regard to globalisation and ensuring that people everywhere have the clean energy needed for comprehensive modernisation. By making fossil and renewable energy sources uncompetitive, ecomodernists also hope that climate risks can be reduced while non-human nature and habitats can be spared. However, for many people in the OECD, the vision of an ecomodernist high-energy future does not resonate with everyday experiences. By most standards, a rich country like Norway is already fairly "sustainable" with a stable baseload supply from hydropower. The problem only arises once one imagines that everyone in a world of 8–10 billion people has a right to live in a similar fashion. While the answer from traditional environmentalism has been that this is simply not ecologically possible and that people in the rich countries have to dramatically reduce their energy consumption in order to accommodate developing world growth, ecomodernists believe that proposing such reductions would trigger a strong anti-environmental populist response and that they, equally important, would end up foreclosing humanity's long-term possibilities as a species (Karlsson 2016). Essentially, ecomodernists suggest that effective climate policies need to work with, rather than against, the transformative forces of the last centuries in order to lessen international inequalities and reduce the risk that developing countries become trapped in a state of premature deindustrialisation due to the lack of reliable energy (Rodrik 2016).

Meanwhile, traditional environmentalists have spared no energy in stoking polarisation and attacking what they consider to be a malign global elite (Klein 2014). In their view, climate change is a moral issue akin to apartheid, and their tactics of visual boycotts and divestment protests also reflect this analysis (Mann and Wainwright 2017, p. 173). Yet, as long as human development remains closely linked to rising carbon emissions, it seems reasonable to doubt the possible success of such tactics. Ecomodernists on their behalf believe that it is crucial to instead restore elite interest in public affairs and reduce overall political polarisation. The Breakthrough Institute in particular has sought to invite conservatives to its dialogue events in the hope of overcoming some of the deadlock surrounding climate policy in the United States. Still, ecomodernist insistence on the role of government in driving innovation inevitably puts it at odds with the dismantling and paralysation of the state that has followed in the wake of the Trump presidency.

Given its progressive and cosmopolitan orientation, ecomodernism is naturally opposed to ethno-nationalist populism but also postliberal visions of ecological austerity. However, in order to offer a coherent alternative to both these ascending forces, ecomodernism in its turn becomes dependent on an emancipatory or even utopian imagination whose realism can only be evaluated *ex-post*. Far from being some passive belief in capitalism's capacity to deliver a "techno-fix", ecomodernism argues that a politics of radical engagement and broad social investments are necessary to facilitate a sufficient advancement of science and technology. In a recent intervention, one of the founders of the Breakthrough Institute, Ted Nordhaus, pointed out that whereas the most successful clean energy initiatives in modern history have all followed a state-led model, most environmentalists today have come to shun big centralised infrastructure and remain captive to a deregulatory neoliberal imaginary in which fossil fuel subsidies and powerful corporate interests are the only things holding back a rapid adoption of renewable energy (Nordhaus 2019). Rejecting this analysis, ecomodernists argue that the innovation of radically new technologies, rather than the deployment of existing ones, must be at the heart of any effective political response to the Anthropocene. In his book-length treatment of ecomodernism, Jonathan Symons writes, "[s]ince the technological metabolism of contemporary civilization is ecocidal, we need an entirely new set of technologies for food production, transport, industry and electricity. Such rapid technological change will only happen if low-carbon innovation becomes a political priority" (2019: 110).

Learning in the Anthropocene

All of this again brings to the fore the temporal disconnect between the ecological stress generated by accelerated economic convergence in the present and the possibilities of ecological redemption in the future. To some extent, this disconnect echoes the older debate surrounding the existence of an Environmental Kuznets Curve (Dinda 2004). Yet, its scope is no longer confined to specific pollutants but pertains to the sustainability of the human enterprise written large.

Considering the current rate of habitat destruction, and rising global temperatures, there is little empirical support for the ecomodernist belief in decoupling of economic growth from environmental impacts. While it is true that demand for a range of natural resources seems to have peaked (McAfee 2019), and that a *relative* decoupling is occurring across numerous indicators, the *absolute* numbers are not encouraging (Wiedmann et al. 2015). Not only does global energy growth continue to outpace decarbonisation (Jackson et al. 2018), but as the oceans fill up with plastics, corals bleach and unprecedented wildfires take their toll, talk of a "Good Anthropocene" may come across as utterly tone-deaf.

Nevertheless, what is under debate is not whether humanity has historically been destructive to its natural environment (it has!) but what long-term planetary trajectory has the greatest chance of ultimately bringing about sustainability. Here, traditional environmentalism has an obvious advantage in that people can easily imagine that their environmental footprint would be smaller if they somehow "harmonised" with non-human nature (or, better yet, if they stopped existing). However, while the Na'vi in *Avatar* may make for excellent science fiction,[8] the political pursuit of such a totalising vision of *Ecotopia* is unlikely to command broad democratic support any time soon. In a world as populous as ours, it is also worth noting that a disorderly "energy descent" would utterly destroy the biosphere as untrained people would go out in nature in search of food and fuel. Thus, it is not surprising that many

[8]The Na'vi is a species on the planet Pandora that embodies many of the ideals of deep ecology.

environmentalists stop far short of deep ecology, arguing that all that is needed is that we all tread a little bit lighter on the planet. However, science lends little support for such an incrementalist belief. According to the latest *Emissions Gap Report*, global net emissions would have to quickly fall back to the levels of the mid-1970s (around 25 $GtCO_2e$ per year) to make possible even a 1.5 °C pathway (Olhoff and Christensen 2018), that is to a time when the global population stood at a mere four billion and neither China nor India were industrialised. Yet, by the same token, there is a little hope of climate stabilisation even if policies slightly more aligned with the ecomodernist vision were to be pursued by a handful of motivated countries.

Again, the problem is one of scale but also what visions (a) can offer scientifically feasible accounts of future socio-natural relations and (b) inspire decisive global action. Given that "radical strategies are called for on either side of the ideological divide" (Arias-Maldonado 2020, p. 12). According to ecomodernism, it is ethically unacceptable to try to solve the ecological crisis on the backs of the poor by denying them the right to a modern life. Instead, humanity must finally make good on the original promise of the Enlightenment and its universal humanism, even if this requires a large amount of proactionary risk-taking and temporarily escalating measures of unsustainability. In order not to only trigger ecological mayhem, such an acceleration would have to be matched by determined political action that would, in turn, require strong civil society advocacy. While some scholars are beginning to accept the ecomodernist view that neither energy efficiency improvements nor renewable energy will be sufficient in the fight against climate change, that does not necessarily translate into support for high-energy innovation but rather into even more radical calls for "deep" energy conservation (Moriarty and Honnery 2019, p. 12), degrowth and a future "defined primarily in terms of frugality, micro-scale localism [and] subsistence" (Trainer 2012, p. 594). What, if anything, in terms of empirical evidence and feedback, would be sufficient to break this impasse?

For long, traditional environmentalists have entertained the belief that humanity would somehow be "saved by disaster" and that, as the material effects of climate change would be felt, support for radical ecologism would grow. However, just as the financial crisis of 2008 did not lead to the demise of neoclassical economics but rather its entrenchment in the form of public austerity, it is equally possible that people will react to the threat of worsening storms by buying sturdier pickups or voting for politicians who promise to keep climate refugees out. Yet, the same problem of political mobilisation applies with equal force to ecomodernism. Of particular concern to ecomodernism is the risk that piecemeal improvements in energy efficiency, especially in combination with overall deindustrialisation, will further diminish rich countries' interest in financing breakthrough supply-side innovation. As rich countries continue to pick the low-hanging fruit of such spontaneous decarbonisation (Pielke 2018), it is possible that they will experience sustained emissions reductions for many years, at least relative to GDP growth, and that this will obscure the fact that the core of their economies remains fundamentally carbon based. Similarly, if ecomodernists are correct that renewable energy is essentially a cul-de-sac with regard to climate change mitigation (Brook et al. 2018), it may still take decades to bring about a sufficient policy re-orientation given how invested many governments and NGOs have become in their support for renewable energy and how insensitive they seem to be to evidence of policy failure (Germany here being a case in point).

Even if only one single death has been attributed to the radiation released after the Fukushima accident while millions die from the burning of fossil fuels every year (Landrigan et al. 2018), public opinion turned strongly against the use nuclear energy following the accident (Kim et al. 2013). This underscores the fragility of a nuclear-centred ecomodernist project. It is likely that a rushed build-out of nuclear energy in response to climate change would lead to accidents and possibly even fatalities. Even if a utilitarian calculation could be used to show how insignificant such local losses would be compared to the devastating global effects of unmitigated climate change, this presents a profound pedagogical problem.

The underlying cause of many of these difficulties is the lack of feedback across scale levels which makes learning in the Anthropocene, which is already inherently difficult due to cultural factors and ideologically motivated cognition (Kahan 2013), even more challenging. Stephen Gardiner famously talked about how the spatial and temporal dispersion of causes and effects invites moral corruption, prompting him to describe climate change as a "perfect moral storm" (Gardiner 2006). This obviously has profound implications for the prospects of an ideal or optimal response to climate change. One does not have to be a cynic to suspect that neither *Ecotopia* nor *StarTrek* is forthcoming any time soon (Karlsson 2013). If the time-scales were more compressed, the sense of urgency would presumably be greater. In his "Vulnerable World Hypothesis", the philosopher Nick Bostrom explores a scenario in which the threat of climate change would be even more catastrophic than it actually is. In this scenario, a doubling of atmospheric CO_2 would lead to rapid warming in the range of 15–20 °C rather than 3 °C (Bostrom 2018, p. 13). In such a "vulnerable world", there would probably be much less hesitation about a war-time mobilisation to quickly build thousands of nuclear reactors in order to fully decarbonise the economy and capture CO_2 from ambient air. However, even in our world, it is still reasonable to expect that increasingly radical policy options will become more salient in the public debate as the environmental crisis gradually worsens.

A common criticism of ecomodernism is that it is hubristic to believe that humanity can somehow detach itself from the natural world. That argument, however, rests on the unexamined premise that it is *not* hubristic to believe that humanity can remain entangled in fragile ecosystems that we have long outgrown, both in our numbers and resource use. As the seminal ecomodernist thinker Martin Lewis put it already in his 1993 article: "when one considers concrete issues regarding the provisioning of our basic material needs, the separation of the human economy from natural systems turns out to offer profound environmental benefits, while the continued immersion of our apparatus of production into the intricate webs of nature is itself highly threatening to the natural world of non-human species" (Lewis 1993, p. 779). In contrast, belief in traditional environmentalism comes with the assumption that humanity can somehow collectively lessen its appetite in an almost ant-like fashion. Historically, such totalising visions have not fared well. If ever realised, they are also likely to undermine the very material conditions that made lifestyle environmentalism and other forms of emancipatory self-expression possible in the first place. A rematerialisation of the economy and a return to labour-intensive forms of agriculture would presumably also be accompanied by a return to survivalist values and a greater emphasis on community self-reliance. On this topic, Stephen Quilley has written about the impossibility of "liberalism in one village" and the associated problem of violence in a degrowth world (Quilley 2011, p. 77), an issue that traditional environmentalism rarely if ever engages with.

Ever Higher Stakes

If an optimal and timely response to the climate crisis is not forthcoming, there are good reasons to believe that political attention will turn to Solar Radiation Management (SRM) or other forms of climate engineering (Heyward 2013). Theoretically, stratospheric aerosol injection of sulphate particles can offer "an escape route against strongly increasing temperatures" (Crutzen 2006, p. 216) should conventional mitigation efforts prove insufficient. While clearly no permanent solution, or a substitute for reducing atmospheric concentrations of greenhouse gases since it would not address ocean acidification, SRM would presumably "buy time" (Wigley 2006) for mitigation by stemming sea level rise and preventing possibly irreversible damages in terms of biodiversity (Long 2016), even as it would give rise to profound governance challenges (Reynolds 2019).

For ecomodernists, SRM represents a daunting predicament. On one hand, it is precisely the kind of risky large-scale intervention in nature that ecomodernism revolts against.[9] In addition, the use of SRM would likely reduce the political urgency of developing new supply-side solutions such as next generation nuclear energy. On the other hand, considering the catastrophic ecological and human losses that SRM has the potential of averting as well as ecomodernism's overriding commitment to human development, it would be unethical to reject SRM simply on the basis that it is "unnatural", especially when considering how deeply entangled humanity already is in nature. As such, most ecomodernists probably agree with Joshua Horton and David Keith in that "given the very strong evidence that SRM would significantly reduce global temperatures and thereby limit climate impacts, particularly in the developing world, we view research on SRM as a moral imperative" (Horton and Keith 2016, p. 90). If nothing else, ecomodernists reckon that such research can be motivated on the basis of ensuring that false hopes are not tied to a technology that ultimately does not work.

---

[9]These risks include the possibility of uneven side effects in terms of altered precipitation and evaporation levels. Similarly, once initiated, a SRM programme would presumably have to continue until concentrations of greenhouse gases in the atmosphere have been sufficiently reduced or there would be a sudden spike in temperature. This so-called "termination problem" (Rabitz 2018) has been much discussed in the literature but the fact that the lifetime of sulphur aerosols in the stratosphere is limited to one to two years could also be seen as beneficial as it makes a SRM programme essentially reversible.

It should not come as a surprise that traditional environmentalists have arrived at a very different analysis of the possible risks and benefits of SRM. Naomi Klein for instance explains how "geoengineering will certainly monsterize the planet as nothing experienced in human history" (2014: 279) whereas Clive Hamilton sees it as "the fulfilment of three and a half centuries of objectification of nature" (Hamilton 2013, p. 200). While environmental arguments against SRM are sometimes invoked, it is clear that the main line of criticism has to do with morality and politics (Anshelm and Hansson 2014). When Justin McBrien writes that "capital now seeks to postpone its demise through planetary geoengineering" (McBrian 2016, p. 135), it is easy to see how the insistence on absolute "system change" drives resistance to SRM and other forms of climate engineering. After all, if SRM can indeed reduce the risk of a climate catastrophe, it would also deprive traditional environmentalism of a key motivational vehicle for its politics.

If the world were to stay on its current path towards 3–5 °C of warming above pre-industrial levels, SRM may quickly emerge as a very real policy option given its short response time compared to emissions reductions (Van Vuuren and Stehfest 2013). If that were to happen, it is likely that an intense political struggle would ensue. With ever rising stakes, the macro-political disputes of the Anthropocene would probably become even harder to settle.

Similarly intractable, ecomodernism has given rise to a fierce debate over the role of intensification and industrial agriculture in land-sparing (Balmford et al. 2018). Echoing many of the arguments above, ecomodernists claim that urbanisation and higher yielding concentrated agriculture have the potential of freeing up land elsewhere for habitat restoration and rewilding (Blomqvist et al. 2015, p. 34). However, for this mechanism to actually deliver meaningful results in terms of conservation, there has to be a strong biophilic ethic in society that ensures that the land spared by intensified agriculture is in fact returned to nature and not converted into for instance energy forestry (as has been the case in much of Europe and the United States). In any case, it seems safe to say that until the hypothetical arrival of truly "game-changing" technologies in the field of synthetic biology or atomically precise manufacturing, the ecomodernist case surrounding agriculture appears comparatively weaker than the arguments advanced in relation to energy and climate. This is especially so when considering factory farming and the unethical treatment of non-human animals in general.

Conclusion

As the world approaches what James Martin once described as the "mid-century canyon" of increasing environmental stress (Martin 2007), there is an urgent need for grand-scale decisions. This chapter has outlined some of the ways in which an ecomodernist vision of such decisions would differ from those of traditional environmentalism. Still, for all their apparent differences, these two visions of the future share some important characteristics, in particular, a highly idealist understanding of history and a belief in human exceptionalism (Mann 2018, p. 455). For ecomodernists, human exceptionalism calls for completing the emancipation from nature that begun with the Enlightenment and realising humanity's axiological potential along an astronomical trajectory in order to make possible planetary-scale rewilding. For traditional environmentalism, it means that humans, unlike any other species, can avoid extinction by collectively self-regulating their reproduction rate and resource use and thus restore a stable equilibrium in relation to its environment. In political terms, ecomodernism promises a democratic and pluralist future characterised by ever greater measures of fluidity whereas traditional environmentalism seeks to restore a common sense of belonging and permanence at the local level.

However, one does not have to look further than the Trump presidency or the SARS-CoV-2 epidemic to realise that the real world does not unfold according to these kinds of ideal types. Often, inertia and irrationality but also sudden urgency seem as important as ideas and visions. Regardless, the twenty-first century does not bode well for non-human nature. Both ecomodernists and traditional environmentalists probably agree that things will get significantly worse before they get any better. As the state of non-human nature continues to deteriorate, all sides will look for empirical evidence to support their respective arguments, including counter-factual arguments about what the world would have looked like if people had only listened to them. However, given the difficulties in agreeing on an appropriate unit of analysis or baseline, science will remain of little help in settling these ideological disputes.

Culturally, the ecomodernist vision of rewilding and universal development runs directly counter to elite romanticisation of Third World agricultural poverty but also to the prevalent fetishisation of locally produced "organic" food. By anticipating a qualitatively different relationship to non-human nature, one in which human sustenance has become increasingly decoupled from the land, ecomodernism points to the long-term liberation of nature from human domination. Attractive as such a future may seem from an ethical point of view, it does leave many difficult questions unanswered, not only about technological feasibility but also the local politics of the envisioned transition and its distributional implications. Still, given the deep unsustainability of the current socio-ecological regime, preserving the status quo may quickly prove to be just as unrealistic as the pursuit of any ecomodernist agenda.

Acknowledgements The author would like to thank Therese Bjärstig, Elias Isaksson, Marcel Wissenburg, Jonathan Symons and Andrew Scerri as well two anonymous reviewers for comments that have been most helpful in revising this chapter. In addition, thanks to the generous support of the Karl Staaff Foundation, it was possible to present a draft version of this chapter at the WPSA Annual Conference in San Diego, California, in April 2019.

References

Alexander, S., & Yacoumis, P. (2018). Degrowth, energy descent, and 'low-tech' living: Potential pathways for increased resilience in times of crisis. *Journal of Cleaner Production, 197*(2:1), 1840–1848.
Alexander, S., & Gleeson, B. (2019). *Degrowth in the suburbs: A radical urban imaginary*. London: Palgrave MacMillan.
Anshelm, J., & Hansson, A. (2014). Battling Promethean dreams and Trojan horses: Revealing the critical discourses of geoengineering. *Energy Research & Social Science, 2,* 135–144.
Arias-Maldonado, M. (2018). Towards a Good Anthropocene? In M. Arias-Maldonado, & Z. Trachtenberg (Eds.), *Rethinking the environment for the Anthropocene: Political theory and socionatural relations in the New Geological Epoch* (pp. 137–150). London: Routledge.
Arias-Maldonado, M. (2020). Bedrock or social construction? What anthropocene science means for political theory. *The Anthropocene Review*. https://doi.org/10.1177/2053019619899536.
Arto, I., Capellán-Pérez, I., Lago, R., Bueno, G., & Bermejo, R. (2016). The energy requirements of a developed world. *Energy for Sustainable Development, 33,* 1–13.
Asafu-Adjaye, J., Blomquist, L., Brand, S., Brook, B. W., DeFries, R., Ellis, E., et al. (2015). *An ecomodernist manifesto*. http://www.ecomodernism.org.
Balmford, A., Amano, T., Bartlett, H., Chadwick, D., Collins, A., Edwards, D., et al. (2018). The environmental costs and benefits of high-yield farming. *Nature Sustainability, 1*(9), 477.
Barry, J. (2016). Bio-fuelling the Hummer? Transdisciplinary thoughts on techno-optimism and innovation in the transition from unsustainability. In E. Byrne, G. Mullally, & C. Sage (Eds.), *Transdisciplinary perspectives on transitions to Sustainability* (pp. 120–137). London: Routledge.
Bastani, A. (2019). *Fully automated luxury communism*. London: Verso.
Baum, S. D., Armstrong, S., Ekenstedt, T., Häggström, O., Hanson, R., Kuhlemann, K., et al. (2019). Long-term trajectories of human civilization. *Foresight, 21*(1), 53–83.
Bazilian, M., & Pielke, R. (2013). Making energy access meaningful. *Issues in Science and Technology, 29*(4), 74–78.
Berger, A., Blees, T., Bréon, F. M., Brook, B. W., Hansen, P., Grover, R. B., et al. (2017). How much can nuclear energy do about global warming? *International Journal of Global Energy Issues, 40*(1–2), 43–78.
Blomqvist, L., Nordhaus, T., & Shellenberger, M. (2015). *Nature unbound. Decoupling for conservation*. Report from the Breakthrough Institute.
Bostrom, N. (2003). Astronomical waste: The opportunity cost of delayed technological development. *Utilitas, 15*(3), 308–314.
Bostrom, N. (2018) *The vulnerable world hypothesis*. Working paper (version 3.21). Oxford: The Future of Humanity Institute.
Brand, S. (Ed.). (1977). *Space colonies*. Whole Earth Catalogue: Sausalito.
Brook, B. W., van Erp, J. B., Meneley, D. A., & Blees, T. A. (2015). The case for a near-term commercial demonstration of the Integral Fast Reactor. *Sustainable Materials and Technologies, 3,* 2–6.
Brook, B. W., Blees, T., Wigley, T. M., & Hong, S. (2018). Silver Buckshot or Bullet: Is a future "Energy Mix" necessary? *Sustainability, 10*(2), 302.
Büscher, B., Dressler, W., & Fletcher, R. (Eds.). (2014). *Nature inc.: Environmental conservation in the neoliberal age*. Tuscon: University of Arizona Press.
Cao, J., Cohen, A., Hansen, J., Lester, R., Peterson, P., & Xu, H. (2016). China-US cooperation to advance nuclear power. *Science, 353*(6299), 547–548.
Cafaro, P., & Staples, W., III. (2009). The environmental argument for reducing immigration into the United States. *Environmental Ethics, 31*(1), 5–30.
Capellán-Pérez, I., de Castro, C., & González, L. J. M. (2019). Dynamic Energy Return on Energy Investment (EROI) and material requirements in scenarios of global transition to renewable energies. *Energy Strategy Reviews, 26,* 100399.

Clack, C. T., Qvist, S. A., Apt, J., Bazilian, M., Brandt, A. R., Caldeira, K., et al. (2017). Evaluation of a proposal for reliable low-cost grid power with 100% wind, water, and solar. *Proceedings of the National Academy of Sciences, 114*(26), 6722–6727.
Crist, E. (2018). Reimagining the human. *Science, 362*(6420), 1242–1244.
Crist, E. (2020). For cosmopolitan bioregionalism. *The Ecological Citizen, 3* (Suppl C): 21–29.
Crutzen, P. J. (2006). Albedo enhancement by stratospheric sulfur injections: A contribution to resolve a policy dilemma? *Climatic Change, 77*(3), 211–220.
Curren, R., & Metzger, E. (2017). *Living well now and in the future: Why sustainability matters.* Cambridge, MA: MIT Press.
Dinda, S. (2004). Environmental Kuznets curve hypothesis: A survey. *Ecological Economics, 49*(4), 431–455.
Dorr, A. (2016). The impact pulse and restoration curves: Going beyond mitigation and stabilization. *Anthropocene, 16,* 61–66.
Frank, A., & Sullivan, W. (2014). Sustainability and the astrobiological perspective: Framing human futures in a planetary context. *Anthropocene, 5,* 32–41.
Fremaux, A., & Barry, J. (2019). The 'Good Anthropocene' and green political theory: Rethinking environmentalism, resisting ecomodernism. In F. Biermann, & E. Lövbrand (Eds.), *Anthropocene encounters. New directions in green political thinking.* Cambridge: Cambridge University Press.
Friedrichs, J. (2013). *The future is not what it used to be: Climate change and energy scarcity.* Cambridge, M.A.: MIT Press.
Gammon, A. R. (2018). The many meanings of rewilding: An introduction and the case for a broad conceptualisation. *Environmental Values, 27*(4), 331–350.
Gardiner, S. M. (2006). A perfect moral storm: Climate change, intergenerational ethics and the problem of moral corruption. *Environmental Values, 15*(3), 397–413.
Goldstein, J., & Qvist, S. (2019). *A bright future–How some countries have solved climate change and the rest can follow.* New York: Public Affairs.
Gott, J. R., III. (1993). Implications of the Copernican principle for our future prospects. *Nature, 363,* 315–319.
Green, C. (2017). Mitigation technology: Half full or nearly empty? *Nature Climate Change, 7*(2), 98.
Hamilton, C. (2013). *Earthmasters: The dawn of the age of climate engineering.* New Haven: Yale University Press.
Heyward, C. (2013). Situating and abandoning geoengineering: A typology of five responses to dangerous climate change. *PS: Political Science & Politics, 46*(1), 23–27.
Horton, J., & Keith, D. (2016). Solar geoengineering and obligations to the global poor. In C. Preston (Ed.), *Climate justice and geoengineering: Ethics and policy in the atmospheric Anthropocene* (pp. 79–92). New York: Rowman & Littlefield.
Jackson, R. B., Le Quéré, C., Andrew, R. M., Canadell, J. G., Korsbakken, J. I., Liu, Z., et al. (2018). Global energy growth is outpacing decarbonization. *Environmental Research Letters, 13*(12), 120401.
Jenkins, J. D., Luke, M., & Thernstrom, S. (2018). Getting to zero carbon emissions in the electric power sector. *Joule, 2*(12), 2498–2510.
Kahan, D. M. (2013). Ideology, motivated reasoning, and cognitive reflection: An experimental study. *Judgment and Decision Making, 8,* 407–424.
Kallis, G., Kostakis, V., Lange, S., Muraca, B., Paulson, S., & Schmelzer, M. (2018). Research on degrowth. *Annual Review of Environment and Resources, 43,* 291–316.
Karlsson, R. (2013). Ambivalence, irony, and democracy in the Anthropocene. *Futures, 46,* 1–9.
Karlsson, R. (2016). Three metaphors for sustainability in the Anthropocene. *The Anthropocene Review, 3*(1), 23–32.
Karlsson, R. (2017). The environmental risks of incomplete globalisation. *Globalizations, 14*(4), 550–562.
Karlsson, R. (2018). The high-energy planet. *Global Change, Peace & Security, 30*(1), 77–84.

Karlsson, R., & Symons, J. (2015). Making climate leadership meaningful: Energy research as a key to global decarbonisation. *Global Policy, 6*(2), 107–117.
Keith, D. W., Ha-Duong, M., & Stolaroff, J. K. (2006). Climate strategy with CO_2 capture from the air. *Climatic Change, 74*(1–3), 17–45.
Kharecha, P. A., & Hansen, J. E. (2013). Prevented mortality and greenhouse gas emissions from historical and projected nuclear power. *Environmental Science and Technology, 47*(9), 4889–4895.
Kim, Y., Kim, M., & Kim, W. (2013). Effect of the Fukushima nuclear disaster on global public acceptance of nuclear energy. *Energy Policy, 61*, 822–828.
Klein, N. (2014). *This changes everything: Capitalism vs. the climate*. London: Allen Lane.
Landrigan, P. J., Fuller, R., Acosta, N. J., Adeyi, O., Arnold, R., Baldé, A. B., et al. (2018). The Lancet Commission on pollution and health. *The Lancet, 391*(10119), 462–512.
Latour, B. (2011). Love your monsters. *Breakthrough Journal, 2*(11), 21–28.
Le Quéré, C., Jackson, R. B., Jones, M. W., Smith, A. J., Abernethy, S., Andrew, R. M., ... & Friedlingstein, P. (2020). Temporary reduction in daily global CO_2 emissions during the COVID-19 forced confinement. *Nature Climate Change*, 1–7.
Lewis, M. W. (1992). *Green delusions. An environmentalist critique of radical environmentalism*. Durham: Duke University Press.
Lewis, M. W. (1993). On human connectedness with nature. *New Literary History, 24*(4), 797–809.
Lilla, M. (2016). The shipwrecked mind: On political reaction. New York: New York Review of Books.
Loftus, P. J., Cohen, A. M., Long, J. C., & Jenkins, J. D. (2015). A critical review of global decarbonization scenarios: What do they tell us about feasibility? *Wiley Interdisciplinary Reviews: Climate Change, 6*(1), 93–112.
Long, J. (2016) Bringing geoengineering into the mix of climate change tools. In C. Preston (Ed.), *Climate justice and geoengineering: Ethics and policy in the atmospheric Anthropocene* (pp. 109–20). New York: Rowman & Littlefield.
Mann, C. C. (2018). *The wizard and the prophet: Two groundbreaking scientists and their conflicting visions of the future of our planet*. New York: Picador.
Mann, G., & Wainwright, J. (2017). *Climate leviathan: A political theory of our planetary future*. London: Verso Books.
Martin, J. (2007). *The meaning of the 21st century: A vital blueprint for ensuring our future*. London: Eden Project.
McAfee, A. (2019). *More from Less. The surprising story of how we learned to prosper using fewer resources-and what happens next*. New York: Simon & Schusters.
McBrian, J. (2016). Accumulating extinction: Planetary catastrophism in the necrocene. Anthropocene or capitalocene? In E. Altvater, E. Crist, D. Haraway, D. Hartley, C. Parenti, & J. McBrien (Eds.), *Anthropocene or capitalocene?: Nature, history, and the crisis of capitalism* (pp. 116–137). Pm Press.
McCray, W. P. (2013). *The Visioneers: How a group of elite scientists pursued space colonies, nanotechnologies, and a limitless future*. Princeton: Princeton University Press.
McMahon, D. M. (2001). *Enemies of the enlightenment: The French counter-enlightenment and the making of modernity*. Oxford: Oxford University Press.
Meyer, W. B. (2016). *The progressive environmental prometheans. Left-wing heralds of a 'Good Anthropocene'*. Cham: Springer International Publishing.
Moriarty, P., & Honnery, D. (2009). What energy levels can the Earth sustain? *Energy Policy, 37*(7), 2469–2474.
Moriarty, P., & Honnery, D. (2016). Can renewable energy power the future? *Energy Policy, 93*, 3–7.
Moriarty, P., & Honnery, D. (2019). Energy efficiency or conservation for mitigating climate change? *Energies, 12*(18), 3543.
Myers, N., & Kent, J. (2003). New consumers: The influence of affluence on the environment. *Proceedings of the National Academy of Sciences, 100*(8), 4963–4968.

Nisbet, M. C. (2014). Disruptive ideas: Public intellectuals and their arguments for action on climate change. *Wiley Interdisciplinary Reviews: Climate Change, 5*(6), 809–823.
Nordhaus, T. (2019). The empty radicalism of the climate apocalypse. *Issues in Science and Technology, 35*(4), 69–78.
Olhoff, A. (Ed.), & Christensen, J. M. (2018). Emissions Gap Report 2018. UNEP DTU Partnership
Pereira, J. C., & Viola, E. (2018). Catastrophic climate change and forest tipping points: Blind spots in international politics and policy. *Global Policy, 9*(4), 513–524.
Pielke, R., Jr. (2018). Opening up the climate policy envelope. *Issues in Science and Technology, 34*(4), 33–40.
Pierrehumbert, R. (2016). How to decarbonize? Look to Sweden. *Bulletin of the Atomic Scientists, 72*(2), 105–111.
Pimentel, D., Marklein, A., Toth, M. A., Karpoff, M. N., Paul, G. S., McCormack, R., et al. (2009). Food versus biofuels: environmental and economic costs. *Human Ecology, 37*(1), 1.
Princen, T. (2010). *Treading softly: Paths to ecological order*. Cambridge, M.A: MIT Press.
Quilley, S. (2011). Entropy, the anthroposphere and the ecology of civilization: An essay on the problem of 'liberalism in one village' in the long view. *The Sociological Review, 59,* 65–90.
Qvist, S. A., & Brook, B. W. (2015). Potential for worldwide displacement of fossil-fuel electricity by nuclear energy in three decades based on extrapolation of regional deployment data. *PLoS ONE, 10*(5), e0124074.
Rabitz, F. (2019). Governing the termination problem in solar radiation management. *Environmental Politics, 28*(3), 502–522.
Reynolds, J. L. (2019). *The governance of solar geoengineering: Managing climate change in the Anthropocene*. Cambridge: Cambridge University Press.
Rodrik, D. (2016). Premature deindustrialization. *Journal of Economic Growth, 21*(1), 1–33.
Shellenberger, M., Nordhaus, T., Navin, J., & Norris, T. (2008). Fast, clean, and cheap: Cutting global warming's gordian knot. *Harvard Law & Policy Review, 2,* 93–118.
Symons, J. (2019). *Ecomodernism: Technology, politics and the climate crisis*. Cambridge: Polity.
Symons, J., & Karlsson, R. (2018). Ecomodernist citizenship: Rethinking political obligations in a climate-changed world. *Citizenship Studies, 22*(7), 685–704.
Trainer, T. (2012). De-growth: Do you realise what it means? *Futures, 44*(6), 590–599.
Trainor, A. M., McDonald, R. I., & Fargione, J. (2016). Energy sprawl is the largest driver of land use change in United States. *PLoS ONE, 11*(9), e0162269.
Umbrello, S., & Baum, S. D. (2018). Evaluating future nanotechnology: The net societal impacts of atomically precise manufacturing. *Futures, 100,* 63–73.
Van Vuuren, D. P., & Stehfest, E. (2013). If climate action becomes urgent: The importance of response times for various climate strategies. *Climatic Change, 121*(3), 473–486.
Wiedmann, T. O., Schandl, H., Lenzen, M., Moran, D., Suh, S., West, J., et al. (2015). The material footprint of nations. *Proceedings of the National Academy of Sciences, 112*(20), 6271–6276.
Wigley, T. M. (2006). A combined mitigation/geoengineering approach to climate stabilization. *Science, 314*(5798), 452–454.
York, R. (2012). Do alternative energy sources displace fossil fuels? *Nature Climate Change, 2*(6), 441.
York, R. (2017). Why petroleum did not save the whales. *Socius, 3,* 2378023117739217.

Elias in the Anthropocene: Human Nature, Evolution and the Politics of the Great Acceleration

Stephen Quilley

Abstract Civilization faces the challenge of reconciling growth with biophysical limits, whilst avoiding devastating geo-political conflict. Radical demands for 'degrowth' are often juxtaposed with realist 'eco-modernist' scenarios. Elias's theory of civilizing processes and his concept of the triad of controls provide an invaluable framework for evaluating the reality-congruence of both trajectories. An Eliasian understanding of ecological conscience formation is synthesized with insights from Ernest Gellner (on exo-education), Walter Ong (on literacy and individuation) and Owen Barfield (on the history of consciousness). Successful navigation of the politics of the Anthropocene implies societal solutions and arrangements that exist in the barely conceivable 'adjacent possible.' Eliasian concepts are invaluable in the exploration of such possibilities, but his rationalism and commitment to greater 'detachment' make him blind to the simultaneous requirement for selectively higher degrees of 'involvement' in the process of both ecological conscience formation and the consolidation of 'imagined community'. Adam Seligman's theory of ritual and sincerity provides a complement to Elias's rationalism. Elias underplays the possibility that more detached scientific and economistic processes of model making and orientation might co-exist with patterns of conscious, creative, cognitive dissonance: modalities of 'paradox', enchantment and participation associated with more relational and reciprocal forms of 'livelihood' economy.

Introduction

Over the next century, our integrated global civilization faces the existential challenge of reconciling continuing socio-economic development with biophysical limits to growth. At issue is the maintenance of societal complexity, i.e. population size; the number of distinct social and occupational roles; topographical and social connectivity and network complexity; the number and diversity of products, goods and services; the continuing trajectory of technical innovation; the number of firms,

---

S. Quilley (✉)
University of Waterloo, Waterloo, Canada
e-mail: squilley@uwaterloo.ca

J. C. Pereira and A. Saramago (eds.), *Non-Human Nature in World Politics*,
Frontiers in International Relations, https://doi.org/10.1007/978-3-030-49496-4_7

institutions, organizations, etc. It is an axiom of ecological economics that such complexity is a function of the throughput of energy and materials and it is intrinsically difficult to reconcile with ecological integrity (Quilley 2017; Odum 2007). This challenge is complicated by the serious risk of devastating geo-political conflict—over land and resources, chaotic migration and spiralling regional and global arms races.

In the field of environmental politics, radical but often rhetorical demands for 'degrowth' are juxtaposed with more self-consciously realist 'eco-modernist' scenarios (Quilley 2017). The former are congruent with biophysical reality, but utopian and unworkable in political terms. The latter are achieving political traction under the guise of the 'green New Deal'—but do not really address the problem of biophysical limits. Norbert Elias's theory of civilizing processes provides an invaluable framework for evaluating the reality-congruence of both trajectories. I will refer back to these broad environmental–political frames in the conclusion. Because Eliasian sociology is replete with a number of oft-recurring technical concepts, where appropriate, the discussion includes short definitions and explanations in boxes.

Box 1 *Civilizing Process*: A long-term process with neither beginning nor an endpoint and involving directional changes in the social structure of society (markets, religion, state organization, etc.). By increasing the intensity and extent of relations of interdependence between individuals and groups, such processes engender more extensive external controls on individual behaviour (laws, rules, social conventions often backed by the power of the state). At the same time, such external constraints are, over time, internalized to become part of an average personality profile or 'habitus.' Characteristic of a particular society in a specific place and historical period, any such habitus generates a distinct 'second nature' that is internalized by individuals and achieves a degree of naturalized 'automaticity', i.e. a set of behavioural patterns that emerge from the interplay of our evolved human nature on the one hand, and culturally specific predispositions on the other. Aside from Elias's interest in controls in relation to interpersonal violence, he refers to numerous other examples including, famously, the increased control over bowel movements that has become a part of 'potty training' since the Early Modern period (Elias 2000).

The Civilizing Process (2000 [On the Process of Civilization (OPC)])[1] centred on a bifocal investigation of psychological and behavioural transformations among the upper and middle classes in Europe on the one hand, and processes of 'pacification' and state formation on the other, is now widely recognized as a rich and complex analysis of European modernization (although he would have been loath to use that word) which rivals the classical accounts of Marx, Weber and Durkheim,

[1] First translated as *The Civilizing Process*, this title has most recently been translated as *On the Process of Civilization* (see detailed citation in Elias 2000). In what follows I will use the latter and the acronym OPC in general discussion.

as well as contemporaries such as Polanyi (1964 [1944]). At the same time, consideration of later works—in areas such as the sociology of knowledge (Elias 2007a), the evolution of language (Elias 2011), the sociology of time (Elias 2007b), the dynamics of individuation (Elias 2010c) and ecology (Elias 2012b); see also Goudsblom (1992)—as the elaboration of core ideas implicit in the structure of OPC, is now engendering a further widening of Elias's reputation and the salience of Eliasian sociology as a contender for the 'central theory' in the social sciences (Loyal and Quilley 2004; Quilley and Loyal 2005). In retrospect, Elias's achievement was to develop a paradigmatic foundation for sociology as the lynchpin in a broader human science encompassing the social, neuro-psychological, physiological, evolutionary and ecological dimensions of human development.

In the long term, the most significant new areas of Eliasian scholarship will focus on the linked problems of human ecology and human nature, i.e. the long-term relationship between the human species and the complex systems of the biosphere, on the one hand, and evolution, ontological and social moulding of human personality and behaviour on the other. In this chapter, I will provide an overview of Elias's contributions in these two areas. My intention is to indicate how an Eliasian sociology of the longue durée can really illuminate the opportunities and risks as well as the wicked trade-offs that are concomitant with the most commonly advocated, and often taken-for-granted, environmental–political scenarios. In drawing attention to these unacknowledged assumptions, the discussion will also highlight a problem arising from the rationalism of Elias's sociology of knowledge, namely that he underplays the role of emotional–affective involvement and ritual in both ecological conscience formation and the formation of 'imagined communities' or the 'we-identities' that are a prerequisite for political action.

The Sociology of Knowledge, Evolutionary Sequence and Big History

Box 2 *Involvement and Detachment*: Modalities of engagement with the world in the process of modelling that characterizes all human cognition. In contrast to the more usual contrast between 'objective' and 'subjective' knowledge, Elias seeks to emphasize that all knowledge creation, and any specific context for human action, involves a balance of both involvement and detachment. This balance is variable over the course of a day (e.g. depending on hunger, agitation, feelings of personal security), a lifetime (e.g. in child development and ageing) and inter-generational history (e.g. with the changing social stock of concepts and knowledge that are available to individuals in particular societies). Some degrees of both involvement and detachment become unavoidable to the extent that they are inscribed in language and narrative conventions. For instance, it's very difficult for modern people to experience a belief in a flat earth or a geocentric solar system, or to recover the simultaneous appreciation of physical 'breath' or 'wind' and 'spirit' that was internal to Anglo-Saxon terminology (Elias 2007a, b; Barfield 1979).

In *Involvement and Detachment* (*I&D*), Elias (2007a) elaborates a theory of knowledge that seeks to establish the relationship across the full array of scientific disciplines within the arc of a 'comprehensive process model' (see Quilley 2004a, b).[2] In this early and prescient example of what has become known as 'big history' (Christian 2011), Elias construes the hierarchy of scientific subjects as a spectrum of empirical domains arrayed by order of complexity and 'levels of integration' (i.e. organizational dynamics relating to nested hierarchies of whole-part relationships, Needham 1937). He underlines also the extent to which this array of disciplines relates to subject matters which themselves emerge in sequence during the course of evolution (Elias 2007a, p. 146). And anticipating the current biological thinking in relation to complexity theory and 'big history' (Kaufman 2000; Christian 2011), Elias argues that there is an unambiguous direction to the great evolution towards increasing complexity: the big bang and star-formation (cosmology, physics), followed by planets (geology, chemistry), life (biology) and eventually, social life and symbolic culture (sociology—see Elias 2011, 2007a, b; Goudsblom 2002).

What is distinctive about this 'model of models' is that Elias integrates it with an historical sociology of knowledge. Since the Renaissance, scientific disciplines emerged from under the aegis of the church with a particular sequence and in the context of new, but now very familiar, academic institutions. But natural science was but the latest in the more general iterative exchange between practical problem solving, increasing social complexity and long-term social development on the one hand, and the expanding repertoire of reality-congruent models of the world on the one other.

Reflecting on this steady expansion of human models and understanding, Elias highlighted the interdependence of two balances or ratios, namely safety and security versus danger and insecurity on the one hand, and emotional 'involvement' versus psychological detachment, on the other. For Elias, these two balances are linked by both positive and negative feedback loops, which can be both 'vicious' and 'virtuous' in their impacts on human well-being and capacities. Dominated by animistic and magical-thinking, early human cognition was fantasy-laden and ego-centric consequent upon heightened emotional involvement. Events and processes in both social and natural domains were understood primarily with regard to a direct meaning they were assumed to have for the self (e.g. 'what does this mean for me?'). But this propensity for ego-centrism was, Elias showed, exacerbated by situations of danger and insecurity—a feedback loop that made more detached and emotionally neutral modelling of the relation between observable and experienced events and processes

[2]This section draws heavily on my exposition of the relationship between biology and sociology in the work of Elias (Quilley 2004a, b)

much more difficult. For much of human pre-history, this cycle of danger and involvement created a permanent impediment to the development of more reality-congruent models of the world. Anticipating the language of Gregory Bateson, Elias argued that it is this 'double bind', rather than any neuro-psychological cognitive failing, that accounts for the painfully slow advance of human knowledge and technology in the hundreds of millennia leading up to the early Palaeolithic. Thus, for example, to collect and transport appropriate stone from one place and to construct tools in another, for the purpose of hunting in yet another location, is a pattern of behaviour that requires planning, psychological restraint and detour behaviours along with the more detached understanding of the qualities of materials in themselves. The foundation of all contemporary human behaviour, social coordination and technical accomplishment took hundreds of millennia to emerge. But over the longest period of time, each incremental extension of detachment consistently expanded the capacity of our Palaeolithic ancestors to control the world around them.

Over the longue durée, concomitant with hundreds of small technological innovations, and in tandem with an expanding lexicon of concepts expressing more 'reality-congruent' models of the relationship between processes and events in the natural world, the danger/safety balance shifts steadily in favour of the latter (at least vis-à-vis non-human nature). In this way, there is a consistent and mutually constitutive relationship between:

(i) the level or degree of detachment that is coalesced into a structure of behaviour and cognition by the public standards of thinking about natural events that inform processes of acculturation and socialization, and
(ii) the level and manner of control of non-human nature represented by public procedures for manipulating them (Elias 2007a, p. 8).

Box 3 *Double Bind*: Gregory Bateson's concept, developed in relation to cybernetics, evolution and schizophrenia, refers to a double-edged 'communication' that points simultaneously to two separate and incompatible responses, each of which having the effect of negating the other. Thus, in evolution, all species are faced with a changing selection environment in response to which they either die out or evolve into a new species—in which case they also die out. Elias uses the term to express a vicious circle or feedback loop in which structural features of both knowledge (models of the world) and material circumstances reinforce each other (Elias 2007a).

Elias also pointed out that over very long periods, the gradual increase in the scale, complexity and patterns of interdependence in social life (I—*socio-genesis*) was associated with both directional changes in the organization of the psyche (II—*psycho-genesis*) and the increasing flows of materials and energy associated with society vis-à-vis the wider biosphere (III—*eco-genesis*[3]). He referred to the co-development of processes on these three levels as the 'triad of controls' (i.e. social controls over people, internal psychological restraints and controls over non-human nature).

Box 4 *Triad of Controls*: The observation that in the course of social development humanity's relation to nature (III—*ecological controls*) changed in tandem with changes in society (I—*social controls*) and personality (II—*psychological controls*). These controls can't be uncoupled. Greater social complexity of any kind (e.g. more specialists and a more complex division of labour or the Internet) necessarily involves much more pervasive ecological controls (e.g. fire culture; agriculture; fossil fuelled industry) but also corresponding changes in the average personality and the kind of internalized psychological controls associated with a particular society. The implication is that one should not expect a society to change radically on one dimension (for instance, a radical process of relocalization associated with 'degrowth') without corresponding changes in the other dimensions (Elias 2012b).

Built into this social–ecological account of long-term development of human culture was a theory of knowledge that linked epistemology to the sociology of knowledge. The development of more reality-congruent, detached models of the world was concomitant with a shift in the safety/danger balance (towards the pole of greater security). The spurt in the reliability and scope of such models associated with the Renaissance and subsequent scientific revolution was made possible by a virtuous cyclical emergence of greater controls over self, others and nature—i.e. increasing social complexity in the context of more effective states and predictable economic contexts (see Fig. 1).

In this way, the gradual shift from a vicious loop or 'double bind' in the relationship between the involvement/detachment and safety/danger ratios to a virtuous loop proceeds according to 'the principle of facilitation.' As the size of this relatively insulated sphere of safety increases (cf. Goudsblom (2002) the 'anthroposphere' within the biosphere), the achievement of more detached understandings becomes progressively easier to achieve. It is for this reason that, at certain threshold moments, the knowledge process is characterized by rapid if not quite exponential acceleration.

Box 5 *Anthroposphere*: Coined by Johan Goudsblom, this term predates the 'Anthropocene' and is cognate with other 'spheres' such as the biosphere, geosphere and lithosphere. It refers to that part of the earth created or modified by human activity (Goudsblom 2002).

[3]Elias didn't use this term but it is concordant with the other two.

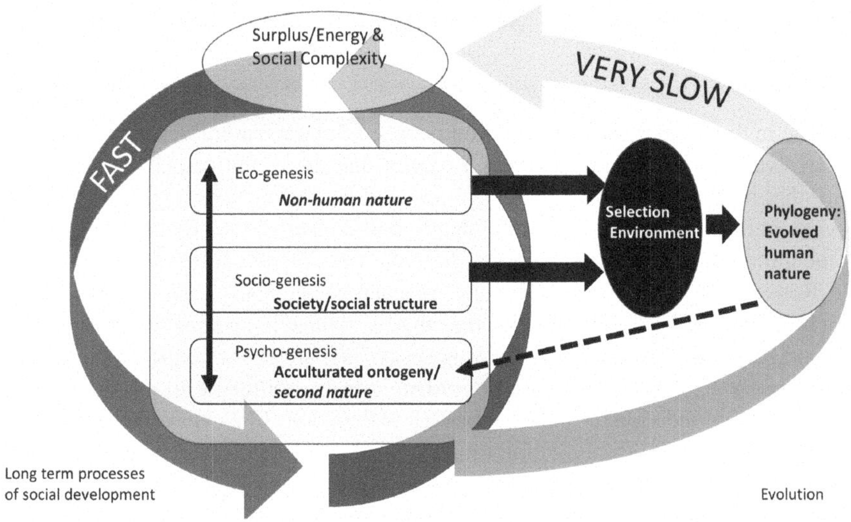

Fig. 1 The triad of controls and the nature of humanity and ecology

Finally, in *Involvement and Detachment,* Elias frequently observes that detachment comes at some psychic and emotional cost. The obstacles inhibiting the move from a geo-centric to a heliocentric image of the solar system had much less to do with empirical data collection and much more to do with the emotional and ontological consequences of moving away from an understanding of humanity at the centre of the universe. As Becker (2014) observed in *The Denial of Death*, this detached cosmological and evolutionary model of the universe in which humanity becomes decentred and potentially accidental undermines critical architectures of meaning and hero/immortality projects which have traditionally insulated human beings from the potentially devastating psychological consequences of mortality awareness. It is for this reason that the social sciences have lagged so far behind the natural sciences in the development of reality-congruent models. It is much harder to develop and propagate detached models of processes in which all human beings are so invested with respect to so many dimensions of life. And even now, the natural sciences are not immune from such dynamics. For instance, highly involved, politicized and unrealistic models that deny any relation between gender and biology are now diffusing from the humanities and beginning to affect even 'hard' sciences such as neuroscience (a colleague at the University of Waterloo has just stopped the chapter from his textbook that deals with neuroscience of sex differences because push-back from activist students is raising the transaction costs of teaching this kind of science, to an unacceptable degree).

Elias on Human Nature, 'Second Nature' and Non-Human Nature

At issue in these conflicts is the relation between 'human nature' as a phylogenetic and evolved attribute of humanity as a species and the cognitive-behavioural plasticity that is consequent upon language and culture. At one extreme, it is possible to view human behaviour both individually and collectively as highly constrained by the tramlines of an evolutionary psychology. Alternatively, following John Locke and Rousseau, a developing individual personality becomes a 'tabula rasa' upon which is imprinted a mental architecture that is malleable and specific to a particular society (Pinker 2002). This view was always favoured by socialists, radicals and most recently gender revolutionaries (too/dangerously) optimistic about the prospect of transforming individuals and societies. How then does Elias view the problem of human nature and 'plasticity within limits'? And how does his understanding of the relationship between phylogenetic species nature and acculturated 'second nature,' relate to the ecological context of non-human nature and to problems of ecological crisis and sustainability?

The Great Evolution

First of all, the propensity for culture evolved from biological processes. Although rudimentary culture is not unique to human beings, the reach and intensity of this new 'level of integration' accelerated exponentially with 'symbol emancipation' (Elias 2011) and the emergence of a species capable, through language, of developing ever more sophisticated models of both the material world and the more abstract universe of meaning frameworks, social relationships and patterns of interdependence.

Elias was always acutely aware of the way in which emerging higher levels of integration and complexity 'canalize' and transform the lower levels from which they are emerging. This was not an original insight but drawn from figures such as Joseph Needham, Julian Huxley, CH Waddington and others associated with the Theoretical Biology Club of 1930s Cambridge (Quilley 2010). So just as the evolution of life (the biosphere) had profound consequences for the chemical and geological evolution of the lithosphere—not least the creation of limestone, the oxygenation of the atmosphere, the transformation of hydrological flows and weather systems—so the emergence of language and culture (the 'anthroposphere'—Goudsblom 2002) have had an enormous and ongoing impact on the evolutionary–ecological dynamics of the biosphere, including mass extinctions, rates of soil formation, deforestation, global warming, the ozone cycle, the emergence of novel ecosystems and even the anthropogenic invention of new species, etc. (Quilley 2011, 2004a, b). From this perspective, the new emergence of any new level of integration so profound as language and culture would make some kind of ecological crisis likely and possibly inevitable. And at the same time, the processual emphasis on disturbance and momentum in the emergence of complexity makes any long-term project of sustainability premised on balance, harmony and dynamic equilibrium unlikely and even implausible. Both the evolution of the biosphere and the long-term dynamics of the anthroposphere tilt consistently towards disturbance, sequential change and developmental ratchets. Complex systems can collapse and shed layers of complexity. Civilizing dynamics can morph to processes of de-civilization (Dunning and Mennell 1998). But evolutionary–ecological and social systems are unlikely to remain in a state of equilibrium for long and cannot be wound back to previous states.

Human Nature, 'Second Nature' and the Rest of Nature

The central idea in OPC relates to the transformation of external social controls into internal, psychological controls or 'social constraint towards self-constraint': i.e. over time, by way of the gradual internalization of more restrained patterns of behaviour and the social expectation to regulate personal conduct, the process of civilisation was defined by a capacity for self-control that becomes so automatic as to appear natural. The expression of deep-rooted evolved behavioural drives and impulses becomes mediated to a greater extent by calculation and foresight but culturally patterned, unconscious behaviour controls. Using the language of Freud, Elias described this 'psychical habitus' of restraint as the formation of a more complex 'superego'. As a technical concept, civilization has no absolute beginning or end, but is understood as a process and not a state or characteristic of 'mind' or 'nature'—and is certainly not an essential attribute of European societies. Rather this progressive internalization of restraints is understood as the unplanned outcome of blind social processes, over many generations. "Civilization is not 'reasonable'; not 'rational, any more than it is 'irrational'. It is set in motion blindly, and kept in motion by the autonomous dynamics of a web of relationships, by specific changes in the way people are bound and live together" (2000 [1939]: 366).

And to the extent that such figurational structures of interdependence, pacification and increasing complexity are transforming societies across the planet in the wake of modernisation and global integration, the theory of civilizing processes would predict concomitant processes of psychological restraint and the internalization of more pervasive and self-monitoring drive controls.

From an Eliasian perspective, there is no tension between the biological and evolutionary universality of the human genome and the manifest variability of human nature. All species exhibit considerable plasticity in the expression of phenotypes in relation to an underlying genotype. This is a function of the ontogenetic growth and development of each individual interacting with a unique environment. Pigliucci (2001) provides a comprehensive account of such 'reaction norms' in biology. In the case of human beings, language and culture function to increase the scope of variation in such 'reaction norms' to such a degree that human nature is often very difficult to define in any prescriptive manner. For millennia, the repertoires of material and symbolic culture (e.g. fire, vocalization, tool making, war, shelter construction, cooking) have expanded. Our nature is to be versatile.

Following Elias, Goudsblom (1992) highlights the interlocking ('hinging') of biological and social processes as a pervasive feature of these earliest stages of human evolution. This 'canalizing' and steering of our evolutionary trajectory by social processes (see Quilley 2010) amounted to the social modification of (biological) norms of reaction.

Since the late Pleistocene, the cultural transmission of models of the world embodied in technical procedures, lexica and categorical systems, etc. has accelerated exponentially consequent upon language. From birth, babies are immersed in language that is not an individual attribute but an emergent property of social groups. Linguistic categories and unreflective epistemological concepts that are implicit in all language use provide an inescapable practical, cognitive and even emotional orientation to the external social and physical worlds. Facilitating more complex and information-laded communication, language greatly elaborated the possibilities for the social coordination of behaviour both in space and time (as with planning a migration or a hunt at a specified time in the future). And in this way, language engendered greater social complexity in social roles and functions and patterns of interdependency not just between individuals within families but between groups. From the outset, such outward social complexity was made possible by, and in turn accelerated, the elaboration of yet more complex social expectations vis-à-vis behaviour. Cultural patterns, which allow for the storage and gradual modification of such formal and informal codes and standards, are an emergent property of social processes—the unplanned outcome of the interweaving plans, actions, emotional and rational impulses of individual people. "From this interdependence of people arises an order *sui generis*, an order more compelling and stronger than the will and reason of the individual people composing it" (Elias 2000 [1939], p. 366).

Bearing this in mind, Elias's argument in OPC is basically that increasing social complexity engendered a progressive ordering of social relations that compelled more and more people to attune their conduct to that of others (Quilley 2004a, b). Over time, such external controls were increasingly complemented by internalized patterns of self-restraint, discernible as a particular pattern of personality. In the context of the relative social and economic stability effected by centralized state monopolies on financial and resource flows, and even more so on physical violence, along with greater consistency in patterns of childhood socialization, "this kind of self-restraint requires a higher degree of automaticity, it becomes, as it were, a 'second nature'" (2000 [1939]: 369). It is this 'automaticity', consequent upon the ontogenetic 'wiring up' of such a psychical habitus during the process of childhood socialization,that makes it second *nature*.

This concept of 'second nature' and his recognition of symbol emancipation as an evolutionary process provide the basis for an accommodation of sociological perspectives with both evolutionary biology and developmental biology.

The Society of Individuals, Survival Units and the Interiorization of the Self: Elias, Lévy-Bruhl and Ong

One thing that emerges very strongly in Elias's work is the deep connection between social structures and the structure of personality. Different societies engender very different kinds of people who think in very different ways. Such ideas had of course been explored in detail by anthropologists such as Frans Boas, Gregory Bateson, Margaret Mead and Ruth Benedict. But developed without reference to the 'culture and personality' school, OPC was the only study to focus on a detailed longitudinal study of societal change over a period of a thousand years. In addition, the conceptual architecture of civilizing processes was explicit in addressing the interaction between phylogenetic and socio/psycho-genetic processes in the moulding of the societal mind. On the other hand, Elias was sympathetic to the controversial ideas of Lucien Lévy-Bruhl (Loyal and Quilley 2013). At the same time as Boas, Benedict and Mead were writing in the United States, the French anthropologist Lévy-Bruhl made a sharp and unfashionable distinction between scientific rationality of advanced societies and 'pre-logical' thought characteristic of primitive societies. Lévy-Bruhl was somewhat vilified in the Anglophone world which was dominated by an anti-colonial relativist sensibility which refused to engage in any kind of comparative evaluation. Evans-Pritchard's (1937) study of witchcraft in the Azande became a benchmark for this hegemonic sensibility and was later famously deployed by Winch (1958) to demonstrate that because the meaning was so dependent on the particularities of context, it was impossible to generate testable and comparative knowledge about different cultures. Winch's book defined the epistemological zeitgeist in the social sciences triggering two decades of interdisciplinary argument about rationality and the conditions for knowledge (Wilson 1970; Hollis and Lukes 1982; Overing 1985). Characteristically obtuse, Elias makes no reference to any of these debates. But Involvement and Detachment should certainly be understood as a subtle response to this long-running controversy about modern versus primitive rationality. Despite Lévy-Bruhl's status as a whipping boy for anthropologists seeking to salve the discipline's colonial conscience, Elias passing defence of the French anthropologist was important and timely (2007a, p. 121, 122, Footnote 10). It was one of the rare occasions where Elias comes close to an overt attack on the burgeoning epistemological relativism that has dogged both sociology and anthropology since the 1970s. Certainly, as with all of his contemporaries, Lévy-Bruhl's work was marred by a philosophical propensity to separate 'reason as an eternal form and knowledge as a mutable content'. But, insisted Elias, Lévy-Bruhl's understanding of primitive 'participation' was an important breakthrough (2007a, p. 121, 122, fn 1).

Like Lévy-Bruhl (1926, 1978a, b), Elias was sensitive to the real gulf that could exist in modes of cognition and thought between more and less complex societies. Unlike anthropologists such as Margaret Mead, Elias was willing to countenance the possibility of long-range directional change. Although personality varied with culture, such variation wasn't random. Rather it was a function of patterns of acculturation and social interdependence (socio-genesis) that were themselves consequent upon broader social–ecological and energy regimes (eco-genesis). Although this emphasis on directional change carried an uncomfortable and unfashionable echo of Victorian progress theory, Elias was careful to distinguish progression in the sense of directional and sequential change, from any normative understanding of progress. Nevertheless, his conception of a 'civilizing process' did shed light on the reality of what Lévy-Bruhl had called 'pre-logical thought'. For Elias, this was better understood in terms of the everchanging balance between processes of involvement and detachment and the degree to which more detached models of the world became woven into taken-for-granted modes of acculturation and public standards of behaviour and so less inflected by individual sensibilities. In this way, Elias was confident in arguing that the thought processes of people in simpler face-to-face societies were likely to be relatively more subject to involved, fantasy thinking, in which models centred on the meaning of events, processes or phenomena for the subject.

In *Orality and Literacy,* Walter Ong drew attention to the enormous impact that literacy has had on the structure of the mind and cognition and on characteristic styles of thinking that are so prevalent in modern societies that it is almost impossible to imagine any other cognitive modality. Literacy, he showed, allows the elaboration of complex clause structures, fine-grained grammatical distinctions and levels of abstraction that are simply impossible in an oral setting. Written texts allow parallel reading, comparison, revision and re-examination to such an extent that they facilitate much greater degrees of abstraction. A significant consequence of such habitual abstraction is that writing engenders an interiorization of cognitive process that is central to the emergence of the modern sense of self—as a free-wheeling homunculus rattling around inside the head and clearly differentiated from other selves 'out there'. What Elias called the *Homo clausus* mentality—the wide-spread but erroneous or misleading idea of the self as contained, as if within a 'thinking statue' (2010a, b, c)—became common sensical and wrongly construed as human universal, in the wake of mass literacy (and of course the wider process of individualization). Acculturated in complex, highly differentiated and mobile societies, modern people have lost sight of the extent to which the selves are, as G. H. Mead argued, relational and interdependent. And yet, as Elias pointed out, this image of sovereign, separate individuals is central to the operation of every modern institution and value-framework from democracy and the law to moral philosophy and the functioning of a modern consumer-oriented, market economy. Gellner (1983) paints a vivid picture of the way in which modern institutions of exo-education are designed to consolidate this self-contained, differentiated, proto-citizen—an individual equipped with a suite of transferable cognitive and language skills, categories and 'route-maps' underpinning unparalleled degrees of social and spatial mobility.

Eventually, the political–economic concomitant of this kind of socio-economic and mental individuation was necessarily a transition in the nature of what Elias (2010a, b, c) referred to as the paradigmatic 'survival unit'—away from extended family, kindred, clan and place-bound community, and towards the institutions of the market (employment, property, shared, pension-funds) and over time the welfare institutions of the (Keynesian) nation-state. In this way what Polanyi (1964 [1944]) referred to as the 'great transformation' centred on the disembedding of newly mobile individuals from the ascriptive structures of traditional society resulting in dangerous social and political instability, and a subsequent 'countervailing movement for societal protection' which saw the consolidation of new survival units associated with market and nation-state (for discussion, see Quilley and Zywert 2019; Quilley 2018).

It can be seen that Elias's account of modernity as a 'society of individuals' and the concept of 'survival units' is highly congruent with Polanyi's account of modernization as disembedding on the one hand, and Gellner's account of the coercive imposition of a national 'high culture' as a prerequisite for the patterns of legitimation underpinning the exclusive solidarity of citizenship (see Loyal and Quilley 2017). And at the same time, the rationalized, literate culture that made possible the kinds of social and spatial mobility that define modern societies (Weber 1968) simultaneously engendered the kind of interiorization of mind and self, described by Walter Ong.

Rationalization, Disenchantment and Problems of Meaning: Elias and Barfield

It is a truism that social and spatial mobility, the freedoms of choice and conscience and the weakening of ascriptive patterns of belief and behaviour associated with the society of individuals have not necessarily made people happier or mentally healthier. In fact, a great deal of nineteenth- and early twentieth-century sociology and philosophy can be understood as an engagement with the specific kinds of unhappiness and spiritual unease associated with modern societies. Consider, for example the young Marx's preoccupation with alienation; Durkheim's concept of 'anomie'; Tönnies distinction between gemeinschaft and gesellschaft; Weber's analysis of disenchantment and the disturbing psychological consequences of rationalization; Nietzsche's disconcerting commentary on the likely consequences of the 'death of God'; R. D. Laing's hypothesis of modern society as engendering schizophrenia; Ernest Becker on the declining efficacy of modern consumer culture to provide ontological insecurity and shield individuals from the disturbing awareness of mortality.

One surprising parallel that may be drawn is between Elias and Owen Barfield—the 'Inkling' interlocutor of C. S. Lewis, famous among a small but discriminating circle for his book *Saving the Appearances,* in which he introduced his concept of 'original participation' and the associated history of consciousness (1965). Barfield's theory centres on the idea that the phenomenal world must be by definition a system of shared representations, and is always constructed in tandem with a participating consciousness. The radical implication of this view is not only that our ancestors thought differently (with Ong, Elias and Lévy-Bruhl) but that the world that they thought about was in fact different also—primarily because they understood and accepted this primary participation, whereas moderns deny the participative nature of the universe and insist on the 'thingy-ness' of the world 'out there'. This is not the place to elaborate Barfield's ideas. It is sufficient to note that in his history of consciousness, he concurs that over the last millennia cognitive processes have changed dramatically. His state of 'original participation' would be described by Elias as a situation of endemic danger and insecurity associated with high levels of 'involvement', a lower capacity for detachment and a predominance of fantasy thinking in which models centred on the relation between phenomena and subjective experience. In Ong's terms, the subsequent systemic shift in the direction of detachment that has been facilitated by the changes in consciousness and a capacity for abstract thinking and the recording and transmission of information associated with literacy.

Barfield concurs with respect to the scope of the cognitive transformation the mode of detachment being described in terms of an elaboration in the capacity for 'alpha thinking'—thinking about the relationship between 'things', 'out there' and construed as separate and detached—an insight very congruent with Elias's critic of process reduction or the construal of dynamic processes as static things (see also Böhm 1980). But, in contrast to Elias, he argues that the detached model making associated with scientific rationalization is misleading and obscures the extent to which modern processes of cognition still depend, just as much as Palaeolithic cognition, on 'participation.' Throughout his work, Elias implies that some degree of alienation or mental unease consequent upon ontological insecurity is an unavoidable concomitant of individuation in complex rationalized societies. Taking a great deal from the architecture of the unconscious mind identified by Freud, Elias was at least implicitly aware that psychoanalysis was an archetypically modern response to modern problems of meaning and integration—and hence his interest in problems of attachment and group analysis (Waldhoff 2007; Penna 2016).

The Eliasian framework suggests that because eco-, socio-and psycho-genesis work in tandem (Fig. 1), any attempt to recover a previous structure of consciousness would only be conceivable with correlative changes at the level of ecology and society, i.e. any diminution of reversal of individuation and recovery of greater integration would likely come at the cost of a loss in social complexity and all this entails—albeit with a possible benefit in the form of a reduction of human pressure on the biosphere.[4] Barfield's state of 'original participation' or what Berman (2000) describes as a diffusely or obliquely focused ontological ease in the face of paradox is, by definition, unavailable to *Homo clausus*—i.e. to mobile, disembedded individuals sustained by complex differentiated and higher energy throughput societies. It is for this reason both Berman and Barfield are adamant in denying the possibility of any kind of 'return' (at least without the total and utter collapse of global industrial society and population levels).

It is in this context that Barfield compares the project of willfully or intentionally recovering a participating consciousness to riding a bicycle. It may be hard to learn to ride a bike, but *unlearning* it, once mastered, is practically impossible because the

[4]Having said this, the fact that Elias was an important figure with Foulkes in the early group-analytical movement underlines that his view was not entirely deterministic and that he recognized considerable lee-way in the manner in which the 'we-less I' typical of later-stage complex societies might be integrated and problems of ontological insecurity and mental instability ameliorated (Mennell 1997; Waldhoff 2007; Maratos 2018).

knowledge and skills involved have moved from the sphere of conscious direction to the domain of muscle memory and 'figurative' thinking in which perception and action are welded and fused into an orientation that has become, in Elias's term, 'second nature'. Placed on a bike, it is impossible not to know 'how' to ride. Regardless of conscious intent, body and back-brain know what to do. In more sociological terms, once disembedded and individuated, the literate self-actualizing citizen who operates as a separate self with a much greater capacity for abstract thought, detour behaviour, etc. may well experience greater unhappiness, self-doubt, narcissistic preoccupation, alienation or ontological insecurity. But they are also unlikely to seek, and nor would they be able consciously to re-embrace, the level of social integration and external constraint necessary to recover the embedded, paradoxical and diffuse sense of one-ness characteristic of a participating consciousness. Unconscious participation can't be willed.

Despite this problem, Barfield does go on to explore the possibility of 'final participation'—a renewed participative mode of cognition and consciousness that is compatible with post-Newtonian scientific procedures and models. As I will argue below, such final participation may depend on a capacity selectively to accommodate cognitive dissonance and, in Elias's terms, creatively and intentionally to juggle modes of involvement and detachment in new ways. This may open up a new trajectory for ecological conscience formation that does not fall foul of the tandem nature of the triad of controls.

The Paradox of Civilization and the Great Acceleration

So how is this account of the relation between human nature, second nature and the nature of consciousness relevant to the twenty-first-century crisis of global ecology? Humanity is a terraforming species par excellence (Quilley 2004a, b). 'Fire stick farming', hunting and later human agriculture, and, with a vengeance, fossil fuelled industrialism, have canalized and transformed the underlying evolutionary-climatic regime, precipitating an accelerating transition from the Holocene into the Anthropocene (Goudsblom 2002).

Although humanity is without doubt creating what appears to be an existential ecological crisis, this is not unprecedented (Goudsblom, 2002, p. 22). As a symbolizing species, with a capacity for the unbroken, inter-generational transmission of an expanding stock of reality-congruent knowledge about non-human nature, our species has carved out a disruptive niche that centres on side-stepping external constraints. These shared and increasingly abstract models of the human environment (including other humans) provide the foundation for all the technical innovation and material construction that continually upgrade our species' capacity to transform the world.

Box 6 *Symbol Emancipation*: Symbol emancipation refers to the evolution of language and a species with the capacity to manipulate symbols (i.e. signs with arbitrary, conventional and culturally specific meanings). Through social stocks of knowledge (models of the world and categories) stored and shared outside of individuals, language allowed the accumulation and transmission of experience over generations and a gradual refinement of culture as a 'means of orientation' in the world (Elias 2011).

This has culminated in an emerging paradigm shift in the relationship between the anthroposphere and the biosphere, with the former now, to a great extent, 'containing' the latter. At the start of the twenty-first century, the greater part of the earth's surface is fenced off, criss-crossed by roads, railways, fences and communications wires.

From the criss-cross of vapour trails, through to the hundreds of orbiting satellites, like every part of the earth, even the outermost layers of the atmosphere are saturated with anthropogenic signatures. Red-neck tycoons and leviathan corporations scramble to exploit mineral deposits on the deepest ocean floors, and, more speculatively, even near-Earth asteroids or the moon. Hunted and fished out, congested with shipping, straddled by pipe lines and cables, the oceans are increasingly construed as farmscapes whose purpose, like agricultural land, is to trap and process carbon and energy. From an evolutionary perspective, this is really significant. The best educated guesses suggest that by the end of the twenty-first century, up to 70% of the terrestrial biomass product will be monopolized by farming, almost certainly using GMOS. This presents a stark 'choice' for organisms that have not thrown in their lot with humanity and are not partners in this terraforming evolutionary regime. If they have a future at all, such organisms will be forced to exist and to evolve at the margins of what Goudsblom called the anthroposphere. No longer the apex of a Tennysonian image of evolution, 'red in tooth and claw', large predators such as tigers or bears have become evolutionary fossils, out of sync with, and maladapted to, a world increasingly created by and for human beings and their allies. Rather than wolves and lions, it is the dogs and cats that have found a permanent niche in a context in which the evolutionary trajectory of the biosphere has become subordinate to the techno-cultural trajectory of the anthroposphere. What Éduoard Le Roy called the 'humanization' of the earth is, however, not unnatural. Rather, it is just as much the outcome of the variation and selection of both human beings and allied species over tens of thousands of years (Budiansky 1994). Eisenberg (1998) likens the process to a hostile corporate take-over. But he also underlines the extensive nature of the coalition involved, which includes not just domesticated plants and animals, but gut bacteria, viruses and thousands of species that have evolved and adapted to the novel ecosystems that have emerged in the wake of anthropogenic change. The resulting anthropogenic web of mutualism and symbiotic cooperation should, according to Coppinger and Smith (1983), be understood as an entirely new evolutionary paradigm—the 'age of interdependent forms' (1983).

The Process of Civilization and Ecological Integrity

Civilization emerged during the benign climatic and ecological circumstances of the Holocene. Other things being equal, the process of civilization described by Elias might be construed (and has been by many) as broadly progressive in a normative sense—a case made most recently by Pinker (2012, 2018). Albeit, at great social cost and with unavoidable levels of violence, the pacification of nation-state societies, the advance of scientific knowledge and the efficiency of price-setting markets have all contributed to a world in which nearly three billion people have been pulled out of absolute poverty, levels of interpersonal violence are declining, medical science makes great strides against crippling diseases, the concept of individual human rights has been sanctified as the centre piece of an emerging global culture and technological innovation proceeds apace.

But, as is often the case, other things are not equal. As Elias showed, the process of civilization depends on controls over non-human nature that are neither neutral nor necessarily sustainable (Fig. 1). The process of eco-genesis that is woven into the human career can be thought of as an escalating tension between the level of integration represented by culture and that of biology. In the twenty-first century, this tension is by most reckonings reaching a crisis point (Rockström et al. 2009).

From the perspective of the co-evolutionary development of the triad of social (I), psychological (II) and ecological (III) controls—and the link between social complexity and energy/material throughput and ecological domination—I would suggest that the possible outcomes of the global ecological crisis are limited. Rifkin (2009) came to the same conclusion, arguing in *The Empathic Civilization* that there was a deep paradox in the extent to which the scientific awareness of human impacts and the global ecological consciousness and emotional identification with the biosphere depend entirely on a scientific and techno-economic advances that are intrinsically unsustainable. To some extent, Pinker's (2018) manifesto for Enlightenment progress is based on the same presupposition—the difference being that for Pinker, the ecological costs are both sustainable and bearable.

An Ecological Civilizing Process?

It is possible that the species with the greatest capacity for destabilizing impacts on non-human nature, may yet prove to be the only species capable of exercising evolutionary self-restraint semi-political and semi-conscious internalization of restraints in relation to nature and environment. Elsewhere, I have discussed the possibility of a trajectory of continuing development and complexification on the basis of an 'ecological civilizing process' (Quilley 2009; see also Aarts et al. 1995). I concluded then that any process of ecological conscience formation involving an internalization of behavioural restraints sufficient to have an impact on the balance between the anthroposphere and the biosphere would imply an intensifying architecture of national and even supranational state controls on behaviour and the emergence of even greater levels of psychological restraint. To the extent that the restrained personality of liberal societies has been associated with significantly higher levels of mental illness and unhappiness, any paradigmatic shift towards an ecological superego would seem to suggest even greater tensions and pathologies.

Any such outcome would depend, to a great extent, on a progressive shift in the involvement–detachment balance. For the notion of sustainable development to have any real traction, it must refer in part to a civilizational spurt towards greater detachment and less fantasy-laden and more reality-congruent models of social relationships between individuals and groups. Such knowledge would be prerequisite for any serious directive remodelling of society. It is possible that the existence of concepts such as 'anthroposphere' is indicative that a deep-seated change in our 'means of orientation' is already in process. The sharp decline in interpersonal violence within modern societies was, for Elias, evidence of a more differentiated superego and drive controls engendered by modern processes of personality formation. This personality structure, Elias argued, is a function of stable state monopolies on violence, a complex array of regulatory structures and incentives and patterns of socialization—all of which are, in turn, made possible by the growth. Energy flows in the form of economic surplus facilitate greater complexity and relations of interdependency which raise the cost of conflict whilst incentivising cooperation and restraint. In this context, it is possible to imagine a concerted regulation and reduction of the ecological footprint of human activities and the emergence of a more 'biophilic' pattern of affective restraint in relation to the ecological impacts of behaviour. But it seems likely that such an outcome would depend on an escalation in the scope and intensity of interdependent relationships between individuals and groups on a global scale, as well as supranational institutions, legal systems and forms of regulatory monopoly accompanied by a significant capacity to sanction rule-breakers. In other words, any putative 'ecological civilizing process' would be still-born without supranational mechanisms of pacification and socialization (Quilley 2009; see also Aarts et al. 1995).

Box 7 *Reality-Congruence*: Elias eschewed terms like 'objective' and 'subjective' because they were static and reduced dynamic processes to things. Rather, he argued, the models that humans use to orient themselves in the world can be more or less congruent with reality, depending on the extent to which more 'detached' categories and theoretical frameworks displaced more 'involved' ones. This terminology did not fall into the trap of positing scientific models that were 'true' (objective) in an absolute sense and retained the insight that human interests and motivations (involvements) are always at play in the creation of knowledge and orienting frameworks. A reasonable objective was to move towards the pole of detachment and greater reality-congruence. Hitherto, this has proved much easier in the natural sciences than in the social sciences where politics, ethics, affect-laden fantasy and normative commitments always reintroduce higher and often unavoidable levels of involvement to bear on models.

From this perspective, the process of civilization is now eroding the ecological pre-conditions for social complexity—a process that left unchecked must engender a countervailing dynamic of 'de-civilization', i.e. reduced scale and intensity of human interdependencies, demographic collapse, a reduction in the capacity of states particularly with regard to their capacity to assert monopolies over taxation and violence and the re-emergence of sub-state survival units organized around clan, family, place, community and tribe. Over the medium term, any concerted socio-genetic movement in this direction would engender a corollary psycho-genetic reversal with the relaxation of internal controls, a shift in the I/We balance towards the latter, more compelling forms of mutual identification and a process de-individualization—all of which could be characterized as the re-emergence of more traditional and gemeinschaftlich forms of society. Overall, such developments would almost certainly lead to a collapse of the global world system. Aside from increasing the likelihood of significant geo-political conflict and war, in the long term, this process of de-civilization would almost certainly be accompanied by the regionalization and localization of both markets, trade and production but also political authority.

Even with the most optimistic appraisal of geo-political opportunities for global environmental governance, it is difficult to imagine any such process of ecological conscience formation having a significant impact on the balance between the anthroposphere and the biosphere, without paradigmatic technical transformations in respect of the metabolic cycling of materials and the underlying energy regime of the global industrial economy (Quilley 2013; Kish and Quilley 2017; Quilley et al. 2016). In the energy-systems language of H. T. Odum, such a scenario implies a reduction in the unit 'transformity' costs of complexity (Odum 2007). 'Transformity' refers to the energy literally embodied in the material infrastructure of a given artefact or system combined with all of the energy lost, transformed and embodied in all the nested subsystems required to engender such a final product. So, a beef burger requires the entire infrastructure of cattle farms, transportation systems, skilled workers, slaughter-houses, food manufacturing sites, packaging plants, regulatory systems and fast-food franchises; and in just the same way, 'equality under the law' requires court houses, university training schools and even translators and 'legal aid.' The concept allows both the overall complexity as well as specific intangible achievements of Western culture (e.g. democracy) to be accorded a notional energy cost measured in embodied solar energy ('SolarEMjoules').

Avoiding the Ecological Sword of Damocles

From an Eliasian perspective, the long-range options for humanity seem limited. With respect to levels of integration and 'the great evolution' (section "The Great Evolution"), the issue turns on the extent to which humanity can stabilize global civilization as a level of integration canalizing and moulding the ecology of the biosphere and all subsequent evolutionary–ecological trajectories (Quilley 2004a, b, 2011). This is the prospect implicit in all versions of eco-modernism. Technological breakthroughs reducing the 'unit costs of complexity' are not inconceivable—although it seems likely that, to the extent that such innovations facilitate ongoing growth, they will generate crisis points and bottle necks with respect to other scarce resources (such as cobalt). With regard to the underlying architecture of the theory of civilizing processes, however, there is a deeper problem. The correlation of social complexity and interdependence with the internalization of restraints suggests possible psycho-social limits to civilization. It seems implausible that the mind of the human animal can be simultaneously individuated—i.e. disembedded and cut loose from the diffuse, immersive pattern of cognition dominated by perception—and continuously further contained and immobilized by a habitus of ambivalence (Bauman 1991) and a super-charged superego. There is something very paradoxical about the formulation of an ecological civilizing process that requires agents of the *Homo clausus* variety to internalize high levels of restraint vis-à-vis consumption or expectations for material standard of living whilst at the same time being construed by the transactional economy as *Homo economicus*—the arbiter of an individualistic, abstract economistic, preference-oriented rationality.

And even more problematic is the extent to which Elias accepts the caustic ontological humanism of the Enlightenment which has proved just as corrosive of meaning and collective cultural capital, or what Macintyre (2007) would understand as shared virtue. For Elias, these problems of individual meaning, social–cultural orientation and democratic legitimation are unavoidable concomitants of the society of individuals dominated and raised up by scientific models and the detached process of model making. The antithesis of detachment for Elias is a social and political world dominated by fantasy thinking; and having lost family and friends to the Nazi regime, Elias had both academic and incontestable personal reasons for foregrounding the dangers of involvement, myth making and the technics of ritual in politics (2013).

Elias was of course aware of the complex interaction of processes of involvement and detachment. Referring to the passion of scientists working to elaborate models of the natural world, he invoked the possibility of 'secondary involvement' (Elias and Dunning 2008). Nevertheless, for the most part, it is an unspoken premise of Eliasian sociology that a state of disenchantment (Weber) involving a concomitant ontological insecurity (Giddens; Laing), anomie (Durkheim) and narcissism (Lasch) is to a large extent an unavoidable function of social complexity.

And yet even this preliminary comparison of his work with the political economy of Karl Polanyi and Owen Barfield's history of consciousness suggests, at least in outline, an alternative possibility that is at least worth exploring. To what extent might it be possible to live in two worlds at once? Adam Seligman's study of ritual and sincerity as two relations to the notion of truth is instructive here. All societies, he argues, have recourse to both modalities. But starting with the protestant reformation, the dominance of ontological individualism in philosophy and the spread of scientific rationality, western-type societies have seen a pronounced shift away from ritual towards the pole of sincerity. And yet it is the technics of ritual that have traditionally allowed the coexistence of radically different understandings of the world without forcing resolution or the collapse of one into the other.

Elsewhere and drawing on Ophuls (2011), I have argued that if the ecological politics of degrowth requires progressives to accept the consequences of some contraction of both state and market, this implies a simultaneous re-expansion of what Polanyi referred to as 'livelihood'—an economy of more place-bound market places, reciprocal and communitarian ties between groups and communities; a greater emphasis on marriage, extended family and kindred, as well as community; a reversal of globalization in favour of national imaginaries, we-identifications and institutional protections (Quilley 2011, 2012, 2013, 2017; Quilley and Zywert 2019; Zywert and Quilley 2018). If the anthropology of pre-modern societies is any guide, it would seem implausible that such communities of reciprocation could be sustained without the re-emergence of a shared and lived understanding of virtue (Drehers 2017; c.f. Macintyre 2007). In essence, this would seem to involve a pronounced process of re-involvement, the re-valorization of non-rational drivers of behaviour, the re-embrace of technics involving ritual, temporal cycles, rites of passage and a place for the rediscovery of what Victor Turner called the anthropology of collective joy or 'communitas' (2012).

If, as Becker (2014) argued, the fear of death is an inextinguishable driver of human behaviour in the form of ontologically insulating 'hero/immortality projects', then the rationalist vision of a culture and politics organized in accordance to ever more detached and reality-congruent models of the social world is likely to be forever receding vision. On the other hand, few would be willing to relinquish the scientific worldview any more than the technological fruits it has brought forth—which raises the question as to whether individually and collectively, it might be possible to live simultaneously in both worlds at once.

There is no room to explore this possibility here except to suggest that Adam Seligman's account of ritual, rooted as it is in decades as a practitioner in the field of mediation, suggests exactly this: the purpose and function of ritual is to stabilize the coexistence of different worlds and worldviews and to mould conscience formation such as to make cognitive dissonance reasonable and even creative. One may get a glimmer of the possibilities of this idea by considering the role of 'elves' in Icelandic planning decisions (see Puolamäki 2012). Roads and developments are cancelled, moved and held up routinely with reference to the interests of communities of elves whose presence is acknowledged obliquely but insistently. The resulting process of accommodation has resulted neither in the disappearance of the elves (as a trivial fantasy) nor the collapse of the scientific rationality embodied in the planning process (the will to mould the external world in response to a singular, anthropocentric vision of rational progress). It was perhaps this possibility to which Owen Barfield alludes when he speaks of 'final participation' as the culminating stage in the evolution of consciousness (1965). In Elias's terms, it might be better understood as an individual and collective capacity to manage the balance between involvement and detachment intentionally, contextually and with deliberation, in relation to stronger and more embedded understanding of virtue, and a more viscous, networked social life, involving the regularization of greater obligations, less (but still considerable) spatial and social mobility.

Conclusion: Eco-Modernization, Degrowth and (Supra)National World Order

Since the 1648 Treaty of Westphalia, the working assumption of those involved in the business of international relations has been that the natural unit of analysis and intervention is the nation-state. This was the vessel through which the vehicle of civic nationalism allowed for the gradual and faltering process of democratization and eventually the partial taming of the market through the welfare state. As Elias (2012b) pointed out, clashing political philosophies notwithstanding, this was in part due to what he called 'functional democratization'—the equalization of power differentials between different groups in more complex economies defined by greater interdependence and opacity. But Westphalia was a significant turning point precisely because it provided an international guarantee for the emergence of small, cellular societies, internally homogenous and with fixed, semi-permeable but controllable borders.

In the wake of forty years of globalization and a concomitant recognition (albeit very gradual) of a global climate emergency, a rather different set of assumptions are beginning to undermine the nation-state as a natural category. De facto mobility of capital, a much more contested mobility on the part of labour, as well as increasing flows of migrants and refugees, are all concatenating factors exposing the shibboleths of national sovereignty, at least in the discourse of globalists, as an unsustainable and unhelpful sham. At the 2019 Liberal Democrat conference in the UK, Brexit negotiator Guy Verhofstadt poured scorn on what he saw as the Little Englander ideas of the Brexiteers. The new world order was, he argued, not driven by parochial nation-states but empires. Pressed by China, the Russian Federation, North America, Europe had, he argued, no option but to coalesce into a supranational federal state—a European Empire.

In the introduction, it was suggested that environmental politics in the Anthropocene was increasingly characterized by a tension between two broad scenarios, namely *degrowth and relocalization* on the one hand and the kind of *eco-modernism* associated with the Green New Deal on the other. I have explored these trajectories extensively elsewhere (Quilley 2011, 2012, 2013, 2017, 2018). How does the Eliasian framework elucidated here help in evaluating the reality-congruence of these models?

Degrowth is not a liberal prospect (Quilley 2013). Systemic relocalization, the reduction in international and even intranational trade, a contraction in the scale of the formal economy would certainly reduce the energy and material throughputs that drive climate change, i.e. a positive development in terms of 'eco-genesis' (III) and the long-term viability of the social–ecological regime undergirding global civilization. But in shedding layers of social complexity, the directional of 'socio-genesis' (I) would be towards lower levels and intensities of interdependence between individuals and groups, both within and between nation-states. In such circumstances, the tandem linkages identified with the triad of controls would suggest a period of 'decivilization', the loosening of internal psychological constraints and a greater prevalence of interpersonal violence (III—Dunning and Mennell 1998). On the upside, in smaller scale more place-bound communities, individuals may well enjoy better mental health and be less prey to processes of alienation, narcissism and anomie associated with highly individualized consumer societies.

With declining trade links, any corollary decline in the superstructure of global governance organizations would see a reversal of the 'functional democratization' (Elias 2012b) that has led to an equalization of power ratios between nation-states and blocs, and diminished the prospect of violent conflict. Together with conflict over scarce resources, supply and distribution problems, such circumstances would be conducive to more frequent wars. Within states, the diminishing resources available to central authorities and an ebbing monopoly over violence would in many places also engender violent conflicts between groups.

Finally, it should also be obvious that there are formidable obstacles to any strategy of intentional degrowth. Clearly, in growth societies, in relation to food, physical well-being and income-security, the combination of state and market has all but completely replaced family, clan and place-bound community as the fundamental survival unit. Any substantial reduction in the securities afforded by the State-Market would engender an immediate legitimation crisis. The drivers of growth are now just as much the ubiquitous reliance on systems of health, employment and policing that are almost impossible to replace overnight, if ever as a matter of policy (Zywert and Quilley 2018). Faced with such realities, even ostensibly radical Green Parties, as they have challenged for electoral representation, have moved away from neo-Malthusian discourses and 'limits to growth'—opting instead for version of ecological modernisation: the green new deal, green growth, etc.

Even more, telling is that Green Parties have without exception embraced the politics of diversity with its emphasis on feminist gender politics, marginality and critical race theory. Whatever their intrinsic merits, such emancipatory positions represent the radicalization of Enlightenment individualism which can only flourish in a high-energy, complex society of individuals (Quilley 2011; Elias 2010a, b, c). Thus, for instance, the 2019 Canadian Green Party manifesto contains a high-profile commitment to universal childcare. By enormously extending entitlements, this policy requires an increase rather than a decrease in fiscal transfers from a growing economy. At the same time, by maximizing female participation in the labour market, it also drives the same economic expansion. This is telling instance of how politically implausible it would be, under any conceivable circumstances, for political parties to proactively embrace a strategy of intentional contraction. Even were such a scenario conceivable with regard to the domestic arena, in the world of climate instability, any individual nation-state or bloc of states would also render itself vulnerable to predatory competition or military pressure from rival states electing to play the strategy of 'last man standing.'

With such dynamics in mind, it should be clear that degrowth would certainly not provide a vehicle for an ecological civilizing process in an Eliasian sense, i.e. spiralling, iterative external (regulatory) and internal (psychic) controls on environmental behaviours (Quilley 2009).

Eco-modernisation, on the other hand, intimates exactly such a process: a steady growth in social complexity, i.e. in the intensity, scale and scope of relations of interdependency between individuals and groups; an expanding architecture of institutional and supranational governance arrangements at the global level (and perhaps even Verhofstadt's green EU empire). In the context of a Green New Deal—unfolding in the context of cooperation between the major blocs—heightened regulation, technical surveillance and social monitoring leveraging processes of shame and embarrassment could conceivably affect a process of ecological conscience formation.

However, from an Eliasian perspective, such a scenario is also problematic for two reasons. First and most obviously, there is an enormous literature debunking the claims of ecological modernists in relation to 'decoupling' of growth from material ecological impacts. Limits to growth are real, and eco-modernisation is predicated on continuing growth by definition. Technological advances notwithstanding, any plausible trajectory of sustainable development must involve some contraction in the metabolic throughputs associated with the formal economy, the state and the market.

The questions really turn on the extent of this contraction, and the degree to which it can be selective. If the earth can support a given metabolic scale with a defined throughput of energy and materials—which is the axiomatic starting position for ecological economics (Daly and Farley 2010)—then how can/should this quotient of social complexity be allocated. The history of central planning in socialist countries is an evidence enough of the folly of moving in this direction. And yet, without a deeply embedded process of ecological conscience formation emerging from patterns of childhood socialization and the social structure of everyday life, it is very difficult to see how a market society can deliver the minimum infrastructures of Market-State society without dissolving the underlying lattice of familial and community obligations that constitute the countervailing domain of Livelihood (see Quilley 2018; Quilley and Zywert 2019).

It is also difficult to imagine a highly individualized society of spatially and socially mobile individuals (Elias 2010a, b, c; Beck 1992, Chap. 5) that did not depend on constantly expanding cycles of consumption. The underlying reasons are complex but stem in the most part from the process of individualization and the disembedding of individuals from place-bound traditional communities. As social psychologists working in the tradition of Ernest Becker have discovered, as the structures of meaning associated with place-affiliations, shared religion and rituals have ebbed away, consumption has become a dominant 'hero/immortality project' for individuals in western societies—creating meaning, undergirding identity, providing ontological security and assuaging fears of death (Arndt et al. 2004).

This problem is severe and to some extent intractable. However, it is at least possible to conceive of a trajectory that combines elements of both the eco-modernist and degrowth scenarios. In theory what is needed is a combination of:

(a) *Eco-modernism*: a technical and infrastructural revolution that delivers cleaner energy and reduces the transformity costs of modern civilization (the Green new Deal). This would be necessary to secure the minimal level of social complexity need to secure stable, peaceful and well-functioning market societies with nation-states capable of delivering internal and external security, as well as sufficient supranational cooperation to deliver geo-political and environmental governance.
And,
(b) *Relocalization and degrowth*: a socially conservative counter-revolution that partially re-embeds individuals into the stream of attachments and family life within place-bound communities and landscapes—or what I have referred elsewhere to as the domain of Livelihood. The latter would be necessary to transform the patterns of socialization and create individual and group lifeways that generate spiritual meaning in the context of shared (religious/spiritual) mythologies, rituals and cosmologies. It is such relational and place-bound attachments that might engender the kind of ecological conscience formation and the internalization of constraints that would be a prerequisite for any post-consumer society.

From an Eliasian perspective, this suggests a more complex and differentiated superego with a more restrained habitus in relation to matters of consumption emerging not on the back of external constraints associated with the State and the Market, but the rather associated with shared ontology and practice woven into the tissue of everyday life.

At first sight, this seems paradoxical. Since, in the poem Dover Beach, Matthew Arnold lamented the 'melancholy long with drawing roar' of the retreating 'Sea of Faith', it has been a staple of modern commentary that modern life must inevitably displace religion, superstition and relational ontologies with a ubiquitous instrumental rationalism and individualism—a firm foundation for consumerism. In this sense, there is an intrinsic relation between the corrosive fungability of the market, which makes everything commensurable in terms of quantity and price on the one hand, and the scientific rationalism which insists on explaining all phenomena and relations in terms of a single unified theory, on the other.

As illuminating as Elias's perspective is, it is fundamentally Weberian in accepting individualisation, rationalization and disenchantment as an unavoidable concomitant of any modern society. The analysis presented here shows that the competing paradigms of eco-modernization and degrowth are both incomplete but, paradoxically, may be complementary. However, this requires that Elias's rationalism is supplemented with an openness to ritual and spirituality as hero/immortality projects (in Ernest Becker's sense):

- Firstly, as a 'technics' by which to allow the lived juxtaposition of scientific rationalism with religio-spiritual forms of participation (in Owen Barfield's sense); and
- Secondly, as a vehicle for allowing market rationality to exist alongside without corroding or absorbing the substantive rationality (in Karl Polanyi's sense) associated with embedded forms of reciprocation in the domains of family and Livelihood.

What makes these possibilities at least conceivable is the way in which, as Seligman (2008) demonstrated, ritual, in particular, allows the stabilization and coexistence of multiple ostensibly incompatible rationalities, worldviews and value-hierarchies—without collapsing or subsuming. As he argues, in contrast to 'sincerity', ritual operates a kind of subjunctive 'as if' moment—of the kind that children are able habitually to slip in and out of, during the course of fantasy play. Although conscious of the teleological dangers of Victorian progress theory, Elias tended to assume that a directional cognitive integration around more and more consistent models was an automatic concomitant of progress—at least to the extent that progress is defined by a social capacity to refine more reality-congruent models of reality. Seligman amends this picture. By stabilizing and making a virtue of, sometimes intentional, cognitive dissonance, ritual can allow strictly reality-congruent and more detached models (in Elias's sense) to operate alongside and even in support of subjunctive ('as if') realities—attachments, identities, relational intuitions—that make it possible for individual human beings behave in ways that are irrational. This is clearly critical for the project of ecological conscience formation. When faced with a question as to why 'I should deprive myself of a car, if 800 million Chinese people don't make the same decision', the rationalized calculus of game theory is of little help. On the other hand, it is religion and prescriptive community values that ensure that old order Mennonites keep driving horse buggies throughout every cold Ontario winter.

It is the simultaneous stabilization and circumscription of such involved patterns of group-thinking alongside rational-scientific modes of cognition and the selective movement between that may allow the tyranny of Elias's 'triad of controls' to be circumvented. And, at heart, this is what is intimated by any combination of traditional and modern, 'small and beautiful' with 'global and high tech' and involvement and detachment. An environmental politics which embraces the technics of ritual and non-rational or more involved drivers of behaviour may be better able to shift in the direction of reversing and loosening the trajectory of *Eco-genesis* (III); whilst in relation to *Socio-genesis* (I) simultaneously tightening some social controls (global governance) whilst loosening others (e.g. much more localized markets and production, community policing) and with respect to *Psycho-genesis* (II) engendering individuals who are able to experience greater participation and psycho-affective involvement in relations with people (face-to-face interaction) and nature, without losing the capacity, selectively to mobilize more detached models and means of orientation in interdependent relation with wider social systems of government, regulation and ecological understanding. This possibility could be characterized as depending on the mobilization of a conscious, creative, cognitive dissonance. And from an Eliasian perspective, this is, I believe, what the project of planetary ecological integrity requires.

References

Aarts, W., Goudsblom, J., Schmidt, K., & Spier, F. (1995). *Towards a morality of moderation. Report for the Dutch National Research Programme on global air pollution and climate change.* Amsterdam: Amsterdam School for Social Science Research.

Arndt, J., Solomon, S., Kasser, T., & Sheldon, K. (2004). The urge to splurge: A terror management account of materialism and consumer behavior. *Journal of Consumer Psychology, 14*(3), 198–212.

Barfield, O. (1965). *Saving the appearances. A study in idolatry.* New York: Harcourt, Brace.

Barfield, O. (1979). *History, guilt, and habit.* Middletown, CT: Wesleyan University Press.

Bauman, Z. (1991). *Modernity and ambivalence.* Cambridge: Polity Press.

Beck, U. (1992). *Risk society.* Cambridge: Polity Press.

Becker, E. (2014). *The denial of death.* London: Souvenir Press.

Berman, M. (2000). *Wandering god: A study in nomadic spirituality.* Albany: State University of New York Press.

Böhm, D. (1980). *Wholeness and the implicate order.* London: Routledge.

Budiansky, S. (1994). *The covenant of nature.* London: Weidenfeld & Nicholson.

Christian, D. (2011). *Maps of time: An introduction to big history* (2nd ed.). Berkeley: University of California Press.

Coppinger, R. P., & Smith, C. K. (1983, Winter). The domestication of evolution. *Environmental Conservation, 10*(4), 283–292.

Daly, H., & Farley, J. (2010). *Ecological economics.* Principles and Applications: Island Press.

Dreher, R. (2017). *The benedict option: A strategy for Christians in a post-Christian nation.* New York: Sentinel.

Dunning, E., & Mennell, S. (1998). Elias on Germany, Nazism and the Holocaust: On the balance between 'Civilizing' and 'Decivilizing' trends in the social development of Western Europe. *The British Journal of Sociology, 49*(3), 339–357.

Eisenberg, E. (1998). *The ecology of Eden.* London: Picador.

Elias, N. (2007a [1987]). *Involvement and detachment* (Ed. Quilley, S., 2nd Revised ed., Vol. 8). Collected works. Dublin: UCD Press [1987 ed., Oxford: Blackwell].

Elias, N. (2007b). An essay on time. In S. Loyal & S. Mennell (Eds.), *The collected works* (Vol. 9). Dublin: UCD Press.

Elias, N. (2010a). *The loneliness of the dying and humana conditio.* (Ed. Scott, A. & Scott, B.) (Vol. 6). The collected works. Dublin: UCD Press.

Elias, N. (2010b). *Mozart and other essays on courtly art* (Ed. Baker E. R., & Mennell, S.) (Vol. 12). The collected works. Dublin: UCD Press.

Elias, N. (2010c [1991]). *The society of individuals* (trans: Jephcott, E., Ed. Van Krieken, R.) (Vol. 10). The collected works. Dublin: UCD Press.

Elias, N. (2011 [1989]). *The symbol theory* (Ed. Kilminster, R.) (Vol. 13). The collected works. Dublin: UCD Press,

Elias, N. (2000 [1939]) 2000 ed. The civilizing process. Oxford: Blackwell; Standard edition (2012a). On the process of civilization: Sociogenetic and psychogenetic investigations (trans: Jephcott, E., Eds. Mennell, S., Dunning, E., Goudsblom, J., & Kilminster, (Vol. 3). The collected works. Dublin: UCD Press.

Elias. (2012b [1978]). *What is sociology* (trans: Morriseey, G., Mennell, S., & Jephcott, E., Eds. Bogner, A., Liston, K., & Mennell, S.) (Vol. 5). The collected works. Dublin: UCD Press.

Elias, N. (2013). *Studies on the Germans: Power struggles and the development of habitus in the nineteenth and twentieth centuries* (trans. and Eds. Mennell, S. & Eric Dunning, D.) (Vol. 11). The collected works. Dublin: UCD Press.
Elias, N., & Dunning, E. (2008) *Quest for excitement: Sport and leisure in the civilizing process* (Ed. Dunning, E.) (Vol. 7). The collected works. Dublin: UCD Press.
Evans-Pritchard, E. E. (1983 [1937]). *Witchcraft, magic and oracles among the Azande*. Oxford: Oxford University Press.
Gellner, E. (1983). *Nations and nationalism*. Cornell University Press.
Goudsblom, J. (1992). *Fire and civilization*. London: Allen Lane.
Goudsblom, J. (2002). Introductory overview: The expanding anthroposphere. In de Vries, B., & Goudsblom, J. (Eds.), *Mappae Mundi: Humans and their habitats in long-term ecological perspective*. Amsterdam: Amsterdam University Press.
Hollis, M., & Lukes, S. (1982). *Rationality and relativism*. Oxford: Basil Blackwell.
Kish, K., & Quilley, S. (2017). Wicked dilemmas of scale and complexity in the politics of degrowth. *Ecological Economics, 142,* 306–317.
Kaufman, S. (2000). *Investigations*. New York: Oxford University Press.
Lévy-Bruhl, L. (1926 [1910]). *How natives think*. London: George Allen & Unwin.
Lévy-Bruhl, L. (1978a [1922]). *Primitive mentality*. New York: AMS Press.
Lévy-Bruhl, L. (1978b [1949]). *The notebooks on primitive mentality (trans: Rivière, P.)*. London: Harper & Row.
Loyal, S., & Quilley, S. (Eds.). (2004). *The sociology of Norbert Elias*. Cambridge: CUP.
Loyal, S., & Quilley, S. (2013, July). Wittgenstein, Gellner, and Elias: From the philosophy of language games to a Figurational sociology of knowledge. *Human Figurations. Long Term Perspectives on the Human Condition, 2*(2). Retrieved from http://hdl.handle.net/2027/spo.11217607.0002.204.
MacIntyre, A. (2007 [1983]). *After virtue: A study in moral theory* (3rd ed.). Notre Dame, IN: University of Notre Dame Press.
Maratos, J. (Ed.). (2018). *Foundations of group analysis for the twenty-first century*. London: Routledge.
Mennell, S. (1997). A sociologist at the outset of group analysis: Norbert Elias and his sociology. *Group Analysis, 30*(4), 489–514.
Needham, J. (1937). *Integrative levels: A re-evaluation of the idea of progress*. Oxford: Clarendon Press.
Odum, H. T. (2007). *Environment, power, and society for the twenty-first century: The hierarchy of energy*. New York: Columbia University Press.
Ophuls, W. (2011). *Plato's revenge: Politics in the age of ecology*. Cambridge, MA: MIT Press.
Overing, J. (Ed.). (1985). *Reason and morality*. London: Tavistock Press
Penna, C. (2016). Homo Clausus, Homo Sacer, Homines Aperti: Challenges for group analysis in the 21st-century. A response to Haim Weinberg's 40th Foulkes Lecture. *Group Analysis, 49*(4), 357–369.
Pigliucci, M. (2001). *Phenotypic plasticity. Beyond nature and nurture*. London: John Hopkins University Press.
Pinker, S. (2002). *The blank slate: The modern denial of human nature*. New York: Viking Penguin.
Pinker, S. (2012). *The better angels of our nature: Why violence has declined*. New York: Penguin Books.
Pinker, S. (2018). *Enlightenment now: The case for reason, science, humanism, and progress*. London: Allen Lane.
Polanyi, K. (1964 [1944]). *The great transformation*. Boston: Beacon Press.
Puolamäki, L. (2012). Individual views and shared landscapes of folklore in Reykholtsdal, Iceland. *European Countryside,4*(2).
Quilley, S. (2004a). Ecology, 'Human Nature' and civilizing processes: Biology and sociology in the work of Norbert Elias. In S. Loyal, & S. Quilley (Eds.), *The sociology of Norbert Elias* (pp. 42–58). Cambridge: Cambridge University Press.

Quilley, S. (2004b, September). Social development as trophic expansion: Food systems, prosthetic ecology and the arrow of history. *Amsterdam Sociologisch Tijdschrift, 31*(3), 321–348.
Quilley, S. (2009). The land ethic as an ecological civilizing process: Aldo Leopold, Norbert Elias and environmental philosophy. *Environmental Ethics., 31*(2), 115–134.
Quilley, S. (2010). Integrative levels and the "Great Evolution": Organicist biology and the sociology of Norbert Elias. *Journal of Classical Sociology, 10*(4), 1–19.
Quilley, S. (2011). Entropy, the anthroposphere and the ecology of civilization: An essay on the problem of "liberalism in one village" in the long view. *Sociological Review, 59,* 65–90. https://doi.org/10.1111/j.1467-954X.2011.01979.x.
Quilley, S. (2012). System Innovation and a new "Great Transformation": Re-embedding economic life in the context of de-growth. *Journal of Social Entrepreneurship, 3*(2), 206–229.
Quilley, S. (2013). De-growth is not a liberal agenda: Relocalization, and the limits to low energy cosmopolitanism. *Environmental Values, 2013*(22), 261–285.
Quilley, S. (2017). Navigating the anthropocene: Environmental politics and complexity in an era of limits. In P. A. Victor & B. Dolter (Eds.), *Handbook on growth and sustainability* (pp. 439–470). Cheltenham: Edward Elgar. Retrieved from https://www.mdpi.com/2071-1050/11/15/4082.
Quilley, S. (2018). Individual or community as a frame of reference for health in modernity and in the anthropocene. In S. Quilley, K. Zywert, & S. Quilley (Eds.), *Health in the anthropocene: Living well on a finite planet.* Toronto University Press.
Quilley, S., & Loyal, S. (2005). Eliasian sociology as a central theory for the human sciences. *Current Sociology, 53*(1), 807–828.
Quilley, S., & Zywert, K. (2019). Livelihood, market and state: What does a political economy predicated on the 'individual-in-group-in-place' actually look like? *Sustainability, 11*(15), 4082. https://doi.org/10.3390/su11154082.
Quilley, S., Hawreliak, J., & Kish, K. (2016, September). Finding an alternate route: Towards open, eco-cyclical, and distributed production. *Journal of Peer Production* (9). Special Issue. Eds. Tréguer, F., Antoniadis, P., & Söderberg J. Retrieved from http://peerproduction.net/issues/issue-9-alternative-internets/peer-reviewed-papers/finding-an-alternate-route-towards-open-eco-cyclical-and-distributed-production/.
Rifkin, J. (2009). *The empathic civilization: The race to global consciousness in a world in crisis.* New York: J.P. Tarcher/Penguin.
Rockström, J., et al. (2009). A safe operating space for humanity. *Nature, 461.7263,* 472–475.
Seligman, A. (2008). *Ritual and its consequences. An essay on the limits of sincerity.* Oxford: Oxford University Press.
Turner, E. (2012). *Communitas: The anthropology of collective joy.* New York: Palgrave Macmillan.
Waldhoff, H. (2007). Unthinking the closed personality: Norbert Elias, group analysis and unconscious processes in a research group: Part I. *Group Analysis, 40*(3), 323–343.
Weber, M. (1968). *Economy and society: An outline of interpretive sociology* (Ed. Roth, G., & Wittich, C.). New York: Bedminster Press.
Wilson, B. (Ed.). (1970). *Rationality.* Oxford: Basil Blackwell.
Winch, P. (1990 [1958]). *The idea of social science.* London: Routledge & Kegan Paul.
Zywert, K., & Quilley, S. (Eds.). (2018). *Health in the anthropocene: Living well on a finite planet.* Toronto University Press.

Empirical Investigations

Anthropocentrisation and Its Discontents in Indonesia: Indigenous Communities, Non-Human Nature and Anthropocentric Political–Economic Governance

Shofwan Al Banna Choiruzzad

Jika negara tidak mengakui kami, kami tidak mengakui negara
(If the state does not recognise us, we don't recognise the state!)
Aliansi Masyarakat Adat Nusantara Declaration, 17 March 1999.

Abstract This chapter discusses the 'anthropocentrisation' of political–economic governance in Indonesia through the establishment, expansion and evolution of the modern state. The process began with colonial state-building in the mid-nineteenth century, following efforts by colonial rulers to exploit their colonies more effectively in order to compete in the global market. The creation of a unified national political and economic governance system with rigidly defined territories gradually displaced ecological governance systems of indigenous communities, and Indonesia's independence led to further institutionalisation of anthropocentric political–economic governance. The authoritarian and developmentalist New Order government (1965–1998) consolidated the power of the state and its control over people and nature, effectively marginalising indigenous communities, despite the formal recognition of *Adat Law*. State transformation in the age of globalisation, fragmentation, decentralisation and internationalisation of state apparatuses has gradually loosened the grip of the state since the 1980s. Indigenous communities, supported by transnational advocacy networks, used this opportunity to create a governance space for themselves. While these initiatives have been partially successful, the loosening grip of the state does not mean the reversal of anthropocentrisation.

Introduction

Human beings are increasingly aware that the changes in their environment, from warmer temperatures (Milman 2019) to more droughts and floods (Schwartz 2018), are consequences of human activities. An overwhelming majority of climate

S. A. B. Choiruzzad (✉)
Department of International Relations, Universitas Indonesia, Jakarta, Indonesia
e-mail: shofwan.albanna@ui.ac.id

J. C. Pereira and A. Saramago (eds.), *Non-Human Nature in World Politics*, Frontiers in International Relations, https://doi.org/10.1007/978-3-030-49496-4_8

scientists, ranging from 90 to 100% according to different studies, strongly believe that the current climate change is human induced (Cook et al. 2016). This realisation led to the introduction of the concept of the Anthropocene, first coined by Paul Crutzen and Eugene Stoermer, which refers to a new geological epoch in which humans (*Anthropos*) are the main factor that shapes the geology and ecology of the earth (Crutzen and Stoermer 2000). The introduction of the concept moves the discussion from a strictly scientific to a political one, as many scholars have argued that the arrival of the Anthropocene is not merely a natural phenomenon but a political–economic one. Some have even argued that the term Capitalocene is more appropriate, since the acceleration of the impacts of humanity on the environment could not be explained without understanding changes in the patterns of human organisation, especially the introduction and evolution of capitalism (Moore 2017).

Focusing on the case of Indonesia, this chapter argues that the arrival of the Anthropocene has been made possible by the 'anthropocentrisation' of political–economic governance through the invention, expansion and evolution of the modern state. Here anthropocentrisation is understood as the gradual process of replacement/displacement or domination of non-anthropocentric political–economic governance, in which the relationship of humans with nature is based on respect and harmony, by anthropocentric political–economic governance, in which humans are at the centre and nature is exploited for the interest of humans. In anthropocentric political–economic governance, humans are considered exceptional, and there is a clear distinction between 'human' and 'non-human'. In this mode of governance, the latter is exploited or preserved for the benefit of the former.

Building upon the State Transformation Framework (Hameiri and Jones 2016; Jones 2018; Jones and Zeng 2019) and Robyn Eckersley's theorisation of the possibility of transforming the state into a non-anthropocentric entity and thus the non-inherent nature of state's anti-ecological nature (Eckersley 2004), this chapter details the anthropocentrisation of Indonesia's political economic governance through the creation and transformation of the modern state since the colonial period. It also looks at how indigenous communities were affected by and responded to this process. Efforts by indigenous communities, supported by a network of academics, lawyers and activists, to push back against this anthropocentrisation to provide 'governance space' for indigenous communities have been partially successful. Despite some formal success, implementation has been impeded by the epistemological domination of an anthropocentric worldview in the state bureaucracy.

The Modern State and Anthropocentrisation

This chapter focuses on the development of the modern state in Indonesia and its anthropocentrising impacts, as the modern state is arguably the most complete human organisation in which anthropocentrism and its zoological politics are institutionalised. Premodern societies may or may not have been anthropocentric, but the establishment of a state often enables and justifies the total domination of the 'non-humans' within a particular territory for the satisfaction of humans, either for people or for profit. The state decides the status of non-humans and how they should be treated, such as when states produce laws defining and mediating animal's agency. A 2013 study pointed out how the state's regulation of dog ownership in the UK and the establishment of protected areas for elephants in India determined the status of those animals (Srinivasan 2013). A similar situation occurred with forests: the state decides which forests are permitted to be logged or cleared and which are to be conserved. With certificates from the state, humans can legally cut the trees and replace them with profitable plants, such as oil palms. Interestingly, the state can even reduce the status of humans to non-humans, as in the case of irregular migrants (Vaughan-Williams 2015, 2016).

As many scholars have argued, many traditional communities hold non-anthropocentric worldviews, seeing non-humans as living and equal counterparts in their traditional ecological knowledge. The distinction between human and non-human is therefore often blurred in traditional communities. These communities are known to live in harmony with nature and observe behavioural patterns that preserve the critical order of nature (Glenn 2014). Acknowledging the value of local traditional knowledge in environmental conservation, modern initiatives to protect the environment have often attempted to incorporate this knowledge into their implementation of conservation policies (Schroeder and González 2019).

In this context, the invention of the modern state in Europe and its expansion into other parts of the world through colonialism can be understood as a process of anthropocentrisation of political–economic governance. The modern state institutionalises an anthropocentric worldview through its bureaucratic machinery, giving it the political technology to control the natural environment and resources. The arrival of the modern state in Indonesia, as we will explain in this chapter, began with surveys of territory and nature as the resources to be exploited for the survival of the state. Forests, often considered sacred by indigenous societies, were mapped, divided and exploited based on legal permits issued by state authorities. Within the rigid and fixed borders of the modern state, the relationship between humans and nature was homogenised.

However, some scholars have reminded us that this anthropocentrising characteristic of the modern state is not inherent to the state itself. While many critical and green political theorists tend to be sceptical or even hostile towards the nation-state that they often consider as inherently ecocidal, Eckersley argued that the state can be transformed to be environmentally responsible (Eckersley 2004, p. 4). Despite its limitations, the state is the most legitimate and at the same time most powerful social institution to act as the 'public ecological trustee' (Eckersley 2004, p. 12). By analysing the social structures of international anarchy, global capitalism and the liberal democratic state, Eckersley argued that the anti-ecological behavioural dynamics generated by these three forces could be reversed. Because the state is not necessarily anti-ecological, it could be transformed into a 'green state' that is animated by ecological democracy (Eckersley 2004, p. 2, 14–15). While none have completely transformed into the idealised 'green state', there are cases where states reversed their anti-ecological behaviour. For example, policy changes by the state successfully reduced the annual rate of deforestation in the Amazon in the 2000s (Thaler 2017).

Thus, it can be argued that while the modern state is often an anthropocentrising force, it only becomes anthropocentric due to its particular historical development. The modern state institutionalises anthropocentrism not because it is an inherently anthropocentric political organisation but because during the early formation of modern states in Europe, as well as during the universalisation of the Westphalian system of nation-states, the modern states relied on the exploitation of humans and non-humans to be able to survive in a highly competitive environment.

Understanding that the anthropocentrising feature of the state is not inherent but rather a historical product necessitates a comprehensive understanding of the forces that shaped the evolution of the state. By relying on the State Transformation Framework, this chapter explains that states continuously evolve and transform in accordance with changes in political–economic relations. In this framework, the domain of national governance is not fixed and is continuously contested. Actors see that they will obtain more privilege if the governance of a particular sector is shifted to another level and will therefore struggle to restructure the system of governance. In this sense, the establishment and transformation of the state is a continuously contested process (Hameiri and Jones 2016). Consequently, while admitting the non-inherent nature of the state's anthropocentrism, this chapter argues that as long as the dominant social forces benefit from the exploitation of humans and non-humans, the state, as well as other governance mechanisms, will continue to be anthropocentric and even ecocidal, though this is not without resistance from other social groups, such as indigenous societies. Thus, to understand the evolution of the modern state and its anthropocentrising consequences, as well as the potential to reverse them, one needs to understand the political and economic competition in each historical phase of a state's evolution.

The historical process of the formation of the modern state in Europe can be understood in this light. European rulers, including the church, monarchs and a rising industrialist class, were involved in bloody struggles for influence resulting in the Westphalian compromise, which gave birth to the concept of the modern nation-state. This historical process in Europe led to the territorial concentration of power in which the modern nation-state asserted sovereign authority over humans and nature inside their clearly defined territories (Linklater 1998). The Westphalian model was then exported to other parts of the world through colonialism. However, unlike in Europe where the consolidation of the modern state, including the importance of territorial sovereignty and the idea of fixed national borders, took hundreds of years, this process was accelerated in many areas under colonial rule (Eilenberg 2012). This is in line with the 'uneven and combined development' argument, stating that early industrialised states often attempted to compress capitalist development in those that industrialised later (Chacko and Jayasuriya 2018). In this case, the development of the modern state, as the political technology that enables efficient exploitation of nature, in Southeast Asia was accelerated by the coercion of colonial rulers.

Colonial powers initiated colonial state-building to more effectively exploit resources in their colonies. This was often done by neglecting existing forms of human organisation, such as traditional communities, and their connection with local geographic features. As explained by many scholars, colonial state-building created, or at least attempted to create, unified political–economic governance based on modern state bureaucracy but at the same time divided traditional indigenous societies with modern borders (Eilenberg 2012). Nationalist movements borrowed and reframed the notion of the modern state to define their national struggle, and after the two World Wars had damaged the foundation of the global colonial order, these nationalist movements successfully carried the momentum for decolonisation and established new states based on the previous colonial states. Thus, many observers argued that the universalisation of the modern state was only achieved after the end of World War II (Hameiri and Jones 2016).

This universalisation of the modern state was not a linear process. Specific patterns of interaction between social forces in the colonies before and after independence shaped the functions and the limits of the newly established states. Since then, states have continued to evolve, mirroring the ever-changing political–economic relations. At some point, this led to the creation of developmental states in some countries in Asia (Vu 2007).

This chapter elaborates on the continuously contested process of anthropocentrisation that arose with the formation and evolution of the modern state in Indonesia (or the Dutch East Indies, as it was known before independence in 1945). It argues that the process of anthropocentrisation began with colonial state-building in the mid-nineteenth century, due to efforts by colonial rulers to exploit the colonies more effectively to be able to compete in the increasingly competitive global market. This process attempted to absorb and integrate traditional communities, many of which were living in a chthonic manner that was harmonious with nature, into a unified national political and economic governance system, which in turn classified nature inside the politically defined territorial boundaries as 'natural resources'. The establishment of the Indonesian state replaced the control of the colonial ruler, but the newly established modern state bureaucracy continued to consider nature as natural resources to be exploited. This time, the exploitation was not for the elites in the colonial metropolis but, at least formally, for the wealth of a specific group of people identified as 'the people of Indonesia'. Since then, the Indonesian state has continuously evolved, reflecting the dynamic political–economic relations that influence it. In those processes, some groups continue to resist the domination of the state and its anthropocentrising consequences, resulting in the formal but often ineffective recognition of *Adat Law*, as well as some other forms of recognition of indigenous societies, including their non-anthropocentric way of governing their relations with nature.

In this, Indonesia's experience is not entirely unique. Scholars have highlighted the ecological impacts of colonial and postcolonial state-building and the political–economic competition surrounding the process in areas such as Latin America. For example, Hecht discussed the emergence of 'insurgent citizenship' in the Amazon, illustrating the continuing struggle of power between Amazonians and the Brazilian state (Hecht 2011). This chapter argues that similar struggles have happened in Indonesia, but in different historical phases and social constellations.

The VOC and the Exploitation of the Non-Human Nature

This chapter does not argue that precolonial Indonesian societies always lived with a non-anthropocentric worldview. Rather, based on abundant studies on societies in precolonial Indonesia and beyond, it argues that before the introduction of the modern colonial state in early and mid-nineteenth century, indigenous or chthonic societies thrived in their non-anthropocentric way of life without much interference from the dominant powers in the archipelago. These dominant powers may or may not have held anthropocentric worldviews or adhered to anthropocentric political–economic governance. While scholars noted that premodern feudal states provided the political technology for elite consumers to coerce communities and purchase commodities to satisfy their needs and wants, which in turn significantly affected the environment far from their immediate domains (Morrison 2018), in Southeast Asia, these precolonial kingdoms often stressed the importance of nature and of maintaining harmonious relations with it in their belief systems and even in the way they governed the state. For example, Javanese philosophy obliged humans, and especially the leader, to maintain the harmony of the universe (*hamemayu hayuning bawono*).

In precolonial Southeast Asia, including in the areas later colonised by the Dutch, the system that governed political–economic relations was predominantly feudal. Different kingdoms ruled different parts of the archipelago, with some kingdoms ruling larger area than others and extracting tributes from smaller ones. While these feudal states were centralised, their degree of control over their subjects and territory differed from the modern state. As many scholars have noted, the feudal system was characterised by 'constant switches in the relative powers of the different layers of society' (Luiten van Zanden 2010, p. 157). The central power was not always able to control its subjects, such as regional elites or even peasants. There were also overlapping claims to land and labour that adjusted flexibly to changes in the relative power between these different groups (Luiten van Zanden 2010).

Furthermore, the feudal model of political–economic governance did not see territory in the same light as the modern state. Kingdoms and sultanates measured their power not in terms of territory but by their ability to earn tributes. These precolonial states prioritised the submission of people, expressed in tributes, rather than total control of territory (Eilenberg 2012, p. 78–79). Regional elites also competed with each other to attract people to be their subjects using various incentives, forming a 'mandala politics' or 'galactic politics' in Java, the island that later was the basis for colonial state-building (Kian 2008). The influence of the central rulers was like a multilayered sphere: strong at the centre but weaker in the outer layers. The Javanese phrase *adoh ratu, cedhak watu* ('far from the king, close to the stones') illustrates that the influence of the king decreased as the distance from him increased. The rulers were also not always interested in subjugating people far from their palace.

The constant variations of relative power between different groups, flexible and overlapping claims to land and labour and the relatively relaxed attitude towards territory created space for less-formal and less-institutionalised organisations of human society which bloomed at the edges of these precolonial kingdoms. Many chthonic or indigenous societies thrived with their non-anthropocentric worldview and coexisted with the relatively more anthropocentric feudal societies. In precolonial Kalimantan, for example, Iban *Dayaks* were never under direct control of the Melayu kingdoms that ruled the island. Some of them did pay tributes (the *serah dayaks*), but their relationship with the feudal state was relatively equal, and cooperation and conflict occurred in different contexts (Eilenberg 2012).

Following Dutch voyagers' 'discovery' of the Indonesian archipelago at the end of the sixteenth century, a mercantile company named the VOC (*Vereenigde Oost-indische Compagnie*, Dutch East Indies Company) was established in 1602 and secured trading posts in the archipelago by 1603. Resistance by local rulers led to the closure of some of these early trading posts, but in 1619, the company successfully captured Jayakarta (now Jakarta), heavily fortified it and renamed it Batavia. This city would later be the centre of Dutch colonial rule (Kian 2008). While the formal status of the VOC was a commercial company, it also acted as a state by raising soldiers, making treaties or declaring wars. Thus, some scholars consider the VOC a 'company state' (Weststeijn 2014).

The VOC's ambiguous position as both a trading company and a political power fit well with the existing system of that time. With its powerful military, the VOC quickly found its place in the messy politics between local rulers (Kian 2008; Luiten van Zanden 2010). It also smoothly adapted to local political–economic structures and even strengthened them, often benefiting from them economically. As it became more entrenched in the local political economy, the company also became increasingly territorial since their profit was tied to their control over land. To maintain this control, the VOC had to maintain a complex relationship with local rulers, as well as with Chinese, Arab and Indian merchants. This situation created an economy with high transaction costs, marked by a highly volatile and non-transparent market (Luiten van Zanden 2010). This, along with external geopolitical factors and the inability of its leaders to effectively manage the financial situation of the company, eventually led to the collapse of the VOC (de Vries and Van der Woude 1997).

Colonial State-Building: Territorialisation and Control Over People and Nature

Following the VOC's bankruptcy, the Dutch state took over possession of the company in 1800. Unlike the VOC, which grafted itself onto the local feudal system, the Dutch state attempted to reshape the political–economic governance of the colony. The Dutch started to develop a modern state bureaucracy in the Dutch East Indies following the Weberian model to effectively 'exploit the East Indies through agrarian production for the European market (coffee, tea, indigo, rubber and so on) and development of mineral resources' (Holleman 1981, p. XXIX). Thus, the introduction of the modern state in the Dutch East Indies was strongly connected to an anthropocentric motive to satisfy the needs of humans in Europe. The expansion of the colonial bureaucracy, centralised within the borders of its 'territory', institutionalised the anthropocentric 'human/non-human nature relationship' as the only mode of relation between humans and nature. Indigenous or chthonic local societies' way of living with nature, which was relatively preserved during the precolonial feudal period, was gradually marginalised. For example, the colonial state tried to control the movement of Iban *Dayaks* in Kalimantan (Eilenberg 2012). While these attempts were not always successful, anthropocentric political–economic governance was becoming the norm.

The development of a modern state with modern bureaucracy began in the early nineteenth century. As van Zanden has explained, it was started by the initiatives of Governor-General Daendels (1808–1811), who attempted to develop a formal Weberian bureaucracy to replace the VOC's feudalistic bureaucracy, which was prone to corruption. A brief takeover by the British, from 1811 to 1816, led by Raffles, did not stop this experiment. Raffles himself continued the initiative by introducing land rent, and when the Dutch returned in 1816, the new Governor-General, Van Der Capellen, continued the initiative by starting a systemic collection of statistical data (Luiten van Zanden 2010).

For colonial state-building, it was important to know the exact extent of the colonial territory. In March 1824, the Dutch and the British agreed to divide the colonial territories in the Malay world and draw up formal borders (Eilenberg 2012, p. 83). Following this development, explorations commenced as a key component of colonial state-building (Tagliacozzo 2010). These explorations became the basis of identifying the fixed territory of the colonial state, which was important for the collection of statistics and taxes, as well as the identification of natural resources. In this sense, the colonial state rigidly defined nature as natural resources to be exploited in the interests of the colonial state.

The identification of natural resources was important not only because the development of the colonial state was aimed at enabling effective exploitation of these resources but also because the colonial state-building process itself needed resources. For example, the concern of the colonial government about its dependence on European coal, as well as a rising demand for fuel, led to the commencement of expeditions in search of coal in the colony. This resulted in the discovery of a coal deposit in an area near Amuntai, Kalimantan, in 1846, causing a rush to find coal, and the creation of a Bureau of Mines in 1852 to be the specific agency responsible for geological explorations (Friederich and van Leeuwen 2017). After 1900, when the modern colonial bureaucracy was in place and the extent of colonial territory was largely identified, the hunt for natural resources became the main motivation for exploration (Tagliacozzo 2010).

Apart from the exploitation of minerals, the non-human nature was also exploited through agriculture. The cultivation system of Van Den Bosch, implemented from 1830 to 1870, brought huge wealth to the Netherlands. Under this system, people in the colony, particularly in Java, were forced to grow and deliver valuable commodities, such as coffee, sugar cane, indigo, pepper and so on, in exchange for meagre payments. These commodities were then sold on the global market through the Netherlands, connecting the colony to the global economy (Bosma 2007; Fasseur 1986). While there was significant criticism of the cultivation system, much of it focused on the *exploitation of humans* (people of the colony) *by other humans* (their colonial rulers). However, this system also significantly affected non-human nature that was 'colonised' by humans to produce commodities to satisfy human needs. Even when the cultivation system was replaced by the Liberal Policy (1870–1900), which opened up the colony for investments, the exploitation of non-human nature continued.

The formation of the colonial state resulted in a leap in the impact of human activities on nature. Previously, various modes of natural resource governance, from feudal to traditional, coexisted in the archipelago. Colonial state-building transformed the Dutch East Indies into a unified national economy integrated into the global economy (Marks 2010). Exploration, control and exploitation of natural resources by humans changed an archipelago with a plural ecological worldview into one anthropocentric entity that was one of the largest exporters of commodities.

However, as state transformation framework would argue, this was a contested process. The expansion of control and exploitation created tension with local communities. The state was never fully able to control all indigenous communities, especially those at the 'edges of the state' (Eilenberg 2012). Bureaucratic centralisation was one of the key initiatives of colonial state-building after Daendels, and many voices in the Dutch state advocated for unification of the legal system to support this effort. However, complex realities on the ground led to a pluralistic legal system (Ravensbergen 2018), with Dutch legal codes, introduced in 1848 and 1849 as part of colonial state-building, only applying to Europeans. While the application of Dutch commercial law was extended to 'foreign Orientals' in 1854, native communities were still regulated according to their own rules (Henley and Davidson 2008).

Following the policy of liberalisation in the late nineteenth and early twentieth centuries, the voices arguing for the unification of the legal system became louder. This development was mostly driven by an anthropocentric desire to control and exploit natural resources. Unification of the legal system, especially the Western system of land ownership and titling, would enable European entrepreneurs to legally take away customary lands (Henley and Davidson 2008). In this period, the introduction of the concept of *Adat Law* (*adatrecht*) by Van Vollenhoven, supported by the famous Orientalist Snouck Hurgronje, provided a strong defence for legal pluralism in the colonial state (Holleman 1981).[1] Van Vollenhoven and Hurgronje argued that forced subjugation to Dutch law would be counterproductive because it would damage the social order and possibly provoke resistance. Rather than forced

---

[1] Van Vollenhoven explained *Adat Laws* as 'the totality of the rules of conduct for natives and foreign Orientals that have, on the one hand, sanctions (therefore, law) and, on the other, are not codified (therefore, adat)'. Van Vollenhoven's unique characterisation of *Adat Law* was explored by various scholars (von Benda-Beckmann & von Benda-Beckmann 2013; Holleman 1981).

legal subjugation by the state, Van Vollenhoven and Hurgronje suggested that recognition would be the starting point for gradual reform of the *Adat Laws* from within the traditional communities themselves, which would lead to appropriate social and economic progress (von Benda-Beckmann and von Benda-Beckmann 2013). This school of thought finally won the debate in the 1920s, as the Great Depression lessened the Western thirst for land in the colonies.

Thus, while there was motivation for the protection of traditional common property against private and state interests (Henley and Davidson, 2008), the recognition of *Adat Laws* had a strong Orientalist ethnocentric bias. Rather than being driven by recognition of the merit of the *adat*, it was driven by an anthropocentric and ethnocentric view of the primacy of the Dutch legal system. However, rather than subjugation, the recognition of *Adat Laws* was designed to absorb local societies into the colonial state by 'fertilizing the Oriental with Western values' (von Benda-Beckmann and von Benda-Beckmann 2013, p. 174). While formally recognising the existence of *Adat Laws*, the colonial state decided when such law would actually be enforced. The colonial state, as the embodiment of anthropocentric political–economic governance, established itself as the main authority within its rigidly defined territories, and various local societies were forced to submit to this authority, including in how they were to interact with their surrounding nature.

The Postcolonial State, Developmentalism and Further Institutionalisation of Anthropocentrism

The consolidation of the modern state in Indonesia continued after the proclamation of independence in 1945, and this consolidation included the anthropocentric worldview of the modern state. The nationalist leaders, who became the elites of the newly established state, framed their struggle as resistance against the exploitation of the people in the colonies (*humans*) by their colonial masters (other *humans*), but had little to say about *the exploitation of the non-human nature by humans.*

While independence was declared as the liberation of the people in the colony, it did not declare the liberation of nature from colonisation by humans. Thus, the Indonesian state replaced the colonial state and took over the authority over people and nature within the defined territory but continued the anthropocentrisation that had begun with colonial state-building. Article 33, point (3), of the original version of the 1945 Constitution of the Republic of Indonesia, mentioned that '*Boemi dan air dan kekajaan alam jang terkandoeng di dalamnja dikoeasai oleh negara dan dipergoenakan oentoek sebesar-besar kemakmoeran rakjat*' ('the land and the waters as well as the natural riches therein are to be controlled by the state to be exploited to the greatest benefit of the people'). The exploitation of people by other people, *l'exloitation de l'homme par l'homme*, was considered morally wrong, but the exploitation of nature by people was not mentioned.

In the early years of independence, military interventions by the Dutch and internal conflicts led to ineffective control of the national bureaucracy. Combined with the anti-colonial view of President Sukarno, which led to the weakening of Indonesia's link to the global economy, this situation caused a decline in the exploitation of natural resources. In the agricultural sector, for example, Indonesia's palm oil export declined, and its global market share shrank from 40% in 1937 to 17% in 1958 (Palm Oil Agribusiness Strategic Policy Institute (PASPI) 2014, p. 11). Nevertheless, the Indonesian state continued to establish itself as the sole authority governing human and non-human nature (i.e. 'natural resources') in its territory. In 1960, the government issued the Basic Agrarian Law, which ruled that all land rights must not be in conflict with the 'national interest' and that the state could lease its lands to the private sector (Henley and Davidson 2008).

The domination of the state over people and nature would be further consolidated with the arrival of the New Order in 1965. Led by a military strongman, General Suharto, who would later be called 'the Father of Development' (*Bapak Pembangunan*), the government transformed Indonesia into an Authoritarian-Developmentalist state. Its policies were based on the ideology of developmentalism, which could be summed up by the 'development trilogy' (*Trilogi Pembangunan*): (1) high economic growth, (2) political stability and (3) equal distribution of the results of development.

High economic growth was to be achieved through the exploitation of Indonesia's nature, which was understood solely as 'resources'. After the fall from power of Sukarno, the new government established laws that enabled extraction of the country's natural resources on a much larger scale. To be able to exploit natural resources through large-scale operations that would deliver high growth, the state invited foreign capital with the enactment of Law No. 1/1967 on Foreign Investment. The government also enacted Law No. 5/1967 on Forestry and Law No. 11/1967 on Mining, which gave the state legal ownership over forest areas and became the basis for legal land-grabbing. The state claimed ownership over land, displacing many people, including indigenous communities, from their traditionally managed lands. The state then gave the rights to exploit these lands and nature to business groups, usually connected to government elites. Corruption and cronyism led to frequent legal and illegal displacements of people from their lands. These lands were then transformed into mining, logging or plantation areas. With these policies, exploitation of Indonesia's natural resources greatly intensified in the New Order Period, leading to average annual GDP growth of 7% from 1965 to 1997 (Thee 2007).

The New Order believed that high economic growth, achieved mostly through the exploitation of natural resources, could only occur under a condition of political stability. This meant that the state also needed to control the population. Thus, the state pacified other social institutions that might influence the relationship between people and between people and their surrounding nature. It 'exoticised' traditions and local cultures to attract tourists but simultaneously marginalised their roles in society (Avonius 2004). 'Adat' (customs, tradition) became an 'ideology of control' (Henley and Davidson 2008). In line with this policy, in 1979, the New Order government enacted Law No. 5/1979 on Village Government, which made the structure of village government uniform all across Indonesia, and replaced traditional village governance (McCarthy 2005). Attempts to control people and natural resources were also conducted through the transmigration policy that brought demographic and land-use changes to Indonesia's rural areas. The central government conducted transmigration to ease the population pressure in Java, further exploit the lands in the outer islands and develop a sense of unified nationhood (Hoey 2003). This policy, supported by the World Bank throughout 1976–1989, reportedly caused deforestation and elimination of indigenous cultures (Connor 2012; Fearnside 1997).

Among the so-called *Trilogi Pembangunan*, equal distribution of the results of development was the most problematic. The New Order government initiated programmes to achieve this aim, such as the Presidential Instruction School (*Sekolah Dasar Instruksi Presiden/Inpres*), and successfully reduced the absolute poverty rate from 40% in 1976 to 11% in 1996 (Thee 2007). However, many people, especially indigenous communities who were displaced and marginalised, never saw the benefits of development. The state's legal control over forests, for example, led to the concentration of wealth in some conglomerates that easily obtained government permits and concessions. Natural resources were exploited, and people were expelled, but the benefit did not trickle down.

The visible results of this kind of development, such as environmental destruction and the marginalisation of traditional societies, provoked resistance to the state, which predominantly responded with force. With its monopoly over the use of violence, the state was generally able to put down these protests, except in a few cases. However, the changing global political economy would transform the power of the state across the globe, and Indonesia was not immune to such change.

Fragmentation, Decentralisation and Internationalisation: Stronger Resistance, Stronger Pressure

Scholars who study state transformation noted that the fall of the Keynes-influenced Bretton Woods compromise, which enabled the consolidation of developmental states in the post-World War era, began with the rise of neoliberalism in the 1970s in the global North, during the administrations of Reagan and Thatcher. This new 'consensus', based on the classical economics idea that the state should not intervene and should be transformed to facilitate the primacy of market, then spread to countries in the global South through various mechanisms, including structural adjustment programmes to global financial markets and the rise of transnational production networks. The spread of this model, in which the role of the state is reconfigured, led to a broadening of the actors involved in global politics and more deregulated transnational flows (Hameiri and Jones 2016, p. 78–81). According to scholars of state transformation framework, the modern state that was universalised in the mid-twentieth century experienced 'fragmentation, decentralization, and internationalization of state apparatuses' (Jones 2018, p. 1–2).

This global trend affected Indonesia in many ways. The transformation of the state in the West gave rise to more influential international advocacy networks, which included those concerned with indigenous rights and the environment, and which were very active in spreading their influence because of the global nature of environmental issues. The attempts of these Western-based global advocacy networks matched the growing discontent of many social groups in Indonesia, including indigenous communities who were marginalised by New Order development policies. In turn, this triggered the birth of organised resistance against the domination of the Indonesian state in the management of humans and nature. Wahana Lingkungan Hidup Indonesia, established in the 1980s, was among the pioneers in advocating for indigenous rights and environmental protection. Scholars have noted that one important factor in the rise of indigenous peoples' rights advocacy in Indonesia was the support from transnational advocacy networks (Henley and Davidson 2008).

Apart from the rise of social movements backed by international actors, another trend which stemmed from state transformation trend was a process of gradual liberalisation from the 1980s. Despite this, the state remained as the most powerful authority.

The situation changed rapidly after the Asian Crisis in 1997. The collapse of the Rupiah, Indonesia's currency, led to a chain of events that caused the fall of President Suharto and ushered in the era of *Reformasi* (Reform). The crisis eroded the legitimacy of the New Order, which was based on the ideology of development, and distrust mounted against the New Order government and the state. Soon, various social forces both inside and outside Indonesia attempted to chip away at the authority of the central government. In order to recover from the crisis, Indonesia requested help from the International Monetary Fund (IMF), which agreed but demanded structural reforms in Indonesia's economy. As President Suharto signed the Letter of Intent between Indonesia and the IMF, much of the state's authority was taken away and shifted to the market with liberalisation, privatisation and deregulation policies. In this way, the control of humans and nature from the state to the market, and the dominance of an anthropocentric worldview in political–economic governance continued to thrive and even became more pronounced.

Article 33 of the Constitution, outlined above, was amended. The amendment included jargon such as 'sustainability' and 'keeping the environment in perspective', but the most important term was 'efficiency'. Furthermore, while ownership legally remained with the state, the management (and thus practical control) of the 'natural riches' could now legally be handed out to the private sector, both domestic and foreign, without being regarded as unconstitutional.

Governance restructuring also happened in a downward direction. Following reform demands, the Post-New Order government enacted regional autonomy Laws No. 22/1999 and No. 25/1999. These laws shifted various spheres of authority of the central government to local governments, including some important aspects regarding control of non-human nature in local administrative territories, such as the issuance of plantation permits.

For marginalised groups, including those who considered themselves *Masyarakat adat* ('customary communities'), the immediate Post-New Order reorganisation provided the opportunity to participate in the renegotiation of the system of governance. During the New Order period, these communities' interaction with and management of human to nature relations in their ancestral lands were repressed and displaced by state organs. Their customary lands were also often appropriated by the state and given to private companies, resulting in the dispossession and displacement of customary communities. Some indigenous communities, such as the *Dayaks* in Kalimantan, supported by international NGOs, had already used participatory mapping as a strategy to resist state development and business initiatives' intrusions into their customary territories since the mid-1990s (Rye and Kurniawan 2017). However, this proved futile in the face of the powerful machinery of the state.

When this political machinery collapsed along with the New Order, many disappointments emerged. With the help of transnational advocacy networks, as well as some academics, *masyarakat adat* movements from all over the archipelago joined forces. In March 1999, the Alliance of Customary Societies in the Archipelago (*Aliansi Masyarakat Adat Nusantara,* AMAN) was established. Their demands were clear: (1) the state must return all customary territories to the customary societies and (2) the New Order organs of village governance must be abolished (Henley and Davidson 2008). While not all customary societies are chthonic in their relationship with nature, many hold non-anthropocentric worldviews. Many of these traditional societies have their own ways of managing the relationship between humans and nature that are not anthropocentric. Thus, shifting governance to the 'local' and 'traditional' levels would provide the opportunity to push anthropocentrisation back by providing indigenous communities some governance space to protect them from the intrusions of state and business.

The efforts to shift the level of governance from the state to traditional or indigenous institutions were partly successful. As discussed in previous sections, since the establishment of the modern state, which was further strengthened during the New Order period, the state controlled forests through agreements on forest use and forestland exchange, which defined who could legally access the forest. In the Post-New Order period, indigenous communities, supported by academics and activists, attempted to take back this control. AMAN declared that approximately 40 million hectares of Indonesian state forestland should be under customary control (Myers et al. 2017). As part of the initiative, AMAN brought a case to the Constitutional Court which challenged an important assumption of state policy. It argued that the assumption that customary forests are part of state forests (*hutan negara*), and thus legally under the direct control of the state, is against the Constitution, which recognises the rights of traditional communities. The Court finally ruled that customary forests are part of titled forests (*hutan hak*), even though they are located within the forest zone. Responding to this development, the government declared that adequately verified customary territories would be released from the designation of the forest zone and would thus be managed by local customary societies (Myers et al. 2017). Similar instances of the state accommodating claims by indigenous communities were made in other areas, such as the recognition of participatory mapping (Rye and Kurniawan 2017), as well as on the larger issue of the recognition of special *adat* rights (Tyson 2011).

Nevertheless, despite some formal successes by indigenous communities and their allies, implementation was often hindered by the epistemological dominance of an anthropocentric worldview in the state bureaucracy. For example, to obtain the recognition of customary territory or customary forest, strict requirements must be fulfilled: '(1) the society is organised as a distinguishable community; (2) there are existing structures and institutional arrangements; (3) there are clear territories and boundaries based on customary law; (4) customary law and customary judiciaries still exist; (5) societies still gather forest products to fulfil their daily needs and (6) the claim has the support of the district government' (Myers et al. 2017). Furthermore, the recognition of the customary forest must consider the 'national interest', as stipulated in the Basic Agrarian Law, which often translates to economic benefits. Since many indigenous societies have limited resources and their structures and institutions have been destroyed by previous state policies, these requirements are difficult to meet.

AMAN activists have noted that these requirements impede the recognition of customary forests and show that the state is not serious about recognising the rights of indigenous communities (Arman 2018). Thus, it is not surprising that by March 2019, only 28,286.34 ha, a total 40 million hectares declared by AMAN, are recognised by the state as customary forests (*hutan adat*) (Arumingtyas 2019). The anthropocentric worldview is also still dominant in the state bureaucracy's accommodation of participatory mapping, which is driven by the need to simplify, classify, administer and measure the people and nature in frontier areas to bring them to the global market for the sake of economic development (Rye and Kurniawan 2017).

Another obstacle to the success of indigenous communities is their small relative strength compared with other social forces that are also striving to protect their own interests. Local elites riding the wave of *Reformasi* have greater control over natural resources vis-à-vis the central government, not to protect nature but to exploit it more freely. Private business also pushed back against the state so that they could expand and exploit nature without much scrutiny. Liberalisation and decentralisation, supported by patronage networks, led to the 'regionalisation' of the exploitation of nature (Tyson et al. 2018; Varkkey et al. 2018). The contentious nature of the governance restructuring process involving different social forces led to an increase in conflicts between the central government and local governments, between local governments and also among different government agencies (Resosudarmo (ed.) 2005). In this messy and crowded arena, the voices of indigenous communities were often side-lined. For example, to obtain recognition for their customary forest, indigenous communities must first get formal acknowledgement from the local administration. However, since the heads of local governments are often more concerned with boosting regional income by conceding the forest to private business for mining or plantation, they often neglect such requests from indigenous communities. During the New Order period, the voices of indigenous communities were silenced. After the *Reformasi*, they can speak louder, but not many ears are willing to listen. Thus, efforts by indigenous communities to push back against anthropocentrisation remain a marginal force in the post-Suharto context.

This is also coupled with the fact that indigenous communities are not homogenous and are continuously evolving. As McCarthy reminded us, we must avoid 'simplifying, reifying or romanticizing *adat*' as a way of living that is always in harmony with nature (McCarthy 2005, p. 80). As his extensive studies in Sumatera show, traditional or indigenous communities are evolving and adapting to anthropocentric political–economic governance and thus are 'neither principally opposed to commercial, economic development, nor has it necessarily supported sustainable forest use' (McCarthy 2005, p. 80). My observation in Riau also shows that some traditional leaders, who have internalised the modern need for material wealth accumulation, used their newly empowered traditional authority to sell the control of some portions of customary lands to companies or *cukong* (rich individuals). In this context, indigenous communities themselves are undergoing an anthropocentrisation from within.

Concluding Remarks: The Contest Continues

By looking at Indonesia's experience, this chapter traced the anthropocentrisation of political–economic governance through the creation and evolution of the modern state and its anthropocentrising consequences. Prior to the creation of the colonial state, many indigenous communities thrived in a non-anthropocentric way of life without much interference from the dominant powers, which did not see territoriality in the way that modern states see it. This chapter argued that the process of anthropocentrisation began with colonial state-building in the mid-nineteenth century, following efforts by colonial rulers to exploit the colonies more effectively to compete in the increasingly competitive global market. The creation of a unified national political and economic governance system with rigidly defined territories gradually displaced the ecological systems of indigenous communities and replaced them with anthropocentric political–economic governance which is designed for the exploitation of humans and non-human nature in the colony. Inside the rigidly defined territory, identified and marked through surveys and agreements among colonial powers in Southeast Asia, 'non-human nature' was treated simply as 'natural resources' to be exploited for the global market and to provide the much-needed finance to sustain colonial state-building. The independence of Indonesia did not stop the anthropocentrisation process, as the postcolonial nation-building process led to further institutionalisation of anthropocentric political–economic governance. The authoritarian and developmentalist New Order government (1965–1998) consolidated the power of the state and its control over people and nature, effectively marginalising indigenous communities despite the formal recognition of *Adat Law*. State transformation in the age of globalisation, including the fragmentation, decentralisation and internationalisation of state apparatuses, gradually loosened the grip of the state after the 1980s. The fall of Suharto in 1998 was followed by brutal competition for control between various social forces. Indigenous communities, supported by transnational advocacy networks, used this opportunity to create a governance space for themselves by attempting governance rescaling.

While these initiatives have been partially successful, such as the recognition of the special *adat* rights, the Constitutional Court's recognition of customary forests as titled forests and the accommodation of participatory mapping by indigenous communities by the state, they are still unable to push anthropocentrisation back. The loosening grip of the state does not mean the reversal of anthropocentrisation, since different social forces, from private business to local elites, are continuously competing for control of humans and nature for the sake of exploitation. Furthermore, indigenous communities themselves, long marginalised and absorbed into the anthropocentric modern state, are also evolving.

References

Arman, M. (2018). *Pengakuan bersyarat hambat implementasi hutan adat* (Conditional recognition hinders the implementation of customary forest). Retrieved March 31, 2019, from https://www.mongabay.co.id/2018/07/02/pengakuan-bersyarat-hambat-implementasi-hutan-adat.

Arumingtyas, L. (2019). *Penetapan hutan adat hanya 1% dari realisasi perhutanan sosial* (Designation of customary forests only accounts for 1% of social forest realization). Retrieved April 4, 2019, from https://www.mongabay.co.id/2019/03/27/penetapan-hutan-adat-hanya-1-dari-realisasi-perhutanan-sosial.

Avonius, L. (2004). Reforming adat: Indonesian indigenous people in the era of Reformasi. *The Asia Pacific Journal of Anthropology, 4*(1–2), 123–142.

Bosma, U. (2007). The cultivation system (1830–1870) and its private entrepreneurs on colonial Java. *Journal of Southeast Asian Studies, 38*(2), 275–291.

Chacko, P., & Jayasuriya, K. (2018). A capitalising foreign policy: Regulatory geographies and transnationalised state projects. *European Journal of International Relations, 24*(1), 82–105 (2018). https://doi.org/10.1177/1354066117694702.

Connor, C. M. (2012). Effects of central decisions on local livelihoods in Indonesia: Potential synergies between the programs of transmigration and industrial forest conversion. *Population and Environment, 25*(4), 319–334.

Cook, J., Oreskes, N., Doran, P. T., Anderegg, W. R. L., Verheggen, B., Maibach, E. W., & Rice, K. (2016). Consensus on consensus: A synthesis of consensus estimates on human-caused global warming. *Environmental Research Letters, 11*(4), 1–7 (2016). https://doi.org/10.1088/1748-9326/11/4/048002.

Crutzen, P., & Stoermer, E. (2000). The Anthropocene. *IGBP Newsletter, 41*, 17–18.

de Vries, J. W., & Van der Woude, A. (1997). *The first modern economy: Success, failure, and perseverance of the Dutch economy, 1500–1800*. Cambridge: Cambridge University Press.

Eckersley, R. (2004). *The Green State: Rethinking democracy and sovereignty*. Cambridge, MA: The MIT Press.

Eilenberg, M. (2012). *At the edges of states: Dynamics of state formation in the Indonesian borderlands*. Leiden: KITLV Press. https://doi.org/10.1163/9789004253469.

Fasseur, C. (1986). The cultivation system and its impact on the Dutch colonial economy and the indigenous society in nineteenth-century Java. In C. A. Bayly & D. H. Kolff (Eds.), *Two colonial empires: Comparative studies in overseas history*. Dordrecht: Springer.

Fearnside, P. M. (1997). Transmigration in Indonesia: Lessons from its environmental and social impacts. *Environmental Management, 21*(4), 553–570.

Friederich, M. C., & van Leeuwen, T. (2017). A review of the history of coal exploration, discovery and production in Indonesia: The interplay of legal framework, coal geology and exploration strategy. *International Journal of Coal Geology, 178*, 56–73 (2017). https://doi.org/10.1016/j.coal.2017.04.007.

Glenn, P. (2014). A chthonic legal tradition: To recycle the world. In P. Glenn (Ed.), *Legal traditions of the world* (5th Edition, pp. 60–97). Oxford: Oxford University Press. https://doi.org/10.1093/he/9780199669837.003.0003.

Hameiri, S., & Jones, L. (2016). Rising powers and state transformation: The case of China. *European Journal of International Relations, 22*(1), 72–98 (2016). https://doi.org/10.1177/1354066115578952.

Hecht, S. B. (2011). The new Amazon geographies: Insurgent citizenship, "Amazon Nation" and the politics of environmentalisms. *Journal of Cultural Geography, 28*(1), 203–223. https://doi.org/10.1080/08873631.2011.548500.

Henley, D., & Davidson, J. S. (2008). In the name of Adat: Regional perspectives on reform, tradition, and democracy in Indonesia. *Modern Asian Studies, 42*(4), 815–852 (2008). https://doi.org/10.1017/S0026749X07003083.

Hoey, B. A. (2003). Nationalism in Indonesia: Building imagined and intentional communities through transmigration. *Ethnology, 42*(2), 109–126.

Holleman, J. F. (1981). *Van Vollenhoven on Indonesian Adat Law*. Dordrecht: Springer.

Jones, L. (2018). Theorizing foreign and security policy in an era of state transformation: A new framework and case study of China. *Journal of Global Security Studies*. https://doi.org/10.1093/jogss/ogy030.

Jones, L., & Zeng, J. (2019). Understanding China's 'belt and road initiative': Beyond 'grand strategy' to a state transformation analysis. *Third World Quarterly, 40*(8), 1415–1439 (2019). https://doi.org/10.1080/01436597.2018.1559046.

Kian, K. H. (2008). How strangers became kings: Javanese-Dutch relations in Java, 1600–1800. *Indonesia and the Malay World, 36*(105), 293–307. https://doi.org/10.1080/13639810802268064.

Linklater, A. (1998). *The transformation of political community: Ethical foundations of the Post-Westphalian era*. Cambridge: Polity.

Luiten van Zanden, J. (2010). Colonial state formation and patterns of economic development in Java, 1800–1913. *Economic History of Developing Regions, 25*(2), 155–176. https://doi.org/10.1080/20780389.2010.527689.

Marks, D. (2010). Unity or diversity? On the integration and efficiency of rice markets in Indonesia, c. 1920–2006. *Explorations in Economic History, 47*(3), 310–324. https://doi.org/10.1016/j.eeh.2009.08.002.

McCarthy, J. F. (2005). Between adat and state: Institutional arrangements on Sumatra's forest frontier. *Human Ecology, 33*(1), 57–82. https://doi.org/10.1007/s10745-005-2426-8.

Milman, O. (2019). 2018 was world's fourth hottest year on record, scientists confirm. Retrieved March 30, 2019, from https://www.theguardian.com/environment/2019/feb/06/global-temperatures-2018-record-climate-change-global-warming.

Moore, J. W. (2017). The Capitalocene, Part I: On the nature and origins of our ecological crisis. *Journal of Peasant Studies, 44*(3), 594–630. https://doi.org/10.1080/03066150.2016.1235036.

Morrison, K. D. (2018). Empires as ecosystem engineers: Toward a nonbinary political ecology. *Journal of Anthropological Archaeology, 52*(March), 196–203. https://doi.org/10.1016/j.jaa.2018.09.002.

Myers, R., Intarini, D., Sirait, M. T., & Maryudi, A. (2017). Claiming the forest: Inclusions and exclusions under Indonesia's 'new' forest policies on customary forests. *Land Use Policy, 66*(May), 205–213. https://doi.org/10.1016/j.landusepol.2017.04.039.

Palm Oil Agribusiness Strategic Policy Institute (PASPI) (2014). *The sustainability of Indonesian palm oil industry: Its role in economic growth, rural development, poverty reduction, and environmental sustainability*. Bogor: IPB Press.

Ravensbergen, S. (2018). Anchors of colonial rule: Pluralistic courts in Java, ca. 1803–1848. *Itinerario, 42*(2), 238–255. https://doi.org/10.1017/S0165115318000323.

Resosudarmo, B. P. (Ed.). (2005). *The politics and economics of Indonesia's natural resources*. Singapore: ISEAS.

Rye, S. A., & Kurniawan, N. I. (2017). Claiming indigenous rights through participatory mapping and the making of citizenship. *Political Geography, 61*, 148–159 (2017). https://doi.org/10.1016/j.polgeo.2017.08.008.

Schroeder, H., & González P., N. C. (2019). Bridging knowledge divides: The case of indigenous ontologies of territoriality and REDD+. *Forest Policy and Economics, 100*(March), 198–206.

Schwartz, J. (2018). More floods and more droughts: Climate change delivers both. Retrieved March 30, 2019, from https://www.nytimes.com/2018/12/12/climate/climate-change-floods-droughts.html.

Srinivasan, K. (2013). The biopolitics of animal being and welfare: Dog control and care in the UK and India. *Transactions of the Institute of British Geographers, 38*(1), 106–119 (2013). https://doi.org/10.1111/j.1475-5661.2012.00501.x.

Tagliacozzo, E. (2010). The Indies and the world: State building, promise, and decay at a transnational moment, 1910. *Bijdragen Tot de Taal-, Land- En Volkenkunde, 166*(2), 270–292.

Thaler, G. M. (2017). The land sparing complex: Environmental governance, agricultural intensification, and state building in the Brazilian Amazon. *Annals of the American Association of Geographers, 107*(6), 1424–1443. https://doi.org/10.1080/24694452.2017.1309966.

Thee, K. W. (2007). Indonesia's economic performance under Soeharto's new order. *Seoul Journal of Economics, 20*(2), 263–281.

Tyson, A. (2011). Being special, becoming indigenous: Dilemmas of special adat rights in Indonesia. *Asian Journal of Social Science, 39*(5), 652–673.

Tyson, A., Varkkey, H., & Choiruzzad, S. A. B. (2018). Deconstructing the palm oil industry narrative in Indonesia: Evidence from Riau province. *Contemporary Southeast Asia, 40*(3), 422–448. https://doi.org/10.1355/cs40-3d.

Varkkey, H., Tyson, A., & Choiruzzad, S. A. B. (2018). Palm oil intensification and expansion in Indonesia and Malaysia: Environmental and socio-political factors influencing policy. *Forest Policy and Economics, 92*, 148–159. https://doi.org/10.1016/j.forpol.2018.05.002.

Vaughan-Williams, N. (2015). "We are not animals!" Humanitarian border security and zoopolitical spaces in Europe. *Political Geography, 45*, 1–10. https://doi.org/10.1016/j.polgeo.2014.09.009.

Vaughan-Williams, N. (2016). The biopolitics of European border security. In S. Prozorov & S. Rentea (Eds.), *The Routledge handbook of biopolitics* (pp. 225–234). London: Routledge.

von Benda-Beckmann, F., & von Benda-Beckmann, K. (2013). Myths and stereotypes about adat law: A reassessment of Van Vollenhoven in the light of current struggles over adat law in Indonesia. *Bijdragen Tot de Taal-, Land- En Volkenkunde/ Journal of the Humanities and Social Sciences of Southeast Asia, 167*(2–3), 167–195. https://doi.org/10.1163/22134379-90003588.

Vu, T. (2007). State formation and the origins of developmental states in South Korea and Indonesia. *Studies in Comparative International Development, 41*(4), 27–56. https://doi.org/10.1007/BF02800470.

Weststeijn, A. (2014). The VOC as a company-state: Debating seventeenth-century dutch colonial expansion. *Itinerario, 38*(1), 13–34 (2014). https://doi.org/10.1017/S0165115314000035.

Ecological Civilization: The Political Rhetoric of "Marxism with Chinese Characteristics"

Justin Heinzekehr

Abstract Critics argue that the Chinese rhetorical concept of ecological civilization has been used to justify an increasingly authoritarian government without substantial improvements in China's environmental footprint. Far from instituting an ecologically responsible socialism, China appears to be continuing on a path of rapid industrialization, growing economic inequality, and repression of individual rights. However, the concept of ecological civilization has played a significant projective, polemical, and appropriative rhetorical role in the Chinese context. An analysis of the function of ecological civilization in Chinese policy shows that it has contributed to a national self-identity based on environmental responsibility by projecting a fictionalized image of itself as an alternative to Western capitalism. The rhetoric of ecological civilization has not always translated into improved environmental policy, nor is its vision for a holistic, socialist society likely to be translated into an objective reality. Still, the concept remains a useful rhetorical tool both for the government and for Chinese citizens attempting to hold the government accountable to its stated ideals. Rather than discounting ecological civilization because of its rhetorical function, its rhetorical effects should be analyzed on its own terms. On the whole, ecological civilization as a rhetorical concept has produced positive political effects with respect to climate change and should be encouraged as a potential tool for environmental progress in China, whether it is used by the Chinese government, Chinese citizens and activists, or international environmentalists.

Introduction: The Potential of Rhetorical Marxism for Environmental Policy

The fundamental question in international relations today is how quickly and comprehensively states and multinational corporations will respond to the threat of climate change. Will most countries continue to prioritize economic growth and production

J. Heinzekehr (✉)
Goshen College, Goshen, IN, USA
e-mail: justinbh@goshen.edu

J. C. Pereira and A. Saramago (eds.), *Non-Human Nature in World Politics*,
Frontiers in International Relations, https://doi.org/10.1007/978-3-030-49496-4_9

over all other considerations, allowing private corporations greater control over the world's natural resources? Or, will states find a way to prioritize the well-being of the majority of their citizens and their non-human constituents?

The answer to this question will have long-term consequences for global political stability. The UN Security Council has already been clear that it considers climate change to be a major factor in East African conflict specifically, and a "threat multiplier" for global conflict in general (United Nations 2019). A 2016 study has shown that, for every one degree Celsius increase, there is a 5% increase in migration, a key metric for political destabilization (Cai et al. 2016, p. 135). The most ambitious study analyzes data from 10,000 BCE to the present and concludes that, for each standard deviation temperature change, there is a 14% increase in intergroup conflict (Hsiang et al. 2013, p. 1212). Climate change may not be enough to cause large-scale violence in otherwise stable countries, but it does exacerbate conflict in countries that are dependent on agriculture, contain minority populations, or have weak public institutions (Koubi 2018, p. 197–202). Based on current trends, we will increasingly come to understand global conflicts and tensions in terms of the localized impact of climate change.

In future decades, China will play a large role in the success or failure of international efforts to mitigate climate change. In 2009, China surpassed the United States as the largest energy user on the planet. "China accounts for 50% of the world's total coal consumption and produces the largest amount of major air pollutants in the world", including greenhouse gases (Ahlers and Hansen 2017, p. 84). As of 2015, China produces 30% of global CO_2 emissions, and its per capita CO_2 emissions are greater than the European Union (7.6 to 6.7, respectively) (Sternfeld 2017, p. 2). As one author puts it, "The blunt truth is that what China decides to do in the next decade will likely determine whether or not [humans] can halt—or at least ameliorate—global warming" (Quoted in Conley 2017, p. 177).

The Chinese response to climate change, governed by the rhetorical concept of "ecological civilization," has taken a different path than that of the United States or European Union. The goal of "ecological civilization" (*shengtaiwenming*) was written into the Chinese constitution in 2012, and points toward national policies that foster a holistic economic system, not opposed to economic growth per se, but committed first to a sustainable and humane society. Ecological civilization is part of what Chinese leaders call "Marxism with Chinese characteristics," which, at least in official rhetoric, is posed as an alternative to the environmentally destructive tendencies of Western capitalism.

Since 2012, however, there has been a good reason to be cynical of China's "ecological civilization." Critics might argue that the term has merely been used to justify an increasingly authoritarian government without substantial improvements in China's environmental footprint. Far from instituting an ecologically responsible socialism, China appears to be continuing on a path of rapid industrialization, growing economic inequality, and repression of individual rights. As I write this chapter, mass demonstrations are taking place in Hong Kong to protest a law allowing extradition to the courts in mainland China.

But part of the discomfort with the concept of ecological civilization comes simply from its obvious function as political rhetoric. It is clear that the concept is not an objective description of economic relations and even contains inherent contradictions in its goals and policies for Chinese society. Especially looking from the outside, there is something that appears disingenuous about using "ecological civilization" as a guiding principle for a country that is so clearly struggling with environmental problems. Partly this suspicion stems from a culturally informed attitude toward language. For instance, Terry Eagleton argues that the whole of Western philosophy, from the Greek sophists to Nietzsche, has conditioned us to be skeptical of rhetorical communication as something done merely for adornment or "effect" and deliberately ambiguous in its content (Eagleton 1981, p. 110).

It would be a mistake, however, to underestimate the role of political rhetoric in environmental policy. If it is true, as Eagleton says, that rhetoric produces "particular discursive and extra-discursive effects without regard for truth or falsity," those effects are nonetheless real. Political rhetoric "fictionalizes the 'real' as to intend a set of effects conducive to certain practices that are deemed, in the light of a particular set of falsifiable hypotheses about the nature of society, to be desirable" (Eagleton 1981, p. 112–13). Therefore, in order to produce specific political results, social actors may use rhetoric as a way to fictionalize reality productively, expose politically undesirable effects, or appropriate useful themes from traditions not specifically associated with the desired effects (Eagleton 1981, p. 113). This productive type of rhetoric is not simply arbitrary but leverages, almost in a literary way, social themes that make it possible to imagine new values and actions.

This chapter argues that ecological civilization has played and might continue to play this "projective, polemical, and appropriative" rhetorical role in the Chinese context. An analysis of the function of ecological civilization in Chinese policy shows that it has contributed to a national self-identity based on environmental responsibility by projecting a fictionalized image of itself as an alternative to Western capitalism. The rhetoric of ecological civilization has not always translated into improved environmental policy, nor is its vision for a holistic, socialist society likely to be translated into an objective reality, at least not in the foreseeable future. Still, the concept remains a useful rhetorical tool both for the government and for Chinese citizens attempting to hold the government accountable to its stated ideals. Rather than discounting ecological civilization *because* of its rhetorical function, its rhetorical effects should be analyzed on their own terms. On the whole, ecological civilization as a rhetorical concept has produced positive political effects with respect to climate change and should be encouraged as a potential tool for environmental progress in China, whether it is used by the Chinese government, Chinese citizens and activists, or international environmentalists.

Ecological Civilization

In 1972, Chinese leaders attended the United Nations Conference on the Human Environment (UNCHE), the first major international conference on environmental issues. It was clear already that developed countries bore the heaviest responsibility for global environmental problems, and China took the opportunity to denounce Western capitalism as causing "worldwide environmental degradation" (Sternfeld 2017, p. 1). In a way, this conference provided the framework that China would use to develop its self-identity as a more sustainable and environmental alternative to Western economies (Tianbo and Meng 2017, p. 19). In line with this self-conception, China began implementing environmental legislation, forming the Environmental Protection Leadership Group, which would evolve into what is now the Ministry of Ecology and Environment.

As China began its process of liberalization and modernization in the 1970s and 1980s, the environmental policy became more and more formalized in the Constitution. "Environmental protection" became a fundamental duty of the state and was written into the Constitution in 1982. Water and air pollution legislation appeared in 1984 and 1987, respectively. Several major energy and resource conservation laws were passed in the 1990s and 2000s. Moreover, during this most recent period, the content of Chinese environmental legislation has matured

> "from attempting to control only the resulting pollution to establishing control of the origin and the whole process of pollution as well; from regulating only the concentration of the pollutants to regulating both the concentration and total amount of pollutants;...and from simply addressing the pollution problem of one corporation to adjusting the entire industrial structure by promoting clean production and developing a cyclical economy" (Tianbo and Meng 2017, p. 24–25).

The most significant development in recent years is the concept of "ecological civilization." The term was introduced by President Hu Jintao in the 17th National Congress of the Communist Party of China in 2007, replacing earlier concepts like "Scientific Outlook on Development". Ecological civilization is now enshrined in the Constitution of the Communist Party of China, and has played a central role in both the 12th (2011–2015) and 13th (2016–2020) Five-Year Plans (FYP), which have articulated climate policy as a key principle for sustainable development. "[The 12th FYP] is widely recognized as marking a turning point in Chinese domestic climate policies because it sets binding targets for carbon intensity and introduced carbon emissions trading for selected pilot regions" (Sternfeld 2017, p. 6). In policy, if not always in practice, ecological civilization has driven the progress that China has made toward environmental regulation.

The new term is meant to communicate not only the fact that China is committed to sustainable development, but also that there is something about Chinese civilization that makes it uniquely suited to lead the world toward a new model of ecological politics. "Domestically it reassures the masses that the government has the determination to solve environmental problems. Internationally, it is Beijing's effort to build the image of a normative and discursive power, capable to contribute new words to the global dictionary" (Wang-Kaeding 2015, p. 45). Ecological civilization is an attempt to connect the ideals of environmentalism to a distinctive culture, history, and politics as the cornerstone of "socialism with Chinese characteristics."

The pillars of ecological civilization are very deliberately connected with principles from traditional Chinese religion. For instance, the Confucian principle of *tianrenheyi*, or harmony between humans and nature, is used to represent the idea that humans must live at a level that natural ecosystems can absorb (Yao 2017, p. 419). Chinese scholars have found important resources in the Yi Ching, which refers to civilization "ordering sky and earth and lighting up the four directions," interpreted as a reference to the natural world as well as the people's internal disposition toward nature. Lao Tze spoke of the law of the Tao filtering through the Earth to humanity. As Jiahua Pan argues, Chinese traditions such as Confucianism, Buddhism, and Taoism contain "such shared values as respect to nature, universal love to everything, as well as harmonious relations between human and nature" (Pan 2016, p. 39). In other words, ecological civilization has involved a re-appropriation by the Chinese government of those religious traditions that were downplayed in the Maoist years. The specific choice of religious terms in the rhetoric of ecological civilization serves to anchor the concept in the specific context of Chinese culture, but also differentiates ecological civilization as a new, modern socialism in contrast to the older version.

Practically, ecological civilization is generally meant to capture the goal of economic development in tandem with an increased quality of life and environment. As opposed to a purely market-based economy that measures value only in terms of purchasing power, the ideal of ecological civilization is that:

> the consumption pattern does not pursue material [accumulation] and a [luxurious] and wasteful life but a green, economic, healthy, rational, and [high-]quality life. …The lifestyle under ecological civilization does not request people to be stingy with daily necessities… Instead, it is on the basis of securing the satisfaction of people's basic material needs, restraining the desire of unnecessary material [accumulation] and consumption (Pan 2016, p. 37).

Thus, the idea of ecological civilization is meant as a direct alternative to the path of market-based capitalism taken in Western history.

Of course, there are tensions inherent in this concept. Even the Central Committee of the Communist Party (CPC) acknowledges:

> The development of ecological civilization is an essential part of the socialist undertaking with Chinese characteristics …From an overall perspective, however, the development of ecological civilization in China still falls behind economic and social development… The contradictions between development and population as well as resources and environment are increasingly prominent, which have become a serious bottleneck hindering sustainable economic and social development (Quoted in Sternfeld 2017, p. 2).

Some critics have argued that, in fact, the attempt to approach environmentalism via any kind of economic development, however sustainable, is inherently misguided. "The catch," as Donovan Conley says, "is that pure economic expansion and pure ecological conservation are not just difficult concepts to reconcile; they are squarely at odds" (Conley 2017, p. 192–193). Under this way of thinking, one might argue that

China is very unlikely to compromise its economic development in any significant way, which means that the entire idea of ecological civilization is "empty" rhetoric.

Others have pointed out that the idea of ecological civilization can be used to support authoritarian governance. Nobel Laureate Liu Xiaobo argues that the idea of harmony between humans and nature has connotations of passivity that undercut the kind of activism necessary for a real environmental movement in China. It suggests that humans are meant to submit to the laws of nature and their leaders, not participate as active citizens (Wang-Kaeding 2015, p. 46). Heidi Wang-Kaeding suggests that ecological civilization relies on a form of ethno-cultural nationalism to create a sense of urgency, but this ethno-cultural nationalism comes at the cost of civic nationalism, i.e., the idea that citizens should have an active role to play in national politics (Wang-Kaeding 2015, p. 54). This makes it difficult for citizens to hold the government accountable to its stated goals of environmental sustainability.

Despite these challenges, the centrality of ecological civilization, even rhetorically, has facilitated significant environmental policy developments in China, which seem to be making at least modest improvements in air pollution and carbon emissions. It is significant that China has become a global leader in environmental policy even as its economy continues to develop and its per capita GDP remains well below its Western counterparts. China is in some ways positioned quite well to contribute to the global conversation on climate change. While rhetoric alone cannot erase the very real challenges that China faces, the rhetorical concept of ecological civilization highlights the ecological opportunities in Chinese society and deepens their potential impact.

From a Marxist perspective, for example, one could argue that any real environmental progress would need to grapple with the existing capitalist economic structures, which make it necessary for corporations to continually expand their production, and therefore, their exploitation of natural resources. China, of course, has a very different relationship to global capitalism than the West does. Although capitalism has certainly shaped China's economy recently, Marxist or socialist ideas are still deeply entrenched in the national psyche, and anti-Western, anti-capitalist rhetoric has long been used as a unifying force in Chinese politics. Whether or not one sees socialist principles at work in the Chinese economy, it is significant that ecological civilization positions itself rhetorically over against global capitalism.

Second, China's status as a world power is still new, which means that political leaders are looking for a way to demonstrate China's leadership on the world stage, including environmental leadership. Because of the desire to demonstrate its cultural superiority, Chinese environmental rhetoric has been couched in national or local terms. On the one hand, excessive nationalism might be alarming if directed against other countries or minority groups within China. However, the basic idea that environmental policy needs to be "China-specific," rooted in local culture and norms, leaves an opening for the development of locally developed ways to approach environmental problems, which naturally encourages activism on the part of individual citizens. On the whole, I would argue, ecological civilization retains enough of a spirit of local activism to counteract the tendency toward passivity in the face of authoritarian governance.

Finally, China's vulnerability to the effects of climate change and other environmental problems increases the urgency with which China has tackled the problem compared to other Western nations. The brute fact of the matter is that China has many internal motivators—air and water quality, widespread flooding, severe drought—that impact citizens' daily lives and tie the rhetoric of ecological civilization back to very concrete, high-stakes issues.

The Legacy of Marxism in China

Since Deng Xiaoping introduced the idea of "One Country, Two Systems" in 1982, the Chinese economy has been complicated and difficult to categorize. In principle and in law, China remains committed to Marxism, but the economy has transitioned to a "socialist market economy" in which overall wealth as well as inequality have skyrocketed. China's Gini coefficient, a measure of social inequality on a scale of 0–100, has risen more rapidly than any other country over the past few years (rising from about 35 in 1990 to 50 in 2015), putting China among the most unequal countries in the world (Jain-Chandra 2018).

When Marxism first took root in China, it was attractive because it promised a route toward prosperity that did not need to follow the Western model, which China had experienced as British and Japanese imperialism. In particular, Marxism spoke to rural farmers who faced oppression from landowners. Mao Zedong wrote,

> "The ruthless economic exploitation and political oppression of the peasants by the landlord class forced them into numerous uprisings against its rule. …It was the class struggles of the peasants, the peasant uprisings and peasant wars that constituted the real motive force of historical development in Chinese feudal society" (Mao 1966, p. 9).

Mao thus developed a unique appreciation for rural workers, as opposed to Marx's focus on the urban working class.

The basic rhetoric and values of Chinese society reflect the same motivations, especially, the development of China on its own terms. Because of its Marxist history and its Marxist Constitution, the favored narrative of the Chinese government is that the unique path of development that China is taking avoids the worst consequences of Western individualism and capitalism, including social inequality. At least in the rhetoric, one hears about Chinese values of social cooperation and social justice as opposed to Western winner-take-all economics. One scholar from the Chinese Academy of Social Sciences, for example, contrasts Western utilitarian, market-based ethics with those of traditional Chinese culture: in utilitarianism, "utility is realized through [the] market. If something has no market value or low market value, it has to be replaced by things with utility or higher utility. The ethical foundations of ecological civilization inherited the ancient Chinese philosophic thinking and sought ecological justice and social justice" (Pan 2016, p. 44). Thus, its history with Marxism means that Chinese nationalist rhetoric resonates with a very different set of values than, say, American nationalist rhetoric.

Practically, this has meant that the Chinese government has continued to encourage Marxist literature and thought in the Chinese education system, even as the economy has moved away from strict Marxist principles:

> Since President Xi took power in 2012, the party has sought to restrict the use of Western textbooks and stop the spread of 'Western values' on campus, referring to ideas about rule of law and democracy that could undermine its hold on power. At the same time, Mr. Xi has demanded that universities expand their teachings on Mao and Marx. In May, he visited Peking University and encouraged students to promote Marxism, saying it was important for the university to "take Marxism as its surname" (Hernández 2018).

But rhetoric always has consequences beyond its cynical usage. At several major universities, for example, student groups dedicated to Marxism and Maoism have emerged, calling for greater social equality, workers' rights, gender equality, and improved conditions for the rural provinces. In 2018, students from some of these university groups organized a protest in Huizhou, where a welding manufacturer had denied its workers the request to form an independent union. One of the protesters, Yue Xin, stated "that she had been inspired by [President Xi's] fight against corruption and his time working in an impoverished village in the countryside as a young man. She added that the campaign in Huizhou had its roots not in foreign ideas, but in the May Fourth Movement of 1919, a student-led uprising in China that the party considers a precursor to the Communist Revolution" (Hernández 2018). The police responded by arresting several of the protesters and putting others under house arrest, further heightening the disconnect between official rhetoric and practice.

The ecological aspects of official rhetoric have also had widespread consequences for the collective Chinese consciousness. Guobin Yang and Craig Calhoun have identified a "green public sphere" emerging among Chinese citizens due to "the proliferation of environmentalist discourse—a green-speak. Contrary to an earlier Maoist and Marxist view of the human conquest of nature, this new discourse warns about the dangers of irresponsible human behavior toward nature and calls for public action to protect the environment" (Quoted in Conley 2017, p. 195). The most prominent example was the cancelation of a dam construction project on the Nu River because of a determined public protest in 2004. But since then, the number of environmental protests has continued to rise; in a country where protests are a common way of expressing discontent (180,000 protests in 2011 alone), the most common point of contention is pollution and other environmental issues (Conley 2017, p. 195).

One could argue whether or not top Chinese officials are sincere in their promotion of ecological civilization as an ideal for the future. The recent reactions to environmental activism have led some to believe that the Chinese government cares more about buttressing its own power than constructing a sustainable society. But it is naïve to expect that any government's self-description would ever be purely altruistic; there will always be a mixture of sincerity and self-preservation in the actions of state leaders. The significant feature in the Chinese case is the possibility of a feedback loop between national rhetoric informed by an ecological Marxism, public awareness of environmental issues, and the implementation of effective policies in response to public demand.

Given the inherent structures of capitalism that work against limits to consumption, China is perhaps in a better position to make deep, long-term policy changes than the next highest contributor to climate change, the United States. The United States has the advantage of better environmental regulation at this point, but its collective values are much more individualistic and tolerant of inequality. China's socialist heritage means that some extreme environmental policies are palatable there, which would be seen as draconian in the United States. It would be difficult to imagine driving restrictions being imposed in a major U.S. city, for example, as they were in Hangzhou in 2014 (Ahlers and Hansen 2017, p. 91–92). The chances that China's economy will develop into a purely Marxist system are negligible, but the reliance on Marxism as a national rhetoric may still drive significant policy changes as a sort of self-fulfilling prophesy.

China's Drive for Political and Cultural Superiority

A similar dynamic occurs due to China's international positioning. Again, one of the connotations of ecological civilization is that China's civilization is the *most* ecological civilization, that China is emerging as the world leader in sustainable, ethical politics. "[Ecological civilization] indicates that China has the historical chromosome and cultural superiority for environmental governance, whereby it can lead the world to a new era of sustainable development" (Wang-Kaeding 2015, p. 38). Angang Hu, a committed nationalist and professor at Tsinghua University, explicitly names the fact that ecological civilization is an overarching strategy for China's global dominance, including an economy that produces the largest GDP, the largest consumer market, the greatest level of high-tech manufacturing, the largest export volume, and the most skilled workforce in the world. (Hu 2018, p. 178–181) The global ascendancy of China, according to Hu, will also include a more benign dominance in terms of sustainability. China will develop the cleanest infrastructure, the most healthy and equitable lifestyle for its people, and the most innovative environmental technology.

China's nationalism, like its Marxism, can be traced back to Mao and the sense of Chinese humiliation at the hand of Western imperialism. Mao's narrative of Chinese history pointed back especially to the First Opium War of 1839 in which the British were able to force the Qing Dynasty to sign a treaty giving the British unlimited trading rights and ceding Hong Kong. "To reclaim its rightful destiny and moral superiority, Mao's China drew from communist ideology to produce a national identity defined by revolutionary Sinocentrism, or Tianxiaism, which argued for solidarity among the Chinese people in opposition to Western imperialism" (Friend and Thayer 2018, p. 38–39). In Mao's view, resistance to Western colonialism required the unification of all of China's population, Han and minority groups, as well as solidarity with other Third World countries. If China was not currently a great world power, it could become the leader of the oppressed countries of the world. One can hear echoes of Mao's version of Sinocentrism in China's response to climate change: this crisis has been foisted onto the rest of the world by the developed West, and China's role is to be the leader of the alternative movement.

In the past few decades, however, some have argued that ethnic nationalism is becoming an increasingly powerful force in Chinese politics. One difference is that China's global position has improved significantly since Mao, so that it can legitimately claim to be a world power already and not simply a leader of the developing world. Because of its changing position, "Chinese nationalism would become less ideological and more Sinocentric, drawing primarily from Confucianism and ideals rooted in nativism and chauvinism" (Friend and Thayer 2018, p. 41). These aspects of Chinese self-identity have always been present to some extent in the collective Chinese consciousness (at least since the time of imperial China), but they have resurfaced to a greater extent as China has transitioned from a developing, Marxist country to a market-based world power.

From an outsider's perspective, this expression of nationalism can be disconcerting. It has raised tensions especially with the United States and increases prospects for military conflict. Probably the more urgent problem with ethnic nationalism in China has to do with treatment of internal minority populations like the Uyghur or Tibetans, who can be seen as undermining a strong, unified China.

On the other hand, especially if its Marxist values can be leveraged to mitigate some of its ethnocentric tendencies, Chinese nationalism has been a strong motivating force toward environmentalism. If ecological civilization continues to be the measuring stick that China's leaders use to demonstrate the nation's superiority, Chinese leaders will be much more likely to implement actions improving the country's environmental impact, or at least make enough of a gesture that they can reasonably claim to be progressing toward the vision of ecological civilization.

Perhaps more importantly, nationalist rhetoric has also had the effect of mobilizing Chinese citizens to work at environmental issues in a more localized way. Some of these are prominent citizens with political, cultural, or economic power, such as Liang Congjie, a historian and founder of the Friends of Earth organization, or Liao Xiaoyi, the environmental activist who founded the Global Village of Beijing. Chinese entrepreneurs or business groups have participated in other market-based solutions to environmental issues (Wang-Kaeding 2015, p. 50–51).

The rhetorical connection between environmentalism and Chinese religions has mobilized religious leaders to participate in grassroots environmental organizing as well. For example, Yang Shihua, a Taoist abbot in eastern China, has built an eco-friendly temple, integrated environmental themes into its rituals, and has been pushing local officials to set aside land for preservation. Other environmental nonprofits have sprung up that are based on Buddhist or Christian value systems (Hernández 2017). As Heidi Wang-Kaeding says, "Nationalism and patriotism, therefore, facilitate the localization of environmentalism in China. It provides fertile soil in which the seed of environmental protection can be planted and nurtured to generate a sense of urgency to call the people into action" (Wang-Kaeding 2015, p. 54). Precisely because Chinese leaders have put forward ecological civilization as a key component of national identity and religion, individual Chinese citizens are more likely to internalize environmentalism in their own communities and contexts.

China's Vulnerability to Climate Change

Aside from the desire to appear superior to Western nations, China has the additional motivation of greater vulnerability to climate change compared to most other countries. A 2011 study from the Center for Global Development measured climate change vulnerability by considering the impact of weather variations on income level and infrastructure. China ranks first in the world for vulnerability to extreme weather events and second, after India, in terms of the greatest number of people at risk from sea level rise (Wheeler 2011). In northwestern China, drought will likely be a major issue, resulting in water shortages, reduced agricultural production, and conflict over water resources.[1] In the coastal regions, especially in the southeast, flooding is the main concern and will likely continue to worsen (Ying et al. 2014, p. 60). Already, China annually evacuates hundreds of thousands of people from flood basins in the southeast. In 2018, the flooding caused about $1.5 billion of damage (Stanway 2018).

In general, environmental threats have become visible and personal to Chinese citizens to a greater extent than in developed nations. Air pollution especially has become part of the common awareness; as Eva Sternfeld observes, "Today, hourly updated data on air quality is available everywhere in China and even children are aware that $PM_{2.5}$ is an air pollutant and a health risk" (Sternfeld 2017, p. 6). And even though public opinion on climate change issues lags behind the more tangible problem of air pollution (in a 2015 poll, only 18% of Chinese participants agreed that climate change was a serious problem) (Stokes et al. 2015), the Chinese government is taking climate change seriously. Last year, China introduced the world's largest carbon trading system to incentivize lower emissions. The market is limited to the energy sector, but covers about 30% of its total carbon emissions (Gardner 2018).

Given the combination of national rhetoric, the current awareness of the importance of air quality and environmental health, and the practical significance of climate change for the average Chinese citizen, it seems likely that in the near future, the energy around environmental activism will begin to be focused on climate change as a significant issue.

The Implementation of Ecological Civilization so Far

The rhetoric of ecological civilization gives China the motivation to make progress toward its environmental vision. However, China faces many material obstacles that cannot simply be wished away, including carbon intensive infrastructure, the tendency to export environmental risks to the provinces, internal inconsistencies in environmental discourse, and inconsistencies in practice and implementation. These problems cannot be solved by rhetoric and place constraints around the vision of ecological civilization.

---

[1] See case study on the Heihe River Basin (Leary 2008, p. 8).

Even though it has put new carbon-emission policies in place, China's infrastructure is still carbon intensive. The country remains the world's largest producer and consumer of coal. At least in the most recent data, greenhouse gas emissions continue at an alarming level (Miller et al. 2019, p. 303). Given the speed with which China's economy expanded over the past half century and the abundance of coal reserves in the country, China's coal dependence has been built into its infrastructure. A recent study from the Carnegie Institution for Science at Stanford estimates that, based on current infrastructure commitments (i.e., with no new fossil fuel plants), China's emissions will grow to about 41% of the global total, compared to 9% for the United States and 7% for the European Union. This is partly because China's coal plants are much younger than those in the U.S. and Europe, and therefore, are likely to be used further into the future (Mooney 2019). Even with the best motivations and some innovative policies, it is unlikely that the country will meet its ambitious goals for reducing carbon emissions in the short-term.

Even when the government does succeed in improving environmental conditions, there has been a tendency to privilege the wealthier urban areas. Sometimes an environmental issue is "fixed" simply by exporting it to the provinces. For instance, urban protests against the construction of power plants or factories may simply result in their being built in rural areas instead. Waste, including hazardous waste, may be collected from large cities and dumped into poorly controlled landfills in less populated areas. "Because of waste fermentation, rain and groundwater soaking in the solid waste, landfills in which this waste is deposited produce leachate, which includes poisonous material. The result is secondary pollution, which has brought a series of problems to local rural society and done great harm to the environment." (Wei and Chen 2016, p. 40). In general, then, the rhetoric of ecological civilization may produce a differentiated motivation where the effort to decrease carbon emissions will be directed primarily toward solutions that impact the urban centers at the expense of less influential regions or populations.

The idea of ecological civilization is broad and encompasses many different, sometimes inconsistent, environmental attitudes. The Chinese government seems to be interpreting ecological civilization as (at least in theory) a form of eco-socialism, with inherent connections to Chinese culture and Marxism. But there are those who interpret ecological civilization under a market-based logic, where the primary solution is to improve green technology and provide incentives to transition to alternative energy. Others interpret ecological civilization as essentially a translation of Western-style environmentalism, including political reform and democratization (Wang-Kaeding 2015, p. 48). Some of these tendencies are not only different but mutually exclusive. Thus, the very concept of ecological civilization may include political actors at cross-purposes with each other.

Finally, Chinese leaders have recently begun cracking down on environmental activism, especially when it seems to threaten the legitimacy or effectiveness of the government. Chinese environmental activists have grown in confidence and are using a wider variety of tactics, including "information politics, intergroup networking, bearing witness, accountability politics, symbolic politics (naming and shaming), undercover investigations, transparency politics, supply chain analysis, blogging, online and in-person petitioning, and the use of the courts to bring class action and public interest lawsuits" (Shapiro 2017, p. 50). It is not necessarily surprising that the growth of environmental activism has attracted greater suspicion, scrutiny, and suppression from officials. But it raises the question of whether ecological civilization can continue to be a motivation for real policy change rather than a mere smokescreen.

However, there are reasons for hope. China's emphasis on ecological civilization has resulted in some significant policy changes, and China has taken a position of leadership in assisting other developing nations toward similar policies (Heggelund and Nadin 2017, p. 107). Some of these policies are having a measurable impact. The air quality measures taken in Hangzhou, for example, reduced the number of passenger cars in the city by 44%, reducing traffic congestion and pollution levels. Similar measures in Beijing and other coastal cities have resulted in modest but measurable improvements in air quality (Ahlers and Hansen 2017, p. 92–93). And, even if it has not met its most ambitious goals, China has succeeded in reducing emissions since 2014, the first time this has happened since emissions tracking began (Sternfeld 2017, p. 13)

Given the political motivation that China has to respond to climate change and other environmental concerns, it seems reasonable to hope that these modest successes signal more significant achievements to come. Hu Tao and Song Peng put it well: "During the last two decades, China created an economic miracle. We hope that, during the next two decades, China will be able to create an environmental miracle and set a prime example for other nations to achieve both economic and environmental sustainability" (Tao and Peng 2017, p. 343–44).

Conclusion

There has been a long history of extreme interpretations of China from Western observers. Relations especially between the United States and China "have been marked for more than sixty years by entrenched misunderstandings, unproductive communication patterns, and simmering rivalries" (Hartnett et al. 2017, p. xi). It is easy to overstate China's promise as a new world leader or to overstate its authoritarianism and social tensions. "[P]opular discussions about our relations swing between two poles: from predictions of a coming Chinese century…to equally exaggerated suggestions of a coming Chinese collapse."

This tendency toward exaggeration has certainly been true in the discourse about global climate change. On the one hand, China's rhetorical emphasis on ecological civilization as a cornerstone of its national policy positions China to make progress on policy to mitigate climate change, particularly in developing economies. On the other hand, China's authoritarian leadership and environmental challenges might lead some to predict a regression toward less political openness and toward accelerating environmental degradation.

Few people, even in the Chinese government, would deny that China faces immense challenges as it works toward a sustainable future, and it is by no means clear to what extent the idea of ecological civilization can be turned into a reality. On the other hand, as argued above, the rhetorical use of ecological civilization as a political concept does help to produce a national interest in mitigating climate change, and there have already been some modest but encouraging results. Rather than dismissing Chinese political rhetoric wholesale, the history of ecological civilization as a concept suggests that it may still be useful as a tool for political change. Environmental policy, in China and elsewhere, needs rhetorical tools that will prepare the social imagination for the necessary transition to a sustainable economy. As such, ecological civilization must function for now as a fictionalization, though hopefully a productive one.

References

Ahlers, A. L., & Hansen, M. H. (2017). Air pollution: How will China win its self-declared war against it? In E. Sternfeld (Ed.), *Routledge handbook of environmental policy in China* (pp. 83–96). New York: Routledge.

Cai, R., Feng, S., Oppenheimer, M., & Pytlikova, M. (2016). Climate variability and international migration: The importance of the agricultural linkage. *Journal of Environmental Economics and Management, 79,* 135–151. https://doi.org/10.1016/j.jeem.2016.06.005.

Conley, D. (2017). China's fraught food system: Imagining ecological civilization in the face of paradoxical modernity. In S. J. Hartnett, L. B. Keränen, & D. Conley (Eds.), *Imagining China: Rhetorics of nationalism in an age of globalization* (pp. 175–204). East Lansing, MI: Michigan State University Press.

Eagleton, T. (1981). *Walter Benjamin, or, towards a revolutionary criticism.* London: NLB.

Friend, J. M., & Thayer, B. A. (2018). *How China sees the world: Han-centrism and the balance of power in international politics.* Lincoln, NE: Potomac Books.

Gardner, D. K. (2018, December 10). Trump is unwilling to tackle climate change. China must step up. *The New York Times.* Retrieved from https://www.nytimes.com/2018/12/10/opinion/china-trump-climate-change.html.

Hartnett, S. J., Keränen, L. B., & Conley, D. (Eds.). (2017). *Imagining China: Rhetorics of nationalism in an age of globalization.* East Lansing, MI: Michigan State University Press.

Heggelund, G., & Nadin, R. (2017). Climate change policy and governance. In E. Sternfeld (Ed.), *Routledge handbook of environmental policy in China* (pp. 97–112). New York: Routledge.

Hernández, J. C. (2017). China's religious revival fuels environmental activism. *The New York Times.* Retrieved from https://www.nytimes.com/2017/07/12/world/asia/mao-mountain-china-religion-environment.html.

Hernández, J. C. (2018). China's leaders confront an unlikely foe: Ardent Young communists. *The New York Times.* Retrieved from https://www.nytimes.com/2018/09/28/world/asia/china-maoists-xi-protests.html.

Hsiang, S. M., Burke, M., & Miguel, E. (2013). Quantifying the influence of climate on human conflict. *Science, 341*(6151), 1212. https://doi.org/10.1126/science.1235367.

Hu, A. (2018). *China's road and China's dream: An analysis of the Chinese political decision-making process through the national party congress.* Singapore: Springer.

Jain-Chandra, S. (2018). Chart of the week: Inequality in China. Retrieved from IMF Blog website: https://blogs.imf.org/2018/09/20/chart-of-the-week-inequality-in-china/.

Koubi, V. (2018). Exploring the relationship between climate change and violent conflict. *Chinese Journal of Population Resources and Environment, 16*(3), 197–202. https://doi.org/10.1080/10042857.2018.1460957.

Leary, N. (2008). *Climate change and vulnerability.* London: Routledge.

Mao, Z. (1966). *Quotations from Mao Tse Tung*. Retrieved from http://www.marxists.org/reference/archive/mao/works/red-book/.

Miller, S. M., Michalak, A. M., Detmers, R. G., Hasekamp, O. P., Bruhwiler, L. M. P., & Schwietzke, S. (2019). China's coal mine methane regulations have not curbed growing emissions. *Nature Communications, 10*(1), 303. https://doi.org/10.1038/s41467-018-07891-7.

Mooney, C. (2019). Existing fossil fuel plants will push the world across a dangerous climate limit, research finds. *Washington Post*. Retrieved from https://www.washingtonpost.com/climate-environment/2019/07/01/existing-fossil-fuel-plants-will-push-world-across-dangerous-climate-limit-research-finds/.

Pan, J. (2016). *China's environmental governing and ecological civilization*. Berlin, Heidelberg: Springer.

Shapiro, J. (2017). The evolving tactics of China's green movement. In E. Sternfeld (Ed.), *Routledge handbook of environmental policy in China* (pp. 48–56). New York: Routledge.

Stanway, D. (2018). East China floods cause over $1 billion in damage this month. *Reuters*. Retrieved from https://www.reuters.com/article/us-china-floods-idUSKCN1LC00W.

Sternfeld, E. (2017). Introduction. In E. Sternfeld (Ed.), *Routledge handbook of environmental policy in China* (pp. 1–13). New York: Routledge.

Stokes, B., Wike, R., & Carle, J. (2015). Global concern about climate change, broad support for limiting emissions. Retrieved from Pew Research Center website: https://www.pewglobal.org/2015/11/05/global-concern-about-climate-change-broad-support-for-limiting-emissions/.

Tao, H., & Peng, S. (2017). Environment and trade. In E. Sternfeld (Ed.), *Routledge handbook of environmental policy in China* (pp. 331–345). New York: Routledge.

Tianbo, Q., & Meng, Z. (2017). Development of China's environmental legislation. In E. Sternfeld (Ed.), *Routledge handbook of environmental policy in China* (pp. 17–30). New York: Routledge.

United Nations (2019). Climate change recognized as 'threat multiplier', UN Security Council debates its impact on peace. Retrieved from UN News website: https://news.un.org/en/story/2019/01/1031322.

Wang-Kaeding, H. N. K. (2015). Fragmented environmental discourse in People's Republic of China: Identity, legitimacy, and local agents. In I. Watson & C. L. Pandey (Eds.), *Environmental security in the Asia-Pacific* (pp. 31–66). New York: Palgrave Macmillan.

Wei, Y., & Chen, G. (2016). Municipal solid waste and environmental equity in China: A case study in South Anhui Province. In Z. Liang, S. F. Messner, Y. Huang, & C. Chen (Eds.), *Confronting the challenges of urbanization in China: Insights from social science perspectives* (pp. 39–56). New York, London: Routledge.

Wheeler, D. (2011, January 24). Quantifying vulnerability to climate change: Implications for adaptation assistance. Retrieved from Center For Global Development website: https://www.cgdev.org/publication/quantifying-vulnerability-climate-change-implications-adaptation-assistance-working.

Yao, X. (2017). The confucian aspiration at the heart of "Tianren Heyi": Eco-centrism, anthropocentrism, and the anthrocosmic relationship in traditional confucian thought. In X. Yao (Ed.), *Reconceptualizing confucian philosophy in the 21st century* (pp. 413–426). https://doi.org/10.1007/978-981-10-4000-9_29.

Ying, X., Bing, Z., Bo-Tao, Z., Si-Yan, D., Li, Y., & Rou-Ke, L. (2014). Projected flood risks in China based on CMIP5. *Advances in Climate Change Research, 5*(2), 57–65. https://doi.org/10.3724/SP.J.1248.2014.057.

America First: The Trump Effect on Climate Change Policy

Sheila D. Collins

Abstract Just when the world's leaders needed to unite to halt the runaway effects of global climate change, the U.S., the largest historic contributor to greenhouse gases, elected a president in 2016 who vowed to withdraw from the Paris climate agreement and to roll back all of the limited environmental regulations that had been put in place by his predecessor, Barack Obama. Donald Trump was elected by a minority of the American people through a peculiarity in the American political system known as the Electoral College. Defying every tradition of democratic rule, and with the assistance of a compliant Republican majority in the Senate, a base of hardcore public supporters and a right-wing media, he has managed to assert his particular brand of authoritarian nationalism on policymaking. Through the use of a variety of sources—U.S. government documents, U.N. reports, the reports of environmental watchdog groups, scholarly and news articles, this qualitative case study lays out the historical context in which Trump's attacks on the environment arose, examines his actions thus far to roll back progress on climate change, and demonstrates that sub-state and non-state actors have become new protagonists in the national and global effort to prevent dangerous climate change. While these efforts are important, much depends on whether Trump can be replaced by a Democratic congress and presidency backed by strong public pressure.

Introduction

Through both Democratic and Republican administrations, the United States government has long dithered on initiating the kinds of changes that could have prevented runaway climate change. If it had taken the steps needed thirty years ago when scientists began issuing alarms, we might not be faced with the kinds of draconian decisions that now have to be made (Rich 2018). The administration of President Barack Obama had begun to take some steps to reverse course, but the Trump administration and reactionary congressional Republicans have already done much to halt

S. D. Collins (✉)
William Paterson University, 65 Rockland Place, New Rochelle, NY 10801, USA
e-mail: sheiladreanycollins@gmail.com; ecointegrityprof@gmail.com

J. C. Pereira and A. Saramago (eds.), *Non-Human Nature in World Politics*,
Frontiers in International Relations, https://doi.org/10.1007/978-3-030-49496-4_10

or reverse these modest changes in environmental policy. Even the climate change related series of droughts, wildfires, tornadoes, and hurricanes that have caused enormous destruction and loss of life in parts of the country over the last few years have failed to dissuade President Trump and his supporters that climate change is real and needs to be addressed.

Trump, however, is not unique. Within the last few years, authoritarian nationalism, representing a repudiation of the post-World War II rules-based global economic order, has become a global phenomenon encompassing countries on every continent. Those rules were established before climate change emerged as a significant global concern, but by 1992, the effects of climate change had begun to become apparent. Consequently, the global community, coming together at the Earth Summit in Rio de Janeiro in 1992, established the U.N. Framework Convention on Climate Change (UNFCCC) whose aim was to set up a process through which greenhouse gas concentrations could be stabilized "at a level that would prevent dangerous anthropogenic (human induced) interference with the climate system … within a time-frame sufficient to allow ecosystems to adapt naturally to climate change, to ensure that food production is not threatened, and to enable economic development to proceed in a sustainable manner" (United Nations Framework Convention on Climate Change, n.d.). In effect, the international community was attempting to write a new set of rules to govern the behavior of nations in the age of the Anthropocene. Although those rules were only haltingly adopted and proved unequal to the increasing urgency of the challenge as defined by the world's scientists (International Panel on Climate Change 2019), Trump's arrival and that of other authoritarian nationalists threatens this process, posing a grave threat to the ability of the human community to limit the rise in global temperatures to 1.5 °C within the next dozen years if we are to avoid dangerous climate change and safeguard both human and non-human nature from its severe effects.

The U.S. Commitment to Curbing Climate Change: The History of a Laggard

To understand Trump's actions, we need to place them within a longer history of the U.S. government's environmental record and shifts in public opinion. As early as 1979, "nearly everything we know about global warming was understood [by scientists]" (Rich 2018, Prologue). Yet despite the efforts of numerous scientists, environmental writers, and activists to call the attention of the American public and policymakers to the urgency of acting to curb greenhouse gases, American public opinion remained stubbornly recalcitrant even as global temperatures rose to historic heights and wildfires, floods and increasingly intense hurricanes affected large areas of the United States (Brulle et al. 2012, p. 165). The earliest poll, conducted in March 1981, showed that most Americans were largely ignorant about the problem. Since then, awareness has grown. Still, awareness did not translate into factual knowledge about the issue nor into a majority agreement that the overwhelming number of the world's scientists understand global warming to be induced by human activity (Egan and Mullin 2017).

The failure of the American public to hold opinions commensurate with the threat articulated by scientists is due largely to the influence of the fossil fuel industry on American politicians and, secondarily, the auto industry. Nowhere, except perhaps in oil dependent states, is the fossil fuel industry as deeply entrenched in political and cultural life as in the United States. And nowhere is the public so attached to the automobile. With deep ties to members of Congress in both parties and with large invested capital in the economies of several politically important states, the fossil fuel industry has worked assiduously to keep climate change from becoming a salient policy issue.[1] Its modus operandi has been to sow doubts among the public about the certainty of the science-based evidence. It has been aided by a mass media that for many years ignored the science of climate change and in recent years by a deeply partisan media environment where facts are disputed, misinformation is cast as the truth and attempts to inform the public about climate change are labeled "fake news."

This level of unconcern wasn't always the case. Prior to 1989, some of the largest oil companies, including Exxon and Shell, made efforts to understand the scope of the crisis and grapple with positive solutions, and many prominent Democratic and Republican politicians recognized the issue and called for immediate and far-reaching policies to deal with it (Rich 2018). In 1980, President Jimmy Carter signed the Energy Security Act which directed the National Academy of Sciences to start a multiyear, comprehensive study that would analyze the social and economic effects of climate change (S. 932 Energy Security Act 1980). But beginning in 1989, a turnaround occurred.

A meeting of experts assembled shortly thereafter to try to outline a policy response to the scientific findings ended with the failure of any coherent policy proposal and, with the election in 1980 of Ronald Reagan to the presidency, the effort to move ahead on climate legislation appeared to be not only stalled, but was set back. After his election, Reagan "appeared determined to reverse the environmental achievements of Jimmy Carter before undoing those of Richard Nixon, Lyndon Johnson, John F. Kennedy and, if he could get away with it, Theodore

[1]Texas with its 34 electoral college votes leads the nation in crude oil and natural gas production, followed by New Mexico and North Dakota. In 2017, Texas accounted for 37% of the nation's crude oil production and 24% of its marketed natural gas production. U.S. Energy Information Administration, Texas State Energy Profile and Energy Estimates, https://www.eia.gov/state/?sid=TX. Pennsylvania (21 electoral votes) is the second largest producer of natural gas after Texas and the third largest coal producing state after Wyoming and West Virginia. Conservative estimates put U.S. direct subsidies to the fossil fuel industry at roughly $20 billion per year; with 20 percent currently allocated to coal and 80 percent to natural gas and crude oil. Environmental and Energy Study Institute, "Fact Sheet: Fossil Fuel Subsidies: A Closer Look at Tax Breaks and Societal Costs," July 29, 2019, https://www.eesi.org/papers/view/fact-sheet-fossil-fuel-subsidies-a-closer-look-at-tax-breaks-and-societal-costs#1.

Roosevelt," appointing anti-environmentalists to head the Environmental Protection Agency (EPA) and the Interior Department, proposing to eliminate the position of science advisor, closing the Council on Environmental Quality, and even removing the solar-thermal panels Jimmy Carter had installed on the White House roof. However, his drastic attacks on the environment alarmed even Republicans at the time, and within a year, both his EPA and Interior appointees were forced to resign (Rich 2018, p. 5–6).

The National Academy of Sciences report, *The Changing Climate*, that had been commissioned during the Carter administration, was completed in 1983. Although it had called for an accelerated transition to renewable fuels, the committee's chairman, who had served on Reagan's transition team, characterized its recommendations as just the opposite in the press briefings on the study. It was better, he said, to be cautious, to wait for American ingenuity to come to the rescue in some future time. The chairman's characterization of that report transformed overnight the state of climate politics as its call for caution and its optimistic belief in American technology received top press billing. There were sporadic attempts after this to renew congressional debate on the issue of climate change, but the Reagan White House blocked any movement, even going so far as to censor the congressional testimony of Jim Hansen, one of the most prominent of the government's climate scientists. Hansen's testimony had struck fear into the fossil fuel industry that its profits would be curtailed if the world moved to limit greenhouse gas emissions. Led by Exxon, the industry then adopted a strategy to emphasize the uncertainty of scientific predictions about the climate. From then on, an intensive lobbying and public relations campaign would seek to persuade lawmakers and the general public of this uncertainty. The only sign of progress on the issue was the U.S.'s signature on the 1987 *Montreal Protocol to Limit Substances that Deplete the Ozone Layer* (Rich 2018, p.1–2).

In the summer of 1988, with record high temperatures recorded, another congressional hearing was held in which the testimony of scientists appeared to finally break through the years of hesitation and inaction that had characterized the debate over the issue in the United States. The press followed with headlines that global warming had begun. At the World Conference on the Changing Atmosphere, held in Toronto, the U. S. delegation proposed a worldwide 20% reduction in carbon emissions by 2000. The conference's final statement agreed on the target but increased the time to 2005.

That fall, global warming became a major subject of the presidential campaign with candidates of both parties giving lip service to the concern. By the end of the year, thirty-two climate bills had been introduced in Congress, led by the National Energy Policy Act of 1988 (Rich 2018, Part 2, 5–7; H.R. 5380 National Energy Policy Act 1988). "It established as a national goal an 'International Global Agreement on the Atmosphere by 1992,' ordered the Energy Department to submit to Congress a plan to reduce energy use by at least 2% a year through 2005 and directed the Congressional Budget Office to calculate the feasibility of a carbon tax" (Rich 2018). Pushed by European countries that had declared even stronger commitments to reducing greenhouse gases, the U.N. established the Intergovernmental Panel on Climate Change to conduct scientific assessments and to develop global policy.

But once again, the progress that had begun would be stalled by forces within the Executive Branch—specifically President George H. W. Bush's chief of Staff, John Sununu–who feared what such a drastic reduction in greenhouse gases would do to the economy and the fossil fuel industry. Once again, Jim Hansen's testimony was censored. This time, the White House censor wanted him to say that the causes of global warming were "scientifically unknown" and might be attributable to "natural processes," and that Congress should only pass climate legislation that immediately benefited the economy, "independent of concerns about an increasing greenhouse effect" (Rich 2018, p. 8).

In November 1989, at the first global diplomatic meeting to decide on whether to endorse a framework for a global treaty on climate change, the U.S. representative succeeded in getting Great Britain, Japan, and the Soviet Union to abandon the commitment to freeze emissions (Rich 2018, p. 11). In 1991, after President George H.W. Bush announced that he would support a treaty limiting greenhouse gas emissions, a new attack on the science of climate change emerged. Charles and David Koch, libertarians who had amassed a $120 billion fortune much of it in the oil and chemical industries, convened a conference whose aim was to initiate a campaign that not only claimed that the science of climate change was uncertain, but that there was no climate change, citing scientists who, for the most part, were employed by the oil industry, to make their case. The company's lobbyists, their network of allied donors, academic centers, think tanks, and sympathetic media outlets like Fox News—a political influence operation "that was without parallel in modern America"–then set about to kill any attempt to legislate action on climate change (Leonard 2019, Chap. 18).[2]

While President George H.W. Bush had signed the United Nations Framework Convention on Climate Change and the Senate ratified it, the president refused to commit to specific emissions reductions. The following year, newly elected President Bill Clinton's proposal for an energy tax, went nowhere—the result of a $1.8 million disinformation campaign by Koch Industries and others. Although Clinton signed onto the Kyoto Protocol, the Senate refused to ratify it, the result of a $13 million lobbying campaign by the fossil fuel industry. There has since been no serious effort on the part of the U.S. to work for a binding global climate treaty (Rich 2018). Although many members of the industry coalition set up to thwart action on climate change resigned in 2002, Exxon continued its campaign for another five years. Although Exxon finally abandoned its strategy, Koch Industries continued the disinformation campaign. Although President George W. Bush had endorsed limitations on carbon emissions in the 2008 campaign, he failed to take any action on climate change during his two terms as President (Farber 2018, p. 98) even refusing to allow his EPA to set fuel efficiency standards for the auto industry (Freeman 2011, p. 349). Despite these setbacks, ever starker warnings from the IPCC had been convincing Democratic members of Congress that they needed to act on climate change. In 2007, majority House speaker, Nancy Pelosi, created the House Committee on Global

[2]For a detailed history of the Koch brothers' efforts to shape public policy, see Christopher Leonard, *Kochland: The Secret History of Koch Industries and Corporate Power in America.*

Warming and Climate Change to try to break down the barriers that had prevented climate change regulation for years. That committee began to hold a series of hearings they hoped would convince more policymakers and ordinary Americans of the seriousness of the threat they were facing.

With the election of President Barack Obama and a majority Democratic congress in 2008, environmentalists were hopeful that after years of inaction, climate change policy could finally move ahead. In an examination of his campaign speeches and press releases, Bailey found that Obama viewed climate change as a significant issue, albeit of less salience than his concern about healthcare. Nevertheless, he outlined a series of broad policy proposals that included proposals to reduce fossil fuel consumption by shifting to renewables (including nuclear power), proposals to promote energy efficiency, including raising fuel efficiency standards for automobiles and building a "smart grid" to distribute energy more efficiently, and a set of proposals to cap or reduce greenhouse gas emissions, including the use of a cap-and-trade system to meet emission targets and promises to work with the international community to address the issue (Bailey 2019).[3] He did not, however, rule out the continued use of fossil fuels during a transition to a new economy. The Koch brothers, however, viewed with alarm Obama's election, seeing his plans to regulate carbon as a threat to their industry. Not only would a carbon-control economy cut into current profits, but it threatened future profits as the world shifted away from fossil fuels (Leonard 2019, Chap. 19).

Obama was faced with several political obstacles to the successful translation of proposals into concrete policy. In order to pass legislation in the United States, presidents must be able to persuade Congress and the public that a course of action is necessary. For a bill to become law, both houses of Congress must agree on it. Despite having a Democratic majority in the Congress during his first two years, he could not rely on every congressional Democrat to support tackling climate change (many represented strong fossil fuel embedded states), and strong Republican opposition (including the Senate majority leader's vow to make Obama a "one term president") made it difficult to construct bipartisan coalitions large enough to pass laws. The year he was elected, the level of Republican anti-environmental voting increased progressively. Thereafter, cooperation between Republicans and Democrats on environmental issues, and the subsequent elite cues provided by the Republican voting record, drove down concern about climate change. Additionally, unemployment increased, and GDP declined, following the 2008 financial collapse (Brulle et al. 2012).

Obama took office during the Great Recession when the economy was the top concern of voters and he was immediately tasked with saving the economy from total collapse (Kuhn 2008). In the economic stimulus plan he enacted, he was able to include a number of green energy initiatives. A cap-and-trade bill had narrowly passed the House in 2009 and environmentalists were hopeful that it would be successful in the Senate. But Koch's political machine went into even more intense action

---

[3]By the time Obama took office, a cap-and-trade bill had already been drafted by the House Committee on Global Warming and Climate Change.

to prevent this from happening, targeting Republicans in their home districts with orchestrated angry "grassroots" protesters who called climate change a hoax, claimed that the cap-and-trade bill would raise consumers' energy prices, and characterized government regulation as tyranny. As the bill was heading toward the Senate, the Koch brothers amplified their alleged "grassroots" efforts with an economic report purporting to show that the cap-and-trade bill would destroy millions of jobs, that electricity prices would jump 50% by 2030, and that $3.1 trillion in economic activity would be lost. The report's conclusions were then amplified by the Koch affiliated media outlets for popular consumption. A third tactic used by the Koch consortium was the creation of a new set of political candidates who would adhere to the Koch line and compete in the next election to defeat members of Congress who believed in global warming (Leonard 2019, Chap. 19). Fighting against this tidal wave of conservative messaging and candidates, Obama and the Democratic Party decided to spend their political capital on getting a health care reform bill and Wall Street regulation passed.[4] And even those two items were an uphill fight. As a result, Obama failed to rally support for the cap-and-trade bill and it died quietly without ever coming to a vote in the Senate. According to many environmentalists, Obama's failure to press for national climate legislation when he had a Democratic majority was perhaps his greatest strategic miscalculation (Lavelle 2016).

The only area in which presidents enjoy almost complete autonomy is in foreign policy, although for an international treaty to come into effect, the Senate must ratify it. Thus, Obama was able to negotiate the nonbinding Paris Agreement in 2015 without congressional approval but could not push for a binding treaty knowing that the Republican Senate would refuse to sign it. With Republicans opposed to any more environmental regulations, with powerful interests like Koch Industries backing an anti-regulatory stance, and with weak public opinion on making climate change a priority, holding the U.S. to its stated commitments in that Agreement would be more difficult.

Backed by heavy campaign spending by the Kochs and fossil fuel industry, Republicans took control of the House of Representatives in the 2010 mid-term elections, reduced the Democratic majority in the Senate, and swept into office a new crop of climate deniers. After this, there was no way Obama was going to be able to achieve significant climate change legislation. Making things even more difficult was that there was no specific statute dealing with climate change. If Obama were to make any progress, he would have to work through the regulatory authority provided to the Environmental Protection Agency (EPA) by existing legislation, the most important being the Clean Air Act of 1963 (modified and amended in 1970 and 1990) (Environmental Protection Agency 1970).

A Supreme Court ruling in 2007 (*Massachusetts v. EPA* [549 U.S. 497, 2007]) had held that greenhouse gases are considered "air pollutants" under the Clean Air

---

[4]Public opinion polls showed health care was a higher priority than climate change, and, moreover, there was less opposition to changing health care policy among Democrats. As it turned out, however, Obama's signature health care legacy, the Affordable Health Care Act of 2010, passed without a single Republican vote.

Act (CAA) and that the EPA must regulate them if they endanger human health or welfare. However, to gain legal standing to any challenge to the EPA's authority, there must be credible evidence that a defendant has suffered or will suffer a concrete or imminent harm. Relying on climate science, the Obama administration was able to show the courts that there was indeed evidence that carbon emissions, estimated to be 36% of U.S. carbon emissions, endangered public health and welfare (Kendall et al. 2018, p. vii). Consequently, the EPA was able to set new auto emission standards for new cars and trucks, the strictest in American history. All of the domestic and foreign auto companies signed letters promising not to challenge this ruling in court. Their compliance was made easier by the fact that the car industry was already under pressure because of stringent regulations that had already been adopted by California (the country's largest auto market) and thirteen other states under waivers granted by the CAA that allowed states to set their own higher standards. The new auto standards were upheld by the court of appeals, and the Supreme Court declined to consider the issue. This ruling was significant because once the EPA sets a binding standard for a pollutant under one section of the CAA, it becomes a so-called "regulated pollutant" for purposes of other CAA programs (Freeman 2011, p. 352). Regulating emissions from stationary sources, such as power plants, factories, oil refineries, and other industrial sources and perhaps even residential sources would be next.

Presidents are given a certain amount of power to exercise what are called Executive Orders—directives to the Executive Branch to do something which can have the force of law. Realizing he was thwarted by a recalcitrant Congress, Obama turned to this unilateral authority to execute his climate plans, issuing an Executive Order which instructed the federal government to prepare for the impact of climate change (Obama 2013). He began his second term by accelerating action on climate change. In August 2015, the Obama administration released the Clean Power Plan under which carbon emissions would be reduced by 32% below 2005 levels by the year 2030 through a combination of coal-powered energy reduction and increases in clean energy sources. The Plan was to be his signature climate accomplishment.[5] Obama's EPA also moved to tackle truck emissions, limit methane leaks from the oil and gas industry, and update energy efficiency standards for home appliances. In addition, he established twenty-three national monuments, more than any other president in history.[6] The U.S. delegation also led the effort to amend the international agreement on CFCs, which will negate the equivalent of 10 years of U.S. emissions (Lavelle 2016).

[5]The Clean Power Plan establishes national carbon dioxide emissions performance rates for existing coal- and gas-fired power plants. Each state then has an opportunity to adopt its own plan—including enforceable emissions limits—for its coal and gas plants.

[6]National monuments are nationally significant lands or waters that are set aside for permanent protection. Presidents can create them under authority given them by the National Antiquities Act of 1906.

The Trump Administration's Attack on Science

Public opinion about climate change is shaped largely by elite cues and structural economic factors (Brulle et al. 2012, p. 169). The bully pulpit that is afforded American presidents is a powerful tool for shaping public opinion and for promoting or discouraging legislation. In Donald Trump the Koch brothers and their supporters found the perfect spokesman for their efforts to prevent any government action to reduce greenhouse gas emissions. Long before he took office in 2017, Trump had tweeted that "the concept of global warming was created by and for the Chinese in order to make U.S. manufacturing non-competitive (Trump@realDonaldTrump 2012)." Since taking office, he has repeated this denialism mantra countless times.[7] Such skepticism about the science of climate change is reiterated by the Right-Wing media of which Trump is a devotee and it is, in turn, echoed and amplified on social media by his base of supporters who make up between 30% and 40% of the American electorate.

Effective national climate change policy requires careful long-term planning based on the best available science and executive coordination across multiple agencies and interests. It also requires a president to educate the public which, as polls show, remains woefully ignorant of the causes and consequences of climate change. Trump's environmental policies, however, are not made after a careful assessment of the science, nor are they focused on the long term. He prefers, instead, to make policy through tweets that reflect his own prejudices and belief system. His tweets can be particularly dangerous as they confuse and obfuscate the issues involved in combating climate change, preventing the public from grasping the nature of the crisis we are facing so that they can make informed decisions about their own lives and public policy. For example, he often confuses momentary changes in weather patterns with climate change, suggesting that if there is a cold spell there is no climate change, as in the following tweet: "Brutal and Extended Cold Blast could shatter ALL RECORDS—Whatever happened to Global Warming?" (Trump@realDonaldTrump 2018). Trump's penchant for playing loose with the facts took a particularly disturbing turn in September 2019 when he contradicted weather forecasters over the path that Hurricane Dorian would take, giving the impression that it would hit Alabama, a state that was nowhere near its path. Rather than correcting his error, he extended the false story for over a week, went on television to show a weather map that appeared to have been doctored with a black sharpie, and had his Commerce Secretary, who oversees the National Oceanic and Atmospheric Administration (NOAA), the scientific agency responsible for weather forecasts, allegedly threaten to fire top officials at the agency if they did not issue a statement agreeing with the President (Flavelle et al. 2019). The resulting about-face by an agency that is supposed to deliver politically neutral scientific information indicates the extent to which scientific integrity is being undermined.

Since coming to office, Trump has reversed the progress that was being made on climate change by the Obama administration, making skepticism about science-based

[7] According to one researcher, by June 2017, he had repeated it 115 times (Matthews 2017).

evidence the centerpiece of his efforts to reshape every executive department that relies on science for its work, as well as politicizing science itself. As of this writing, 100 regulations seen as overly burdensome to the fossil fuel industry, including major Obama-era policies aimed at fighting climate change, have been or are in the process of being eliminated (Popovich et al. 2020; Sabin Center for Climate Change Law 2019). Over a dozen organizations concerned about this attack on climate change and other forms of environmental protection as well as major newspapers like the *Washington Post* and the *New York Times* are in the process of tracking these changes (Box. 1).

Box 1 Organizations tracking Trump administration's rollback of climate change and other environmental protection policies

Brookings https://www.brookings.edu/interactives/tracking-deregulation-in-the-trump-era/.

Center for American Progress https://www.lawofthelandproject.org/.

Center for Western Priorities https://medium.com/westwise/government-shutdown-oil-and-gas-permitting-tracker-69f25dfa4dc0.

Coalition for Sensible Safeguards https://sensiblesafeguards.org/.

Columbia University, Sabin Center http://columbiaclimatelaw.com/resources/climate-deregulation-tracker/.

Columbia University, Sabin Center http://columbiaclimatelaw.com/resources/silencing-science-tracker/.

Earthjustice https://earthjustice.org/features/two-years-overruling-trump.

Environmental Data and Governance Initiative https://envirodatagov.org/.

Environmental Integrity Project https://www.environmentalintegrity.org/.

Global Energy Institute (U.S. Chamber of Commerce) with support from **Beveridge and Diamond** https://www.globalenergyinstitute.org/energy-tracker.

Harvard Law School Environmental and Energy Law Program https://eelp.law.harvard.edu/epa-mission-tracker/https://eelp.law.harvard.edu/regulatory-rollback-tracker/.

Institute for Policy Integrity https://policyintegrity.org/deregulation-roundup.

National Geographic https://www.nationalgeographic.com/news/2017/03/how-trump-is-changing-science-environment/.

NYU State Energy and Environmental Impact Center https://www.law.nyu.edu/centers/state-impact/ag-actions.

National Resources Defense Council https://www.nrdc.org/trump-watch.

Reuters https://www.reuters.com/trump-effect/energy-environment.

Union of Concerned Scientists' Center for Science and Democracy https://www.ucsusa.org/center-science-and-democracy/attacks-on-science#.W5Ad7ehKi72.

Source: List compiled by Sheila D. Collins

Effective government depends on a reliable process for filling senior government positions with qualified professionals who are dedicated to doing the public's work (National Task Force on Rule of Law 2019). In pursuit of his goals, President Trump has appointed climate science deniers and representatives of the fossil fuel industry to head the agencies most responsible for the environment, namely, the Environmental Protection Agency (EPA) and the Departments of Energy (DOE) and the Interior (DOI), the latter of which is responsible for managing the nation's public lands. Two of his appointees have since had to resign over alleged corruption but have been replaced with others equally as committed to the same agenda. The result, according to the Environmental Data and Governance Initiative (EDGI), is that the EPA's major enforcement metrics against environmental polluters–"inspections, civil cases, criminal cases, compliance costs, penalties, defendants, sentencing—are some of the lowest recorded in decades" (EDGI 2018, p. 4).

The Suppression of Science and Science-Based Findings

The Trump administration has diminished climate change as a priority factor in environmental management. "Altering descriptions of what climate change science is and what climate change researchers do, agencies have restricted access to key informational resources, either by removing links to these pages, removing the pages themselves from the public domain, or by modifying descriptions of what those resources are" (EDGI 2019a, p. 29). The administration has also attempted to undermine public discourse, specifically around climate impacts, mitigation, and adaptation (EDGI 2019a, b). Such changes can have critical regulatory implications (Atkin 2017). While all of the actions taken by the Trump administration to undermine the use of science in policymaking reach into almost every area of the government's functions, I have chosen to highlight a representative sample of only those directly related to the science of climate change. These actions—which represent only a part of the administration's total war on science—fit into two broad categories of abuse: (1) The withdrawal of scientists and suppression of scientific information and (2) the politicization of science. Among them are the following:

- "The contents of several websites, including the EPA's Climate Change website; Climate and Energy Resources for State, Local, and Tribal Governments website; and the Clean Power Plan website were all removed from public view thus depriving the public of critical scientific information" (EDGI 2019b; Friedman

2017a, b). Use of the terms "climate change," "clean energy," and "adaptation" dropped by 26% between 2016 and 2018 (EDGI 2019a, b).

- "Over half of all pages where 'climate change' was completely removed from public access (73/136) were U.S. Environmental Protection Agency (EPA) pages. Such removals are significant given that EPA.gov was the 1,750th most visited website in the U.S. in early 2019, giving it more reach than even WhiteHouse.gov" (EDGI 2019a, p. 2).
- The EPA withdrew an Obama request for more detailed information on oil and natural gas facilities that had aimed to better track the industry's methane and volatile organic compound emissions and proposed eliminating federal requirements that oil and gas companies control leaks of methane from new wells, storage facilities, and pipelines (Lavelle 2019).
- The EPA withdrew guidance for federal agencies to include greenhouse gas emissions in environmental reviews (Federal Register 2017).
- The EPA removed its Clean Power Plan website by redirecting it to a newly created webpage called "Energy Independence" (Union of Concerned Scientists (UCS) 2019a).
- The EPA's Office of Science and Technology removed the word "science" from its mission statement, instead emphasizing "economically and technologically achievable performance standards" (Atkin 2017).[8]
- The Trump administration has ignored the findings of a major scientific report issued in November 2018 by thirteen federal agencies that present the starkest warnings to date of the consequences of climate change for the United States, including dire threats to its economy, health, and environment (Davenport and Pierre-Louis 2018).
- Both the EPA and Department of the Interior (DOI) have been prevented from using many kinds of scientific studies vital to their decision-making (UCS 2019a).
- The Department of State, Department of Energy (DOE), and the EPA removed information about the federal government's international obligations regarding climate change, downplaying U.S. involvement.
- Half of the scientists who serve on a key review board that affects major policy decisions at the EPA have been fired and replaced by executives from the industries the agency is supposed to regulate (Milman 2017; Eilperin and Dennis 2017).
- The EPA has stopped collecting data on pollutants released from farms. Methane from industrial farms is a potent contributor to global warming (UCS 2019b).
- The US Department of Agriculture (USDA) prevented the release of a plan for how the agency can effectively respond to the impacts of climate change. It has also refused to publicize peer-reviewed studies by USDA scientists that show the impacts of climate change (UCS 2019a).

---

[8]The Union of Concerned Scientists (UCS), a science advocacy organization, shares EDGI's concerns. "The role of the EPA is to protect public health and safety," said Andrew Rosenberg, Director of UCS's Center for Science and Democracy. "So what you want a science office to do is make sure you're using the best science available, and what's safe for the public. That's a pretty critical role".

- The EPA has stacked scientific advisory committees with industry representatives rather than adhere to scientific protocols and has dismantled one-third of 1,057 advisory committees (UCS 2019).
- Trump officials delayed for several months a press release from the U.S. Geological Survey announcing a new study on climate change and then released a highly edited version that removed references to the study's main findings (UCS 2019).
- The Director of the US Geological Survey (USGS 2019) "ordered that scientific assessments produced by USGS can only use climate models that project the impact of climate change through 2040, rather than through the end of the century" as has been done previously, thus suppressing worst case scenarios (Davenport and Landler 2019).
- The DOI and two of its agencies have carried out policies that block or restrain federal scientists from attending or presenting at scientific conferences (UCS 2019).
- The DOI quietly rescinded multiple science-based policies on climate change and conservation. These changes undercut the Department of Interior's (DOI) ability to adhere to its mission to conserve and manage the Nation's natural resources (UCS 2019).

The Politicization of Science

Previous governments have been guilty of manipulating science for political ends. The George W. Bush administration went farther than others, politicizing science "to a degree otherwise unseen in modern times" (Goldman et al. 2017, p. 4). In his inaugural address, President Obama vowed to "restore science to its rightful place" (Obama 2019). During his fraught administration, he made several efforts to restore scientific integrity to federal policymaking including recognizing climate science, acknowledging the risks associated with it, and allowing federal climate science to be communicated and to inform policymaking. He also put forth a framework for addressing losses of scientific integrity in government and furthering data access and scientific advisory committee independence (Goldman et al. 2017, p. 13). But progress slowed over the course of the administration, leaving important scientific integrity initiatives unfinished (Goldman et al. 2017).

With the election of Trump, the politicization of science has reached a new high. "We are now at a crisis point, with almost weekly violations of previously respected safeguards" (National Task Force on Rule of Law 2019). According to *Washington Post* journalists, the federal government is undergoing a "rebranding." "Climate change" is out, replaced by "resilience." "Clean energy investment" has been changed to "energy investment." And the mission of key agencies has shifted to reflect Trump's fossil fuel focused agenda (Mooney and Rein 2017). Below are a few examples of such politicization as cited by watchdog groups.

- Scientists at the US Geological Survey (USGS) are currently required to obtain advance permission before speaking to reporters, despite the fact that the agency's scientific integrity policy specifies that the communications office need only be "notified" (UCS 2019).

- White House officials stopped a senior analyst at the State Department's Bureau of Intelligence and Research from submitting scientific information on climate change in a written testimony to the House Intelligence Committee (UCS 2019).
- The Department of Energy (DOE) prevented climate scientists from mentioning "climate change," "emissions reductions," or "Paris Agreement" in their communications (UCS 2019).
- Both the EPA and DOI have directed political appointees to review discretionary grants to make sure that they align with the priorities of the Trump administration. Such grants are those that would be made to institutions of higher education or non-profits that can legally advocate for certain policies. While it is not uncommon for political appointees to get involved in the grants process, their involvement has generally been limited to broadening solicitations for grant proposals. It is clear that this new process was specifically set up to undermine science and scientific experts, especially those working on climate change related issues (UCS 2019).
- The Interior Department reassigned its leading climate scientist to an accounting role after he stressed dangers posed by climate change (Clement 2017; Eilperin 2017).
- Agency officials at the EPA, the DOI, and the USDA have blocked agency scientists and contractors from presenting research about climate change and related ecological issues at professional conferences (Negrin 2017; Patterson 2017; Kaplan 2017).
- After Joshua Tree National Park's Twitter account published a series of tweets based on climate change science, former Secretary of the Interior Ryan Zinke reprimanded the park's superintendent and ordered the park to cease posting tweets on climate change (UCS 2017).
- Former EPA head Scott Pruitt "attacked the independence of the Science Advisory Board (SAB), by ordering that no scientists receiving EPA grant funding could serve on the SAB. Industry scientists faced no additional restrictions on service, and the new board more than doubled the number of industry representatives" (Reed 2017).

Trump's Reversal of Obama-Era Climate Change Policies

One of Trump's first vows as president was to withdraw from the Paris Climate Agreement that had been orchestrated by President Obama charging that it was unfair and infringed on U.S. sovereignty. The Agreement requires countries to put forward their best efforts to reduce carbon emissions through Nationally Determined Contributions (NDCs) and to make progressively more ambitious pledges to reduce emissions following global stocktaking every five years, including regular reporting on their emissions and implementation efforts in an effort to limit global warming to no more than 1.5 °C above pre-industrial levels, the threshold beyond which the IPCC has determined would cause irreversible risks to human and natural systems. While a majority of the other parties to the Paris Agreement have remained committed to the Agreement's goals and, in December 2018 in Katowice, Poland agreed to a set of guidelines to operationalize the reporting requirements, the opposition of the United States is certain to have an effect, especially, on those countries now headed by similar authoritarian leaders whom Trump has been embracing. For example, Jair Bolsonaro, the right-wing populist president of Brazil has opened the Amazon to rapid deforestation and development. But even countries that have committed themselves to the Paris Agreement may be unwilling to take the most radical steps needed to halt CO_2 emissions if they believe the United States—the second largest producer of greenhouse gases—is unwilling to curb its own appetite for fossil fuels.

Trump cannot formally withdraw the U.S. from the Paris Climate Accord until November 5, 2020, just days after the next presidential election, so his threat is still somewhat theoretical, but his decision is not without negative consequences. According to Curtin (2018, p. 1), Trump "has created a powerful countervailing force acting against the momentum which the Paris Agreement on climate change hoped to generate." The "Trump Effect," according to Curtin, has applied a brake in three ways: (1) U.S. Federal rollbacks have increased the attractiveness of fossil fuel investments globally; (2) The U.S. decision to withdraw from the Agreement has created moral and political cover for others to follow suit; and (3) Goodwill at international negotiations has been damaged. He cautions that in the short-run, the Paris Agreement can withstand the Trump Effect, "but in the medium and longer-term, it will continue to be assailed by instability and uncertainty until broader and deeper structural factors within the U.S. political economy can be addressed" (Curtin 2018, p. 1).

Under the rubric of promoting energy independence and economic growth, the Trump administration has reversed Obama-era policies that had sought to move away from the use of fossil fuels in the transition to a renewable energy economy and has engaged in attempts to undermine research on renewable energy. Executive orders can be overruled by Congress and the courts as well as by succeeding presidents. The Clean Power Plan has been fiercely fought by the fossil fuel industry and congressional conservatives. While Congress was not able to overrule it during Obama's tenure, Trump did just that in June 2019.

In pursuit of further oil and gas exploration on public lands, the Trump administration has engaged in the largest reduction of federally protected lands in U.S. history. Slated for such exploration are a number of national monuments that Obama had created during his last years in office as well as Alaska's Arctic National Wildlife Refuge (ANWR), a 19.6 million acre tract of pristine wilderness in eastern Alaska that is home to gray wolves, musk oxen, caribou, and imperiled polar bears; a resting place for migratory birds from all 50 states and six continents; and a land that is vital to the culture and survival of the indigenous Gwich'in people. In support of Trump's policy, the DOI failed to consider 18 memos from staff scientists who had raised scientific and environmental concerns about proposed oil and gas operations in the ANWR. These documents were excluded from public view and from the DOI's draft environmental assessment and were not released during Freedom of Information Act (FOIA) requests filed by advocacy groups.

The Trump administration rescinded a 2015 Bureau of Land Management (BLM) rule that set standards for well construction, wastewater management, and chemical disclosure around hydraulic fracturing on public lands. "Without such a rule in place, fracking on federal lands has less oversight and transparency, putting the public at risk" (UCS 2017).

In an attack on renewable energy, the DOE withheld almost $100 million of appropriated funds from the Advanced Research Projects Agency-Energy (ARPA-E) which typically funded researchers exploring alternative energy and improvements to the efficiency of existing energy technologies and in the 2020 federal budget reduced its funding by 70% (Natter 2019).

Mr. Trump has also taken aggressive steps to allow more planet-warming pollution from vehicle tailpipes and power plant smokestacks, reversing actions the Obama administration had taken and has begun a lawsuit against the state of California and 13 other states that had set emission standards higher than his proposed standards (Davenport 2018; Friedman 2018).

Trump is not alone in his attempts to undermine environmental regulation. A pro-business, anti-regulatory Republican majority in the Senate is also responsible. Since achieving ascendancy, this relatively new Republican cohort has sought to tear apart the edifice of environmental law constructed largely under a Republican president, Richard Nixon, and formerly sustained by bipartisan support. While in the majority in both houses of the legislature, these Republicans voted nearly 200 times to undercut clean air and water laws while blocking efforts to limit global warming, protect public lands, and guard against future oil spills. Having lost control of the House of Representatives in the 2018 election but still in control of the Senate, Republicans have been trying to attach amendments called "riders" to dozens of must pass appropriation bills that would significantly reduce the country's commitment to providing clean air and clean water, saving pristine public lands from oil and gas drilling, and protecting endangered species (*New York Times Editorial Board,* 2018).

However, public opinion, which heretofore had tended to put climate change down low on a list of concerns, has now begun to change. A nationally representative survey from Yale University and George Mason University reported that seven in ten Americans (73%) now think global warming is happening, an increase of ten percentage points since March 2015. 69% of Americans say they are "somewhat worried" about climate change and 291% are "very worried." "These are the highest values since the surveys began in 2008, and the 'very worried' category shows an 8% jump compared to the previous survey published in April 2018" (Leiserowitz et al. 2018). Another national poll conducted by Reuters/Ipsos between November and December 2018 found that 72% of Americans consider climate change to be a moderate, serious, or imminent threat, and 69% said that the United States should work with other nations to curb climate change, including 64% of Republicans and 80% of Democrats (Caspani 2018). According to Pew research, the rise in concern, however, masks deep partisan divisions. Most of that rise has come from Democrats, not Republicans whose opinions remain largely unchanged (Kennedy and Hefferon

2019). The sentiment that dealing with climate change should be a top priority for Congress has grown within the last four years from 34 to 44%, yet it is still a minority sentiment, and the increased interest in prioritizing it comes from Democrats, not Republicans (Kennedy and Hefferon 2019).

American public opinion on climate change will have to move more decisively, however, if we are to see dramatic policy change in the U.S. on the issue. Despite being concerned about climate change, many are still reluctant to support specific policies that would address it. Several key climate policies on the ballot in the 2018 election, including a carbon tax in Washington State and an aggressive renewable power target in Arizona, were defeated and too many Americans are still purchasing large gas guzzling cars and refusing to move out of flood and wildfire zones, even after they have been hit by disaster (Kormann 2018). The mainstream media aided by a fossil fuel industry disinformation campaign is largely responsible for the American public's still tepid response to the imminence of climate change. In the Reuters/Ipsos poll, only 35% of respondents believed that climate change is an *imminent* threat. During the 2017 elections, the media failed to ask the presidential candidates a single question about climate change. Except for the *New York Times* and *Washington Post* which now regularly cover climate change, most U.S. mainstream media coverage of several weather disasters over the summer of 2018 failed to cite climate change as a driver (Hymas 2018). The Democratic Party elite is also partly responsible. When the Democratic presidential primary debates began in the summer of 2019, the Party announced that there would be no debate focused on climate change, fearing that it was not a priority for voters and that their candidates should stick to more immediate concerns like health care and the economy. Despite this rejection, climate change did enter into the debate topics, fueled by a backlash among young climate activists and a group of progressive candidates who do take the issue seriously.

Sub-National and Civil Society Actors as Protagonists in the Service of Protecting the Planet

While there has been no movement at the national level to halt climate change, a series of sub-national governments—cities, counties, and states along with businesses–have been acting on their own to reduce greenhouse gases, employ renewable energy, and retrofit their infrastructures for resilience. Two of the most ambitious states are California and New York. California, the fifth largest economy in the world and the victim of devastating climate-related drought and wildfires, has taken the lead on climate change. Under a Democratic governor and legislature, it has developed an integrated climate change strategy, with a goal by 2030 of reducing greenhouse gas emissions to 40% below 1990 levels, procuring 50% of its electricity from renewable sources, reducing today's petroleum use in cars and trucks by up to 50%, doubling its energy efficiency in existing buildings, and increasing carbon sequestration in its forests, wetlands, rangelands, and farms among other measures (California Climate Change, n.d.). In October 2019, over 600 veteran EPA officials called for a congressional investigation into Trump's threatened punishment of California for its effort to set its own auto emission standards by withholding federal funds for other state priorities (Environmental Integrity Project 2019).

After the election of a Democratic legislature and intense lobbying pressure from climate activists, New York State passed in the summer of 2019 the most ambitious climate change bill in the country and possibly the most ambitious of any legislature in the world, pledging to slash its planet-warming pollution 85% below 1990 levels by 2050, and offset the remaining 15%, possibly through measures to remove carbon dioxide from the atmosphere (NY State Senate 2019; McKinley and Plumer 2019).

In addition to individual state-level efforts, three multi-state efforts are now providing coordinated responses to climate change. For several years now, the Regional Greenhouse Gas Initiative, a cap-and-trade system among nine Northeastern and Mid-Atlantic states (with two more about to join), has been auctioning a steadily dwindling supply of carbon pollution permits to power plants, using the revenue to invest in energy efficiency, renewable energy, and other clean energy programs (Plumer 2019; Regional Greenhouse Gas Initiative). The United States Climate Alliance, a relatively new bipartisan coalition of the governors of 25 states is committed to reducing greenhouse gas emissions consistent with the goals of the Paris Agreement. Each state in the alliance commits itself to implement policies that advance the goals of the Paris Agreement, with the goal of reducing greenhouse gas emissions by at least 26–28% below 2005 levels by 2025; track and report progress to the global community, including when the world convenes to take stock of the Paris Agreement; and accelerate new and existing policies to reduce carbon pollution and promote clean energy deployment at the state and federal level (United States Climate Alliance, n.d.). Altogether these states account for over one-third of all U.S. greenhouse gas emissions and nearly half its population (Plumer 2019). These states have vowed to resist the Trump administration's attempt to revoke the waiver that allows states to set their own emission standards (United States Climate Alliance).

Just days after Mr. Trump's announcement that he was withdrawing from the Paris Climate Agreement, these states, and an even broader group of American mayors, businesses, investors, regional associations, faith institutions, and universities declared "We are still in," pledging to meet the goals of the Paris agreement in spite of Trump's refusal (*New York Times* Editorial Board 2017). Co-chaired by former New York City mayor and philanthropist, Michael Bloomberg, and former California governor, Jerry Brown, the commitments of more than 3,000 signatories are now formalized as "America's Pledge," which will aggregate and quantify the actions of states, cities, businesses, and other non-national actors to drive down their greenhouse gas emissions consistent with the goals of the Paris Agreement, report their efforts to the international body, and catalyze further climate action by providing detailed roadmaps for similar business-level, city, and state action in the U.S. and potentially in other countries (America's Pledge, n.d.). These actors represent more than half the U.S. economy—economic activity that is equivalent to that of the third largest country in the world. If followed through, current state and city climate policies would get the United States about two-thirds of the way toward its commitment under the accord (Fulfilling America's Pledge 2018). In December 2018, nine Eastern states and the District of Columbia announced they would work together to put a price on emissions from transportation fuels and invest the revenue in lower carbon projects.

This is an extraordinary initiative. In effect, these sub-national players have become protagonists in the global negotiations process that until now was the exclusive domain of national governments in what some have called a "complex, multilevel, or polycentric governance system" (Betsill 2017, p. 189). Their initiative could become a model for sub-national and non-state actors in other countries whose national governments are moving too slowly or not at all to reduce greenhouse gas emissions. National governments are still needed for the heavy lifting required to keep greenhouse gases below 1.5 °C above pre-industrial levels, but states, counties, and cities can serve as laboratories, testing which policies work and which don't, sharing best practices and perhaps catalyzing others to follow suit.

But more than these efforts are necessary. It will take a massive international people's movement to push the world's leaders to take the radical steps that are needed to curb greenhouse gas emissions. If there is to be a massive people's movement on climate change in the United States and elsewhere, it will come from millennials and those younger, who are facing the consequences of their leaders' failure to act. Polls show that U.S. millennial voters, the largest voting demographic, consistently rank climate change as an issue of top concern. These young people along with their counterparts in Europe and other parts of the world have been at the forefront of a series of marches, rallies, student strikes, and lobbying in statehouses and in Congress to demand that their governments act on climate change and do so quickly. In the U.S., they have become active in politics and can be found in two movements, 350.org and a more recent organization called the Sunrise Movement whose mission statement says "We are building an army of young people to make climate change an urgent priority across America, end the corrupting influence of fossil fuel executives on our politics, and elect leaders who stand up for the health and wellbeing of all people" (Sunrise Movement, n.d.).

These young activists have already succeeded in electing a number of young progressive Democrats to Congress who have begun to lobby for the creation of a Green New Deal, modeled on Franklin Delano Roosevelt's New Deal of the 1930s which was both an economic and an environmental recovery program (H. Res. 109 2018; Collins and Goldberg 2014). Their proposal, in the form of a general set of principles introduced in the House of Representatives in 2019 as a nonbinding "resolution," calls for a massive effort comparable in scope to what went into the New Deal during the Great Depression and the war effort in the 1940s to completely restructure our economy and infrastructure so as to end the U.S.'s reliance on fossil fuels. Like the original New Deal, it combines both socioeconomic justice principles with environmental sustainability. Trump and his Republican supporters have derided the idea of a Green New Deal, calling it unrealistic, unaffordable, ill-conceived, a threat to American jobs and the economy and have tarred it with what for Republicans is a pejorative label, "socialism." In fact, the proposal is eminently realistic if we are to meet the carbon threshold set by the IPCC. So powerful has this grassroots movement become that the outlines of the Green New Deal they have proposed have been supported by at least six of the Democratic candidates who were running for president in 2020. While many of the older, more moderate Democrats are unwilling to be as bold as the House of Representatives resolution submitted by these new Democrats, the party leadership in the Senate has nevertheless stated that it will make combatting climate change a central issue of the 2020 campaigns—a striking shift on an issue they have shied away from for the past decade (Davenport and Stolberg 2019).

It remains to be seen how fast a growing climate movement can move national policy in the right direction before it is too late. Trump has already done irreparable harm to institutions, and any future Democratic administration will have to work doubly hard to make up for the loss of momentum on climate change incurred under President Trump. Some of what Trump has tried to do may be thwarted by the courts, but litigation is slow and the timeline for action is short. Moreover, Trump has been able to stack the courts with conservative, business-oriented judges. His renewed support for coal, however, is running up against the reality that renewable energy—particularly solar and wind—is now more cost-effective than coal. And there is some encouragement in the fact that the other major emitters have all signed onto the Paris Agreement and many world leaders have reaffirmed their Paris commitments after Trump's withdrawal announcement (Betsill 2017). Yet, if the world is to achieve the Paris Agreement's goal of remaining below a 2 °C temperature increase, there is no room for any country or region, especially as large as the United States, to remain a high-carbon reservation (Sprintz et al. 2017). There will be little change in national policy unless the U.S. can elect a Democratic president and Congress and/or public opinion, perhaps faced with ecological collapse, engages in a massive political movement unlike anything the U.S. has yet seen.

Conclusion

A president who denies climate change and places the interests of the fossil fuel industry above environmental sustainability and the long-term integrity of the planet's socio-ecological systems was elected just when what is needed is international cooperation to halt the runaway climate change that is threatening the peace and sustainability of human civilization. As a leader of one of the world's major powers, Trump's withdrawal from the Paris Climate Agreement leaves the world with an increasingly dangerous and uncertain future. In the absence of a federal commitment to reducing greenhouse gas emissions, sub-national governments, businesses, and civil society actors have stepped into to do what they can to maintain U.S. commitments to the Paris Accord, in essence becoming new protagonists in the international arena that was once the province of sovereign states. It is these actors, along with a growing international movement of young people who are taking to the streets and running for office to press for swift action on climate change, who provide some hope that the world may yet avoid climate disaster.

References

Atkin, E. (2017, March 7). The EPA's 'Science' Office removed 'science' from its mission statement. *The New Republic*. Retrieved September 18, 2019, from https://newrepublic.com/article/141174/epas-science-office-removed-science-mission-statement.

Bailey, C. J. (2019). Assessing President Obama's climate change record. *Environmental Politics, 28*(5), 847–865. https://doi.org/10.1080/09644016.2018.1494967.

Betsill, M. M. (2017, July 3). Trump's Paris withdrawal and the reconfiguration of global climate change governance. *Chinese Journal of Population Resources and Environment, 15*(3), 189–191. https://doi.org/10.1080/10042857.2017.1343908.

Brulle, R. J., Carmichael, J., & Jenkins, J. C. (2012, September). Shifting public opinion on climate change: an empirical assessment of factors influencing concern over climate change in the U.S., 2002–2010." *Climatic Change*, 169–188, https://doi.org/10.1007/s10584-012-0403-y.

California Climate Change (n.d.). California Climate Change. http://www.climatechange.ca.gov.

Caspani, M. (2018, December 13). More Americans view climate change as 'imminent' threat: Reuters/Ipsos poll. Retrieved March 7, 2019, from https://www.rE.U.ters.com/article/us-climate-change-usa-poll/more-americans-view-climate-change-as-imminent-threat-rE.U.ters-ipsos-poll-idUSKBN1OC1FX.

Center for Science and Democracy, Union of Concerned Scientists (2017). Attacks on science. Retrieved October 6, 2019, from https://www.ucsusa.org/center-science-and-democracy/attacks-on-science.

Clement, J. (2017). I'm a scientist. I'm blowing the whistle on the trump administration. *Washington Post*. Retrieved September 6, 2019, from https://www.washingtonpost.com/opinions/im-a-scientist-the-trumpadministration-reassigned-me-for-speaking-up-about-climatechange//07/19/389b8dce-6b12-11e7-9c15-177740635e83_story.html.

Climate Deregulation Tracker (n.d.). Sabin Center for Climate Change Law, Columbia Law School. Retrieved March 4, 2019, from http://columbiaclimatelaw.com/resources/climate-deregulation-tracker/.

Collins, S., & Schaffner Goldberg, G. (2014). *When government helped: Learning from the successes and failures of the new deal*. New York: Oxford University Press.

Curtin, J. (2018). The Paris Climate Agreement versus the Trump effect: Countervailing forces for decarbonization. *IIEA European Affairs*. Retrieved March 22, 2019, from https://www.iiea.com/energy/the-paris-climate-agreement-versus-the-trump-effect-countervailing-forces-for-decarbonisation/.

Davenport, C. (2018). Trump administration unveils its plan to relax car emission rules. *New York Times*. Retrieved August 2, 2018, from https://www.nytimes.com/2018/08/02/climate/trump-auto-emissions-california.html?module=inline.

Davenport, C., & Gay Stolberg, S. (2019). Pressed by climate activists, Senate Democrats plan to 'go on offense.' *New York Times*. Retrieved March 4, 2019, from https://www.nytimes.com/2019/03/04/us/politics/senate-democrats-climate-change.html?searchResultPosition=1.

Davenport, C., & Landler, M. (2019, May 27). Trump administration hardens its attack on climate science. *New York Times*, A1.

Davenport, C., & Pierre-Louis, K. (2019, November 23). U.S. climate report warns of damaged environment and shrinking economy. *New York Times*. Retrieved March 4, 2019, from https://www.nytimes.com/2018/11/23/climate/us-climate-report.html.

Egan, P. J., & Mullin, M. (2017). Climate change: U.S. public opinion. *Annual Review of Political Science, 20*, 209–227.

Eilperin, J. (2017). Senate Democrats call for an investigation of climate scientist whistleblower complaint. *Washington Post*. Retrieved September 25, 2019, from https://www.washingtonpost.com/news/energy-environment/wp/2017/07/24/senate-democrats-demand-ig-probe-of-interiors-reassigning-top-career-officials/.

Eilperin, J., & Dennis, B. (2017). EPA dismisses half of key board's scientific advisers; interior suspends more than 200 advisory panels: Interior secretary Ryan Zinke "is committed to restoring trust in the department's decision-making," a spokeswoman said. Washington: WP Company LLC d/b/a *The Washington Post.* Retrieved from http://ezproxy.cul.columbia.edu/login?url=https://search-proquest-com.ezproxy.cul.columbia.edu/docview/1896208293?accountid=10226.

Environmental Data and Governance Initiative (2018). The first 100 days and counting, Part 3, changing the digital climate. Retrieved October 10, 2019, from https://100days.envirodatagov.org/changing-digital-climate/#h.5apjn836y9u6.

Environmental Data and Governance Initiative (2018, November). A sheep in the closet: the erosion of enforcement at the EPA. Retrieved October 8, 2019, from https://envirodatagov.org/wp-content/uploads/2018/11/Sheep-in-the-Closet.pdf.

Environmental Data and Governance Initiative (2019a). The new digital landscape: How the trump administration has undermined federal web infrastructures for climate information. Retrieved October 7, 2019, from https://envirodatagov.org/wp-content/uploads/2019/07/New_Digital_Landscape_EDGI_July_2019.pdf.

Environmental Data and Governance Initiative (2019b). Examples of changes to the EPA website that undermine public access to scientific information. Retrieved October 7, 2019, from https://envirodatagov.org/4372-2/.

Environmental Data and Governance Initiative (n.d). Website monitoring. Retrieved October 10, 2019, from https://envirodatagov.org/website-monitoring/.

Environmental Integrity Project (2019). Almost 600 EPA veterans call for investigation into Trump's political abuse of EPA authority. Retrieved October 11, 2019, from https://www.environmentalintegrity.org/news/almost-600-epa-veterans-call-for-investigation-into-trumps-political-abuse-of-agency-authority/.

Environmental Protection Agency (1970). Summary of the Clean Air Act. https://www.epa.gov/laws-regulations/summary-clean-air-act.

Farber, D. (2018). U.S. climate policy: Obama, trump, and beyond. *Revista De Estudos Constitucionais, Hermenêutica e Teoria do Direito, 10*(2), 95–108. Retrieved from http://ezproxy.cul.columbia.edu/login?url=https://search-proquest-com.ezproxy.cul.columbia.edu/docview/2134276631?accountid=10226.

Federal Register. (2017). Withdrawal of final guidance for federal departments and agencies on consideration of greenhouse gas emissions and the effects of climate change in national environmental policy act reviews. Retrieved October 18, 2019, from https://www.federalregister.gov/documents/2017/04/05/2017-06770/withdrawal-of-final-guidance-for-federal-departments-and-agencies-on-consideration-of-greenhouse-gas.

Flavelle, C., Friedman, L., & Baker, P. (2019). Commerce chief threatened firings at NOAA after Trump's Dorian tweets, sources say. *New York Times.* Retrieved September 13, 2019, from https://www.nytimes.com/2019/09/09/climate/hurricane-dorian-trump-tweet.html.

Freeman, J. (2011). The Obama Administration's national auto policy: Lessons from the 'car deal.' *Harvard Environmental Law Review, 35*(2), 343–374. http://www.law.harvard.edu/faculty/freeman/HLE209-2.pdf.

Friedman, L. (2017a). E.P.A. scrubs a climate website of 'climate change.' Retrieved September 28, 2019, from https://www.nytimes.com/2017/10/20/climate/epa-climate-change.html.

Friedman, L. (2017b). E.P.A. cancels talk on climate change by agency scientists. *New York Times.* Retrieved September 28, 2019, from https://www.nytimes.com/2017/10/22/climate/epa-scientists.html.

Friedman, L. (2018). Cost of new E.P.A. coal rules: Up to 1,400 more deaths a year. *New York Times.* https://www.nytimes.com/2018/08/21/climate/epa-coal-pollution-deaths.html?module=inline.

Fulfilling America's Pledge (2018). How states, cities and businesses are leading the United States to a low-carbon future. Retrieved September 28, 2018, from https://www.bbhub.io/dotorg/sites/28/2018/09/Fulfilling-Americas-Pledge-2018.pdf.

Goldman, G., Reed, G., Halpern M., Johnson C., Berman E., Kothari, Y., & Rosenberg, A. (2017). Preserving scientific integrity in federal policymaking: Lessons from the past two administrations and what's at stake under the Trump administration. Center for Science and Democracy at the

Union of Concerned Scientists. Retrieved October 6, 2019, from https://www.ucsusa.org/sites/default/files/attach/2017/01/preserving-scientific-integrity-in-federal-policymaking-ucs-2017.pdf.
H. Res. 109. (2018). Recognizing the duty of the Federal Government to create a Green New Deal. Retrieved February 28, 2019, from https://www.congress.gov/116/bills/hres109/BILLS-116hres109ih.pdf.
H.R. 5380. (1988). National Energy Policy Act of 1988. Retrieved from https://www.congress.gov/bill/100th-congress/house-bill/5380.
Hymas, L. (2018). Climate what? Media kept on chasing Trump, not climate change, in 2018. Retrieved March 7, 2019, from https://grist.org/article/climate-what-media-kept-on-chasing-trump-not-climate-change-in-2018/.
Kaplan, S. (2017). Government scientists blocked from the biggest meeting in their field, *Washington Post*. Retrieved September 6, 2019, from https://www.washingtonpost.com/news/speaking-of-science/wp/2017/12/22/government-scientists-blocked-from-the-biggestmeeting-in-their-field.
Kendall, A., Hanjiro A., Mahoney, E., & Huijing, D. (2018). Regulatory reform: Assessing life-cycle-based greenhouse gas standards. National Center for Sustainable Transportation. Retrieved September 8, 2019, from https://rosap.ntl.bts.gov/view/dot/36892.
Kennedy, B., & Hefferon, M. (2019). U.S. concern about climate change is rising, but mainly among Democrats. Pew Research Center. Retrieved October 11, 2019, from https://www.pewresearch.org/fact-tank/2019/08/28/u-s-concern-about-climate-change-is-rising-but-mainly-among-democrats/.
Kormann, C. (2018). Efforts to fight climate change had a tough election day. *The New Yorker*. Retrieved October 12, 2019, from https://www.newyorker.com/news/news-desk/efforts-to-fight-climate-change-had-a-tough-election-day.
Kuhn, D. P. (2008). Exit polls: Economy top issue. *Politico*. Retrieved September 4, 2019, from https://www.politico.com/story/2008/11/exit-polls-economy-top-issue-015270.
Lavelle, M. (2016). Obama's climate legacy marked by triumphs and lost opportunities. Retrieved October 6, 2019, from https://insideclimatenews.org/news/23122016/obama-climate-change-legacy-trump-policies.
Lavelle, M. (2019). Trump tries again to roll back methane rules for oil and gas industry. Retrieved October 13, 2019, from https://insideclimatenews.org/news/29082019/methane-regulation-oil-gas-storage-pipelines-epa-rollback-trump-wheeler.
Leiserowitz, A., Maibach, E., Rosenthal, S., Kotcher, J., Ballew, M., Goldberg, M., & Gustafson, A. (2018). Climate change in the American mind. Retrieved March 7, 2019, from http://climatecommunication.yale.edu/wp-content/uploads/2019/01/Climate-Change-American-Mind-December-2018.pdf.
Leonard, C. (2019). Kochland: the secret history of Koch Industries and corporate power in america. New York: Simon & Schuster.
Massachusetts v. EPA, 549 U.S. 497. (2007). Retrieved from https://supreme.justia.com/cases/federal/us/549/497/.
Matthews, D. (2017). Donald Trump has tweeted climate change skepticism 115 times. Here's all of it. *Vox*. Retrieved October 3, 2019, from https://www.vox.com/policy-and-politics/2017/6/1/15726472/trump-tweets-global-warming-paris-climate-agreement.
McKinley, J., & Plumer, B. (2019). New York to approve one of the world's most ambitious climate plans. *New York Times*, Retrieved June 18, 2019, from.https://www.nytimes.com/2019/06/18/nyregion/greenhouse-gases-ny.html.
Milman, O. (2017). EPA removes half of scientific board, seeking industry-aligned replacements. Retrieved October 16, 2019, from https://www.theguardian.com/us-news/2017/may/08/epa-board-scientific-scott-pruitt-climate-change.
Mooney, C., & Rein, L. (2017). Don't call it 'climate change': How the government is rebranding in the age of Trump. Retrieved October 7, 2019, from https://www.washingtonpost.com/news/energy-environment/wp/2017/05/26/just-dont-call-it-climate-change-rebranding-government-in-the-age-of-trump/.

National Task Force on the Rule of Law and Democracy, Brennan Center for Justice (2019). Proposals for reform volume II. Retrieved October 12, 2019, from https://www.brennancenter.org/our-work/policy-solutions/proposals-reform-volume-ii-national-task-force-rule-law-democracy.

Natter, A. (2019). Trump again seeks deep cuts in renewable energy funding. *Bloomberg News*. Retrieved October 9, 2019, from https://www.bloomberg.com/news/articles/2019-03-07/trump-said-to-again-seek-deep-cuts-in-renewable-energy-funding.

Negrin, E. (2017). Energy department scientists barred from attending nuclear power conference. *Huffington Post*. Retrieved October 9, 2019, from https://www.huffpost.com/entry/energy-department-scientists-barred-from-attending_b_597f7e2ee4b0cb4fc1c73b8e.

New York State Senate (n.d.). Senate bill S6599. Retrieved from https://www.nysenate.gov/legislation/bills/2019/s6599.

New York Times Editorial Board (2017). States and cities compensate for Mr. Trump's climate stupidities. Retrieved March 9, 2019, from https://www.nytimes.com/2017/06/07/opinion/climate-change-cities-states.html?module=inline.

New York Times Editorial Board (2018). The dirty little deals that would foul the environment. Retrieved March 9, 2018, from https://www.nytimes.com/2018/02/19/opinion/republican-environment-policy.html.

Obama, B. (2013). Executive order-preparing the United States for the impacts of climate change. Retrieved February 18, 2019, from https://obamawhitehouse.archives.gov/the-press-office/2013/11/01/executive-order-preparing-united-states-impacts-climate-change.

Obama, B. (2019). President Barack Obama's inaugural address. Retrieved October 5, 2019, from https://obamawhitehouse.archives.gov/blog/2009/01/21/president-barack-obamas-inaugural-address.

Patterson, B. (2017). Govt. scientist blocked from talking about climate and fire. *E&E News*. Retrieved October 9, 2019, from https://www.eenews.net/stories/1060065143.

Plumer, B. (2019). A 'green new deal' is far from reality, but climate action is picking up in the states. *New York Times*. Retrieved March 9, 2019, from https://www.nytimes.com/2019/02/08/climate/states-global-warming.html.

Popovich, N., Albeck-Ripka, L., & Pierre-Louis, K. (2020). The Trump Administration is reversing 100 environmental rules. Retrieved May 29, 2020, from https://www.nytimes.com/interactive/2020/climate/trump-environment-rollbacks.html.

Reed, G. (2017). Scott Pruitt deals yet another blow to independent science advice at the EPA. Retrieved September 8, 2019, from https://blog.ucsusa.org/genna-reed/scott-pruitt-deals-yet-another-blow-to-independent-science-advice-at-the-epa.

Regional Greenhouse Gas Initiative (n.d.). Retrieved from https://www.rggi.org/.

Rich, N. (2018). Losing earth: The decade we almost stopped climate change. A tragedy in two acts. *New York Times Magazine*. Retrieved March 1, 2019, from https://www.nytimes.com/interactive/2018/08/01/magazine/climate-change-losing-earth.html.

S. 932 Energy Security Act (1980). Retrieved from https://www.congress.gov/bill/96th-congress/senate-bill/932.

Sabin Center for Climate Change Law, Columbia University (n.d.). Retrieved October 5, 2019, from http://columbiaclimatelaw.com/resources/climate-deregulation-tracker/.

Sprintz, D.F., Saelen, H., Underdal, A., & Hovi, J. (2017). The effectiveness of climate clubs under Donald Trump. *Climate Policy, 18*(7). https://doi.org/10.1080/14693062.2017.141009.

Sunrise Movement (n.d.). Retrieved from http://www.sunrisemovement.org.

Trump, D. J. (2012). DonaldJTrump@realDonaldTrump. Retrieved October 3, 2019, from https://twitter.com/realDonaldTrump/status/265895292191248385.

Trump, D. J. (2018). DonaldJTrump@realDonaldTrump cited by Nick Visser, Scientists slam Trump's clueless climate change tweet: "He's a clown.'" *Huffington Post*. Retrieved October 3, 2019, from https://www.huffpost.com/entry/donald-trump-climate-change-tweet_n_5bf6019ae4b0eb6d930b6d25?ncid=APPLENEWS00001.

Union of Concerned Scientists (2017). Joshua Tree National Park superintendent reprimanded for climate change science tweets. Retrieved September 6, 2019, from https://www.ucsusa.org/resources/attacks-on-science/joshua-tree-national-park-superintendent-reprimanded-climate-change.

Union of Concerned Scientists (2019a). The state of science in the Trump era. Retrieved October 1, 2019, from https://www.ucsusa.org/resources/state-science-trump-era.

Union of Concerned Scientists (2019b). EPA stops collecting date on pollutants released from farms. Retrieved September 8, 2019, from https://www.ucsusa.org/resources/attacks-on-science/epa-stops-collecting-data-pollutants-released-farms.

United Nations Framework Convention on Climate Change (n.d.). Retrieved from https://www.iucn.org/theme/global-policy/our-work/united-nations-framework-convention-climate-change-unfccc.

United States Climate Alliance (n.d.). Retrieved from http://www.usclimatealliance.org.

Rights of Nature in the European Union: Contemplating the Operationalization of an Eco-Centric Concept in an Anthropocentric Environment?

Hendrik Schoukens

Abstract A global movement to give legal rights to nature is slowly gaining momentum in the face of the ongoing biodiversity crisis that is hitting our planet. At its core, the concept of rights of nature presupposes a novel template for ecological governance, which is aimed at prioritizing nature's right to exist and to flourish through a societal and legal reform. This chapter makes two separate arguments. First, it argues that rights of nature constitutes a powerful new paradigm for ecological governance. Second, it demonstrates the difficulties and opportunities encountered when operationalizing this emerging concept within the existing legal order through a case study on the interplay between nature's rights and the existing EU legal order. In spite of the identified alignment between a more rights-based approach to nature protection and the existing EU environmental law, several long-standing defects of the existing EU legal order seem to block a further operationalization of rights of nature. Against the backdrop of the relatively strict case law at EU level with respect to standing in environmental cases before EU courts and the persisting anthropocentric nature of many EU environmental directives, this chapter concludes that the recognition of rights of nature at the national level, either through strategic litigation or legislative amendments, is to be approached as the most realistic pathway to the short-term operationalization of ecocentrism in the EU.

Introduction

The past decades have seen the adoption of a growing number of environmental treaties, laws and regulations aimed at halting ecological decline. While the scheme of the existing environmental regulations appears impressive at first sight, an increasing number of authors posit that these models of protecting nature and the environment are fundamentally flawed and therefore no longer apt to achieve their objectives (Wood 2014; Boyd 2017). This failure is most glaring when it comes to climate

H. Schoukens (✉)
Ghent University, Ghent, Belgium
e-mail: hendrik.schoukens@ugent.be

J. C. Pereira and A. Saramago (eds.), *Non-Human Nature in World Politics*, Frontiers in International Relations, https://doi.org/10.1007/978-3-030-49496-4_11

change.[1] A similar grim picture emerges when it comes to biodiversity loss.[2] Even the European Union (EU), which often portrays itself as an environmental frontrunner, is no exception to that general rule, with more than 70% of its natural habitats and 60% of its protected species currently in a dire conservation status.[3] Taking stock of the limited results on the basis of modern environmental law, a push for more systematic and transformative changes of the existing legal frame is becoming increasingly noticeable within the existing literature (Cullinan 2011). For one, it is by no means surprising to read into the 2019 report from the Intergovernmental Science-Policy Platform on Biodiversity and Ecosystem Services (IPBES) a call for 'transformative change', which is further defined as a 'fundamental, system-wide reorganization across technological, economic and social factors, including paradigms, goals and values'.[4] It is increasingly entertained that our existing legal frameworks poorly reflect the eco-centric shift that is needed to curb and, if possible, reverse the ecological decline our planet is witnessing (Gaines 2014; Bosselman 2016). A frequently echoed critique regarding the ineffectiveness of modern environmental laws is that such regulations are based on the omnipresent Cartesian, dualistic approach to nature (Boyd 2017), and often primarily focused on regulating and legalizing the destruction of our natural ecosystems (Wood 2014).

Granting legal personhood or, as some put it, 'rights' to nature has recently emerged as a promising new movement in the legal discourse, with various jurisdictions having incorporated 'rights of nature' in existing statutes or laws.[5] According to the proponents of more eco-centric approaches to law, a legal system in which the environment or nature is a legal person with rights will eventually lead to a more coherent and effective approach to certain environmental problems (Shelton 2015). Instead of granting rights *to* nature, the more rights-based approach assumes that nature *has* certain rights—thus leading to the concept 'rights *of* nature'. To use the words of Professor Stone (1972)—whose groundbreaking and influential article 'Should trees have standing' lay at the root of the rise of rights of nature movement—nature as a 'rights holder' would 'have a legally recognized worth and dignity in its own right and not merely to serve as a means to benefit us'.

---

[1] UN Environment, Emissions Gap Report 2018, https://www.unenvironment.org/resources/emissions-gap-report-2018 (accessed 15 September 2019).

[2] S. Diaz et al. (ed.), Summary for policymakers of the global assessment report on biodiversity and ecosystem services of the Intergovernmental Science-policy Platform on Biodiversity and Ecosystem Services (2019), advance unedited version.

[3] European Environment Agency, State of nature in the EU. Results from reporting under the nature directives 2007–2012, EEA Technical Report, No. 2/2015.

[4] See supra, footnote 2.

[5] For an extensive overview, see: Living Law, Giving Nature a Voice, Legal rights and personhood for Nature—an analysis of progressive legal developments in other jurisdictions where Nature Rights have been recognized in national constitutions, other domestic laws and litigation proceedings, 2018, https://www.livinglaw.co.uk/single-post/2018/07/06/Giving-Nature-a-Voice—Granting-Nature-Legal-Rights-a-progressive-path-to-balance-between-people-and-the-nature-world (accessed 15 September 2019).

The revolutionary idea of granting legal rights to nature generated mixed reactions in the legal literature of the past decades. Some authors discarded the idea as fundamentally inapt to deal with environmental challenges (Elder 1984) or posit that—since nature has no traceable 'voice'—the focus of granting rights is wrong-sided (Giagnocavo and Goldstein 1990), while others, such as Nash (1993), took up the idea and translated it into a plea for what he dubs 'biotic rights'. That said, it remains undisputed to say that this novel governance paradigm has remained dormant for several decades at the international forum, with the adoption of the World Charter of Nature in 1982 by the General Assembly of the United Nations being a notable exception.[6] Yet, remarkably so, recent years have witnessed a remarkable surge in the number of jurisdictions in which nature has been granted explicit rights and personhood. More than four decades after the publication of Stone's groundbreaking article, the country of Ecuador was the first to implement rights of nature in its governing laws (Tanasescu 2016),[7] with Bolivia quickly following suit (Villalba 2013).

As of today, however, the EU remains a notoriously absent actor in this debate. At times when rights of nature is emerging even in the United States (Boyd 2017), albeit at local level,[8] it is remarkable to note that no concrete implementation of rights of nature is to be encountered within the EU. Even so, the ongoing ecological crisis is also a European one. With the exception of one modest local ordinance adopted in a Dutch municipality, which contained a plea for the explicit legal recognition of the rights of the Wadden Sea,[9] no tangible example is to be found of the rights of nature within the EU. The EU's self-image as an environmental frontrunner—which appears to be justified in view of the remarkable array of progressive environmental directives that have been passed during the past decades—might partly help to explain why the EU is currently lagging behind in the global movement towards an explicit recognition of rights of nature. This begs the question to what extent granting legal personhood to nature might be an effective pathway to cure the existing flaws of the EU legislative framework with respect to environmental protection. More sceptical authors like Bétaille have argued that implementing rights of nature in the EU legal context is unlikely to further the cause of environmental restoration seeing that its main substantive tenets are already present in the existing legal order (Bétaille 2019).

This chapter makes the case that the latter, more sceptical position does not hold ground. First, it is argued that granting rights of nature not only ties in with several prominent Western ideas, championed by renowned and prominent natural scientists and ecologists such as Darwin and Leopold, but also presents a unique opportunity to move away from the human exceptionalism which has dominated the debate

[6]World Charter of Nature, A/RES/37/7, United Nations General Assembly (1982).

[7]More information can be found at https://therightsofnature.org/wp-content/uploads/pdfs/Rights-for-Nature-Articles-in-Ecuadors-Constitution.pdf (accessed 15 September 2019).

[8]See, for instance, the granting of legal personhood to Lake Erie by the City of Toledo. For more information, see https://www.theguardian.com/us-news/2019/feb/28/toledo-lake-erie-personhood-status-bill-of-rights-algae-bloom (accessed 15 September 2019).

[9]See for more background: De Waddenzee heeft ook rechten en die horen in de Grondwet te staan', https://www.trouw.nl/samenleving/de-waddenzee-heeft-ook-rechten-en-die-horen-in-de-grondwet-te-staan~a874d8b0/ (accessed 15 September 2019).

regarding legal personhood. Second, I will submit that granting rights to nature has the potential to effectively bolster environmental protection in the EU. Through a comprehensive assessment of the precise interplay between the existing EU legal order and the prevailing rights of nature template—taking into account the recent jurisprudence of the Court of Justice of the EU—I will outline the potential obstacles and opportunities in this respect.

Rights of Nature: What, Where, When and How?

Not Such a Novel Idea, After All?

Over the past decades, granting substantive rights and legal personhood to nature has emerged as a novel practice aimed at breaking down the hierarchical relationship between mankind and nature and, as a result, allowing for a more balanced approach to economic developments. It is closely linked to the concept of 'Earth Jurisprudence' (Schilmoller and Pelizonn 2013), which entails that the whole Earth Community has 'fundamental rights', including the right to exist and to flourish (Cullinan 2011). Human acts that infringe upon those rights are deemed 'unlawful'. To an unexpected audience this rights-based approach might sound outlandish. For, as of today, Western law almost exclusively treats the environment, nature and non-human animals as property and mere 'things' or 'commodities', which can be purchased, sold and used by humans to their own liking (Stone 1972). Already back in the 1970's, Singer (1975) contended that rights ought to be accorded to animals on the basis of sentience, which he defines as the capacity for pain and happiness. In the meantime, leading scientists, like the Dutch-American primatologist Frans de Waal, have demonstrated that animals, such as primates, are at least capable of proto-moral behaviour (de Waal 2013).

In other words, simply maintaining that animals lack moral agency and therefore cannot be regarded as a contracting party to the social contract no longer seems to be a conclusive argument against granting certain rights to animals (Driessen 2017). However, the rights of nature rhetoric goes one step further than the animal rights movements (Staker 2017). While the latter, like human rights lawyers, prioritize the individual sentient being, rights of nature has a more broad ambit (Chapron et al. 2019). To be more precise, it seeks to explicitly acknowledge the intrinsic value of ecosystems and nature itself. This is revolutionary, especially when put against the benchmark of the existing environmental legislation. As Emmegegger and Tschentscher (1994) succinctly put it: 'one cannot truly speak about protecting the intrinsic value of the environment without granting it explicit substantive rights'.

To be sure, advocating for rights for nature is not a recent phenomenon. Already back in 1897, Salt (1894) claimed that 'if we are ever going to do justice to the lower races (i.e. animals), we must get rid of the antiquated notion of a 'great gulf' fixed between them and mankind, and must recognize the common bond of humanity that unites all living beings in one universal brotherhood'. At the same time, Darwin's scientific findings dealt a significant blow to the prevailing view on the role of Mankind: Man should no longer be regarded as a special creature endowed with a divine blessing by an omnipresent God (Darwin 1874). This ultimately culminated in Aldo Leopold's famous writings in A Sand County Almanac (1949), in which he advocated a different land ethic, 'which changes the role of Homo sapiens from conqueror of the land community to plain member and citizen of it'. In this narrative, Leopold (1949) held that life forms that share the planet with people should be allowed to live 'as a matter of biotic rights, regardless of the presence or absence of economic advantage to us'.

One of the first 'legal' appearances of the idea of turning nature into a rights-bearing subject took place in the context of the renowned infamous *Sierra Club versus Morton*-case, decided by the U.S. Supreme Court in 1972. The renowned lawsuit revolved around the permits granted by the Forest Service, which allowed Disney Enterprises to build a complex of recreational facilities in the Mineral King Valley in the Sierra Nevada of California. During the proceedings, Christopher Stone authored his famous article, in which he posited that nature should be granted legal personhood, which entailed that it could be represented in court through a system of guardians. Ultimately, Stone's view was rejected by the U.S. Supreme Court, which decided to deny the environmental organization standing. However, Justice Douglas paid lip service to Stone's thesis in his dissenting opinion.[10]

Rights, Standing and Legal Remedies

It is instructive to go back to Stone's original thesis in order to identify the most recurring features of the rights of nature rationale. At the time, Stone put forward the following three core elements that are to be observed when implementing rights of nature: *first*, the rights of nature rationale entails granting legal personhood to nature, which implies it has the necessary standing to launch proceedings in its own right; *second*, in determining the granting of relief, courts are required to take into account its own (ecological) injury, which goes beyond the personal damages incurred by private individuals; *third*, that relief must run to the benefit of the damaged nature. A common denominator in terms of substantive rights is the steadfast emphasis that is put on preserving the *integrity* of a specific ecosystem (e.g. a river) or nature as a whole, as is the case in Ecuador and Bolivia, for instance. As was also underlined by Stone, rights of nature does not necessarily entail that nature should be granted 'absolute' rights in every single instance. As is the case with several human rights, such as the right to free speech, rights of nature are not necessarily 'absolute' either (Stone 1972). Rights of nature is often aligned with a more liberal approach to standing before court, which might fundamentally alter the assessment courts need to make when issuing an injunction or granting damages. Since nature lacks

[10]Sierra Club v. Morton, 405 U.S. 727 (1972).

a 'voice', virtually all of the recent manifestations of rights of nature entail some sort of guardianship approach, which ensures an effective voice for the environment even when no authorities or individuals are willing to start legal action on their own behalf (Shelton 2015).

Guardians at Work

Ecuador was one of the first countries to explicitly implement rights of nature in its legal order. When revising its constitution in 2008, the country chose to place the concept of *Buen Vivir*—which translates as 'harmonious coexistence'—at the heart of its novel constitutional legal order, thereby reasserting the interdependency between nature, people and society. From the preamble of the revised Constitution alone, one can infer that nature no longer is to be approached as a commodity. Since the much-heralded revision of 2008, the Ecuadorian Constitution also explicitly includes a section articulating the rights of nature. Pursuant to Article 71 of the Constitution, 'Nature, or Pacha Mama, where life is reproduced and occurs, has the right to integral respect for its existence and the maintenance and regeneration of its life cycles, structure, functions, and evolutionary processes'.

Along similar lines, the Ecuadorian Constitution now grants nature the right to be restored when damaged.[11] It also requires the government to prevent or restrict activities leading to species extinction, ecosystem destruction and permanent alteration of life cycles.[12] In addition to recognizing the rights of indigenous people to remain on their ancestral lands, the provisions grant 'all persons, communities, peoples, and nations' the right to call upon the public authorities to enforce the rights of nature.[13] This entails that there is no requirement to prove a personal interest when relying upon nature's rights before court. Interestingly so, Ecuador is also the first country in which the legally recognized rights to nature were applied before court (Kotzé and Calzadilla 2017). The Vilcabamba River case, where two plaintiffs sought legal relief on behalf of the river in order to stop the detrimental impact it suffered from an authorized road project, was dubbed the 'first rights of nature lawsuit' of the world. And, most importantly, nature 'won'. Ultimately the Court, with reference to the precautionary principle, upheld the claim reasoning that injuries to nature are 'generational damage', which will not only impact present generations but also future people.[14]

In other lawsuits, rights of nature were invoked by plaintiffs as well and, in some instances, even by the judges themselves. A preliminary case-law analysis showcases that not all these attempts were successful (Kauffman and Martin 2017). The lack of a

[11]Ecuadorian Constitution, Article 72.

[12]Ibid, Article 73.

[13]Ibid, Article 71.

[14]R.F. Wheeeler and E.G. Huddle v. Attorney General of the State of Loja (2011), Judgment No. 11121–2011 (2011, March 2011).

clear hierarchy within the constitutional framework, which grants significant leeway for authorities when balancing rights of nature with other economic considerations, is to be singled out as a major shortcoming (Whittemore 2011). The latter is starkly demonstrated by the unsuccessful lawsuits aimed at halting the construction and operation of an open pit copper mine—dubbed 'Mirador'—at a vulnerable ecological site.[15] Neither the Ecuadorian nor the Bolivian government has fundamentally altered their resources-based economic policies in view of the right of nature-based legislation. Yet Ecuadorian judges are increasingly prone to unilaterally applying rights of nature, even when contested from a governmental point of view (Kauffman and Martin 2017). In addition, the 2015 ruling of the Ecuadorian constitutional court, in which the precedence of the rights granted to nature over economic interests was upheld, appears to be slowly trickling down to other jurisprudence, further highlighting the importance of the new paradigm and its legal teeth (Kauffman and Martin 2018).[16] At best, the prevailing picture regarding the concrete application of rights of nature in the field appears to be ambivalent, with the emergence of several progressive court decisions that are contrasted with the blatant lack of a wholesale paradigm shift towards eco-centrism.

In sharp contrast to the Ecuadorian model, the New Zealand model opts for a different approach, namely, granting legal personhood to specific ecosystems. The Whanganui River Claims Settlement Act, which was passed by the New Zealand Parliament in 2017, arguably represents one of the most inspiring and comprehensive efforts to operationalize rights of nature.[17] It is most renowned for explicitly recognizing the Whanganui River, a sacred river for the Whanganui Iwi, the Maori people who see the river as their ancestor, as bearer of rights and responsibilities equivalent to a person (Hsiao 2013). The Act itself is to be approached as the result of a long-standing physical and legal battle by Maori people against the colonial domination of the river and its water, which lasted more than 150 (!) years. The wording of the Act reflects a clear concession to the Maori view, which depicts the river as a whole, living, spiritual being (Iorns Magallanes 2013). The Act now explicitly recognizes the river and all its tributaries as a single legal entity dubbed *Te Awa Tupua*, which has rights and interests, and is the 'owner' of its own river bed.[18] Key to the management, which is aimed at preserving the integrity of the river system as a whole, is the appointment of two custodians—one by the Crown and the other by the Whanganui Iwi—which are mandated to perform the guardianship role on behalf of the river.[19] They are expected to act in a similar manner as legal guardians represent children and place the interests of the river central to their actions.

---

[15]On the latest development regarding the Mirador project, see https://www.reuters.com/article/us-ecuador-mining/ecuador-begins-large-scale-mining-at-mirador-copper-project-idUSKCN1UD36F (accessed 15 September 2019).

[16]Ecuadorian Constitutional Court, 2015. Sentence NO. 166-15-SEP-CC, case no. 0507-12-EP.

[17]Te Awua Tupua (Whanganui River Claims Settlement) Act 2017, New Zealand Act No. 7.

[18]Ibid, section 14.

[19]Ibid, section 15.

As of today, however, no specific actions have been brought before the national courts in New Zealand on behalf of the nature which has been accorded legal personality through specifically tailored legislation (O'Donnel and Talbot-Jones 2018). Likewise, the new legislation has not led to massive expropriation programmes: the property rights of corporations, farmers and other inhabitants still remain valid. Pollution still exists. This being the case, the recognition of rights of nature does not necessarily require the adoption of specific legislation but can also involve the judiciary. In some instances, plaintiffs might file a lawsuit in order to seek legal personhood for a natural entity, such as a river. In 2007, this strategy was followed by some plaintiffs in the U.S. State of Colorado, seeking a declaratory ruling which would hold that the Colorado River was a person possessing rights, among others the 'right to exist, flourish, regenerate, be restored and naturally evolve'. However, the plaintiffs ultimately decided to drop the case.[20] Similar lawsuits have yielded interesting rulings in countries such as India[21] and, more recently, Colombia. In a 2015 decision, the Colombian Constitutional Court decided to grant the Atrato River, its tributaries and its basin 'the right to be protected, preserved and restored by the State and the communities'.[22] Interestingly, the court was also moved to designate two custodians for the management of the river, thereby reflecting the recent developments in New Zealand. A similar pattern is noticeable in another legal procedure relating to the Colombian Amazon Rainforest, where the Supreme Court was persuaded to declare the area of the Amazon rainforest within its national borders an 'entity subject to rights'.[23] While the recognition of the Amazon rainforest as a legal person was not totally alien to the plaintiffs' claims, the eagerness with which the Colombian judges made a bridge between a human rights-based litigation strategy and rights of nature is remarkable.

Inherent Obstacles, Limited Effectiveness and Recurring Criticism

The increased popularity of rights of nature has also triggered some criticism and scorn by several scholars.[24] However, Stone had already anticipated some of the most perennial criticism by stating that '(t)hroughout legal history, each successive extension of rights to some new entity has been, thereto, a bit unthinkable (…) each

[20]For more background information, see https://www.aspenjournalism.org/2017/12/05/colorado-river-personhood-case-pulled-by-proponents/ (accessed 15 September 2019).

[21]Uttarakhand High Court, Mohd. Salim v. State of Utturakhand & Others (2017); Uttarakhand High Court, Lalit Miglan v State of Utturakhand & Others (2017). As indicated above, these rulings were provisionally halted by a ruling of the Indian Supreme Court.

[22]Constitutional Court of Colombia, case T-633/16, Center for Social Justice Studies et al. v. President of Colombia et al., Judgment of 10 November 2016.

[23]Supreme Court of Colombia, judgment STC 4360-2018 of 5 April 2018, Sect. 13.

[24]For an overview, see Burdon and Williams (2016).

time there is a movement to confer rights onto some new 'entity', this proposal is bound to sound odd or frightening or laughable' (Stone 1972). In a similar vein, he posited that even the fact that only humans are moral agents and, as a result, nature should not be granted rights since it is—evidently—incapable of voicing its own 'feelings' or interests, does not present a definite obstacle (Stone 2012). For infants, incompetents, terminally comatose and corporations have also been granted certain rights and can claim standing in legal procedures.

I strongly concur with Stone's view. It is relatively easy to push aside rights of nature as an exercise in empty rhetoric and advocate for a better enforcement of the existing rules. Yet rights of nature is not to be approached as an all-encompassing alternative for the existing environmental legislation. Rather, it is to be seen as a bedrock for a shift towards a novel normative approach to nature protection, which goes beyond the well-entrenched dualistic view on nature and mankind (Taylor 1998). Moreover, it is indeed widely accepted that personhood is not limited to people, who remain just one of the two main types of persons. Besides human beings, who the law tags as 'natural (or physical) persons', the law also acknowledges the existence of legal (artificial) persons. The main difference between both categories is that the latter requires a deliberate decision by the lawmaker to artificially create legal personhood (Pietrzykowski 2018). And whereas supporters of the so-called 'will or choice theories' assume that 'rights' should in principle be aimed at the protection of the right's holders free will (Hart 1982), champions of the so-called 'interest theories' link the usage of the concept of rights to 'mere' interests that the lawmakers regard as deserving legal protection against third parties (Raz 1986). Decades of research into animals' abilities to communicate have yielded some groundbreaking results and underline the moral capabilities of many animals (de Waal 2013; Driessen 2017). Surely, intelligent and sentient animals such as Elephants, Killer Whales and Primates possess interests that merit legal protection. But even at a more principal level, I see no real difference between granting certain rights to ecosystems and corporations. While we consider corporations and companies to constitute the self-evident pillars of our current economic system—which explains why case law has acknowledged that corporations can be accorded certain human rights and responsibilities that are enjoyed by natural persons (La Folette and Maser 2017)—a similar move can equally be envisaged for ecosystems. That is, if we are prepared to acknowledge their primary importance for society.

Even so, critics insist that it remains impossible to know whether a river or mountain really prefers to be 'preserved' or exploited (Sagoff 1974; Warnock 2012). Along similar lines, one might argue that the notion of rights is not the appropriate instrument to bring about the shift in environmental governance needed to protect the environment. As stated above, this view only holds ground within the framework of the so-called 'will theories'. It presupposes that we restrict the notion of personhood to human individuals, based on our unique moral capabilities. A similar rationale was also upheld in the dismissal of a *habeas corpus* action on behalf of Tommy, a chimpanzee held in captivity, by the New York Supreme Court in 2014. The court noted that law is the result of a social compact.[25] Humans can enter into a 'social compact' whereas animals and *a fortiori* nature are barred from doing so since they cannot assume certain duties. However, this reasoning equally remains challengeable in view of the mounting criticism regarding the notion of human 'free will'. For one, an increasing number of thinkers call into question the artificial premise underlying the liberal belief in feelings and the free choices of the individual (Harari 2018). Against this backdrop, which is built upon a rationale that is openly debunking the notion of 'human exceptionalism', it is no longer unthinkable to argue for a third category of non-personal subjects of law for certain animal species (Pietrzykowski 2018) or, when it comes to natural entities such as rivers and mountains, to deliberately accord them the status of subjects with certain rights, just as we have done with companies and corporations.

Another line of criticism discards rights of nature as a 'quick legal fix', which will bring about little result if not accompanied by social reform (Burdon and Williams 2016). Evidently, enforcement remains a key challenge to be addressed when effectively implementing rights of nature (Bétaille 2019). Even so, as demonstrated above, the number of lawsuits in which rights of nature is successfully being invoked is steadfastly rising in Ecuador, attesting to the potency of rights of nature if effectively relied upon before national courts (Kauffmann and Martin 2017). That being said, if one were to follow the above-mentioned line of reasoning, one could also easily dismiss the usage of the concept of inalienable human rights in view of the persistent human rights violations that still occur in many places in the world.

To be fair, the disruptive effect of rights of nature will evidently give rise to legal challenges, which is starkly illustrated by the recent lawsuits that were launched by the farmers against the local ordinance granting legal personhood to Lake Erie. Such cases are to be framed as an illustration of the (often) limited local competences regarding the protection of ecosystems, which might turn local legislation into mere rhetorical instruments for the furtherance of rights of nature. However, they are also indicative of the uphill battle that is to be fought in order to translate rights of nature into our anthropocentric legal order. In other words, they underline the transformative potential of rights of nature and should thus not be used as an argument to dismiss the novel paradigm altogether.

Interim Conclusion

In this section, it has been established that rights of nature clearly has the potential of creating a novel normative groundwork which allows for a fundamental shift towards a more eco-centric-based governance. Evidently, the simple implementation rights of nature will not constitute a definitive and all-inclusive answer for the impending

---

[25] State of New York Supreme Court, Appellate Division (Third Judicial Department), Non-Human Rights Project on Behalf of Tommy v. Lavery and Circle L Trailer Sales, Judgment (Opinion) of 4 December 2014.

ecological crisis if not accompanied by a major shift in decision-making and administrative practice. However, the majority of the criticism directed at rights of nature can be easily refuted, either by referring to the human exceptionalism it presupposes or by referring to the fact that corporate personhood is also to be seen as a human construct.

The Interplay Between the EU Legal Order and Rights of Nature: An Environmental Frontrunner, in Word and Practice?

The EU as a Self-Declared Environmental Frontrunner?

Having thoroughly demonstrated why rights of nature could indeed be approached as a crucial new template for addressing the current economic plight, the second question to be addressed in this paper is the potential interplay with existing EU law. To begin with, on the European Commission's website it is stated that the '(t)he EU has some of the world's highest environmental standards. Environment policy helps green the EU economy, protect nature, and safeguard the health and quality of life of people living in the EU'.[26] At first glance, this self-praise seems justified. The progressive march forward of EU environmental law cannot be negated, from the adoption of the first EU environmental action programme over the adoption of several progressive environmental directives to the introduction of an environmental title in the EC Treaty by the 1986 Single Act. In 1993, the Maastricht Treaty continued this progressive evolution by making the environment an official EU policy area, whereas the Treaty of Amsterdam (1997) laid down the duty to integrate environmental protection into all sectoral policies with a view to promoting sustainable development.

This resulted in several—albeit often half-hearted—attempts to further 'green' policy areas, such as the common agricultural policy (CAP) and the common fisheries policy (CFP). During the past two decades, combatting climate change has become a top priority, which has also led to the inclusion of this specific policy goal in the Lisbon Treaty of 2009 (Vedder 2010).

As of today, however, no explicit reference to rights of nature can be found in primary EU law. Nonetheless, Article 3 of the Treaty on the European Union (TEU) puts 'sustainable development' at the centre of the EU's internal market, specifying that the 'internal market (...) shall work for the sustainable development of Europe based on balanced economic growth (...) and a high level of protection and improvement of the quality of the environment'. In turn, Article 11 of the Treaty on the Functioning of the European Union (TFEU) fleshes out the 'integration principle', pursuant to which environment protection requirements are to be integrated into the

---

[26]https://europa.eu/european-union/topics/environment_en (accessed 15 September 2019).

definition and the implementation of the EU's policies and activities, 'in particular with a view to promoting sustainable development' (Hovden 2018).

Of course, a mere reference to 'sustainable development' does not guarantee that, when balancing economic, social and ecological considerations, the latter will ultimately prevail. In other words, it does not guarantee a fundamental rethinking of our current economic, political and legal paradigms in light of the damage that these approaches have done to the ecosystems. In addition, most authors have accepted that—in view of its very broad and vague wording—the provisions linked to sustainable development are not enforceable before court (Verschuuren 2016). Yet, in spite of their lack of legal teeth, these provisions have led to the adoption of numerous secondary legislation—which is enforceable—and action programmes. For instance, the EU's progressive implementation of the precautionary principle, which entails that in case of uncertainty, instructions may still be given to remove harmful products, such as certain pesticides, from the market, is to be singled out as one of the major achievements of the EU environment policy of the past decades (Schoukens 2018).

It is, of course, impossible to treat all relevant pieces of EU environmental legislation in the context of this chapter. Given the focus of rights of nature advocates on rivers, nature and pollution, I will briefly focus on the main pieces of EU legislation which address these very sub-topics. As a general remark, it has to be borne in mind that the EU has not chosen to grant legal personhood to rivers, nor to protected sites. Where water protection is concerned, most international rivers and waterways within the EU are covered by the existing international agreements, which detail the cooperative management of international watercourses. More broadly speaking, however, reference has to be made to the EU's Water Framework Directive, adopted in the year 2000,[27] which obligates the Member States to achieve good qualitative and quantitative status of all water bodies (including marine waters up to one nautical mile from shore) through the adoption of river basin management plans by 2015. In turn, the SEA[28] and EIA Directives[29] require authorities to integrate environmental considerations into decision-making procedures with respect to plans, programmes and projects liable to generate significant effects on the environment. A further attempt to operationalize the polluter pays principle is offered by the 2014 Environmental Liability Directive, which requires preventive and remedial action in the context of 'pure' ecological damage.[30] Lastly, the EU has also adopted two pieces of legislation explicitly aimed at halting the biodiversity loss within its territory. The 1979

[27]Directive 2000/60/EC of the European Parliament and of the Council of 23 October 2000 establishing a framework for community action in the field of water policy [2000] OJ L 371/1.

[28]Directive 2001/42/EC of the European Parliament and of the Council of 27 June 2001 on the assessment of the effects of certain plans and programmes on the environment [2001] OJ L 197/30.

[29]Directive 2011/92/EU of the European Parliament and of the Council of 13 December 2011 on the assessment of the effects of certain public and private projects on the environment [2011] OJ L 26/1.

[30]Directive 2004/35/CE of the European Parliament and of the Council of 21 April 2004 on environmental liability with regard to the prevention and remedying of Environmental Liability [2004] OJ L 143/56.

Birds Directive,[31] which was later complemented by the more cross-cutting Habitats Directive,[32] not only explicitly protects hundreds of endangered species within the EU but also obligates the Member States to designate the most ecologically valuable parts of their territory as protected sites (Natura 2000). Stretching over 18% of the EU's land area and almost 9.5% of its marine territory, the EU hosts the largest coordinated network of protected areas in the world.

Beyond explicit legislation, the so-called multiannual Environment Action Programmes, which put forward legislative proposals and goals for the EU environment policy, are continuously setting a high water mark in the realm of environmental protection as well. The 7th Environment Action Programme, which was adopted by the Council and Parliament in 2013 under the title 'Living well, within the limits of our planet' lay down the main objectives for the environmental policy and governance within the EU for this decade[33] Underpinning this vision is the recognition that Europe's economic prosperity and well-being is intrinsically linked to its natural environment. Whereas such claims do not explicitly pay lip service to rights of nature, they at least recognize the intrinsic interdependency between the long-term survival of our economic system and the proper conservation of our natural ecosystems.

Pertaining to the pressing challenge of combatting climate change, the EU adopted a 2030 climate & energy framework in 2014 that puts forward a 2030 target to reduce greenhouse emissions by 40% of 1990 levels.[34] Likewise, in its Biodiversity Strategy, the EU has committed itself to halting the loss of biodiversity and ecosystem services as well as restoring 15% of its degraded ecosystems by 2020.[35]

A More Ambivalent Practice?

The picture painted above appears impressive at first glance. In 2013, the European Environment Agency identified a staggering number of 63 legally binding targets and 68 non-binding objectives set out in EU policy covering the 2010–2050 period (Hovden 2018). However, in spite of its lofty policy ambitions and its progressive environmental directives and regulations, the EU's environmental outlook still remains bleak. Admittedly, looking back on the last 40 years, one has to acknowledge that the local environment is in some places arguably in a better state than

---

[31]Council Directive 79/409/EEC on the conservation of wild birds [1979] OJ L 103/1 (further referred to as 'Birds Directive'). The initial Birds Directive was codified in European Parliament and Council Directive 2009/147/EC on the conservation of wild birds [2010] OJ L 20/7.

[32]Council Directive 92/43/EEC on the conservation of natural habitats and of wild fauna and flora [1992] OJ L 206/7.

[33]Decision No 1386/2013/EU of 20 November 2013 on a General Union Action Programme to 2020 'Living well, within the limits of our planet' (2013) OJ L 354/171.

[34]For more background: https://www.consilium.europa.eu/en/policies/climate-change/2030-climate-and-energy-framework/ (accessed 15 September 2019).

[35]European Commission, Report from the Commission to the European Parliament and the Council. The Mid-Term Review of the EU Biodiversity Strategy to 2020, COM(2015) 478 final.

in the 1970s, at least in some respects (e.g. water quality). Yet the European State and Outlook Report (SOER) 2015, compiled by the European Environment Agency (EEA), provides for a grim reading when zooming in on the global outlook.[36] While significant improvements have been made in reducing ecosystem exposure to excess levels of acidification, the majority of the protected nature within Western Europe is still exposed to air pollution levels that exceed the so-called critical loads for eutrophication, i.e. the upper limit that an ecosystem can absorb without significantly damaging its structure or function. These bleak conclusions were subsequently repeated in the 2015 Mid-Term Review of the EU Biodiversity Strategy.[37] Notwithstanding recent 'greening' efforts of the EU's agricultural policies, most studies still point out that farmland nature is in steep decline.[38] Likewise, a lot of progress still has to be made when it comes to sustainable fishing practices, with more than 50% of the EU's fish stocks being subject to catch regimes which are above the maximum sustainable yield.[39]

Even the EU's track record when it comes to combatting climate change, arguably the most pressing environmental challenge for the twenty-first country, is far from perfect. Even though the EU is meeting its objectives under the Kyoto Protocol, many of the goods needed for maintaining our current western lifestyle are imported from elsewhere around the globe, which means that part of the EU's CO_2 emissions (i.e. polluting activities) are outsourced to other parts of the world, where legal standards are more relaxed in terms of environmental protection. European businesses, consumers and policymakers have relatively limited knowledge and influence to bring about sustainable changes in the production processes.[40] As to the new reduction pledges, several NGOs argue that they are not ambitious enough in view of the goals set out by the Paris Agreement.[41]

Procedural and Substantive Obstacles: No Standing for Nature and Climate, After All?

No Explicit Recognition of Rights of Nature in the EU Treaties

The lack of express acknowledgment of rights of nature does not necessarily mean that the main underlying principles of the rights of nature discourse described above are not implicitly endorsed by or reflected in the existing EU legal order. Seminal

[36] European Environment Agency, The European Environment: State and Outlook 2015: Synthesis Report, 2015, at pp. 8–9.

[37] Ibid.

[38] See https://www.theguardian.com/environment/2018/mar/21/europe-faces-biodiversity-oblivion-after-collapse-in-french-bird-populations (accessed 15 September 2019).

[39] European Environment Agency, supra note 37, p. 74.

[40] Ibid, pp. 93–95.

[41] See https://euobserver.com/environment/142591 (accessed 15 September 2019).

concepts, such as the precautionary principle, are well rooted in the current EU legal order. However, a more detailed analysis that aptly reveals that the EU is incapable of adequately addressing the systematic challenges arising in the context of environmental policy becomes all the more pressing. The EEA itself concluded in its 2015 report that '(t)he transition to a green economy is a long-term, multidimensional process that will require a move away from the current linear economic model of 'take-make-consume-dispose' which relies on large quantities of easily accessible resources and energy. This will necessitate profound changes in dominant institutions, practices, technologies, policies, lifestyles and thinking'.[42]

Interestingly, the EEA itself proposes to 'recalibrate' the existing policy approaches in order to allow for such a transition. As of today, little evidence is put forward that the EU institutions are willing to effectively implement the shift towards ecological law. In line with the 2012 United Nations Conference on Sustainable Development, the focus within the EU is still on 'green economy', while topics such as 'degrowth', which calls into question one of the dominant paradigms of our current economic system, remain notoriously absent in the current discourse. The underlying premise still appears to be that 'infinite economic growth' can be aligned with an improved environmental performance. In itself, this is not surprising if one takes into account the wording of Article 3 TEU, which links sustainable development to 'balanced economic growth', without specifying any requirements for a balancing exercise needed in scenarios in which ecological and economic interests might clash. Even the presence of Article 11 TFEU, which merely requires the Union's actual environmental protection standards to be integrated into all relevant policy sectors, is incapable of ensuring adequate environmental protection through the other sectoral policies (Hovden 2018).

According to Somsen (2017), 'the absence in EU environmental law of an ecological equivalent to human dignity (ecological integrity) implies boundless institutional discretion to preside over and acquiesce in the destruction of the environment'. Bosselmann, in turn, argues that a rule prohibiting harm to the integrity of ecological systems is fundamental, which is currently absent in modern environmental law (Bosselman 2016; Bosselman 2008). Granting substantive rights to nature has precisely emerged as a novel practice aimed at breaking down the hierarchical relationship between mankind and nature and, as a result, allowing for a more balanced approach to economic development, banning or at least phasing out practices that are fundamentally at odds with an ecosystem's resilience.

Therefore, the preliminary conclusion appears to be that the EU's environmental instruments are falling short of the above-mentioned rationale underlying rights of nature. Some authors go as far as to qualify the EU as 'ecologically illiterate' (Hovden 2018). As I understand, there are two main issues here. *First*, EU environmental law is characterized by a piecemeal and reactive approach. Crucial policy areas, such as the CAP, have not consistently been made subject to the environmental principles laid down by the EU environmental title in the TFEU. No hierarchical relationship can be observed here, which renders the achievement of the overarching goals very unlikely.

[42]European Environment Agency, supra note 37, at p. 155.

Second, the EU's environmental policy in itself, as well as the primary principles upon which the EU legal order is built, is predominantly anthropocentric, to the extent that they refuse to explicitly acknowledge the increasing interdependency between nature and mankind.

Procedural Impediments and Unexpected Opportunities: Can EU's Endangered Nature Be Represented Before Court?

This being said, it is interesting to analyse how the rights of nature rationale could be further implemented and/or operationalized within the existing EU legal order. For, as is demonstrated by the above-mentioned recent manifestations of rights of nature, it is not uncommon to implement rights of nature within the context of anthropocentric legal schemes. This was, for instance, the case in New Zealand and Colombia. And the EU's comprehensive legal framework in respect of environmental protection provides an excellent case study to outline the most seminal issues to be addressed in this context.

In this respect, a first necessary stop relates to the provisions on access to justice in environmental cases at EU level. As demonstrated above, liberal standing for nature in court was one of the principal reasons why Stone pitched his revolutionary idea of granting rights to nature in 1974. Stone's plea did not stop at simply advocating a more liberalized approach to standing for environmental NGOs in cases where nature is at peril. He went one step further and claimed that natural objects should be granted the benefit of standing. This would give a host of potential guardians, such as environmental NGOs, the possibility to enforce nature's rights. Stone differentiated between two situations: bringing a claim as a custodian on behalf of a protected species is to be regarded as completely different from filing a lawsuit as an environmental NGO in order to promote their own statutory objectives (Stone 2010).

Arguably Stone's narrative has not managed to create a shift in the discourse on standing for nature in the United States. Granted, in the famous 1991 *Palila*-case, an endangered bird which is protected under the U.S. Endangered Species Act, a plaintiff was seeking relief on behalf of the protected bird. And surprisingly, the Ninth Circuit Court seemed to concur, holding that the *Palila* 'also has legal status and wings its way into federal court as a plaintiff in its own right'.[43] However, in most of the subsequent cases in the United States in which the standing of protected species was upheld, more traditional plaintiffs were present, such as environmental NGOs.[44]

This is no different in the EU legal order, where nature also has no direct standing. As already alluded to, much has changed in the international framework regarding environmental accountability and justice since the 1960s. With the adoption of the 1998 Convention on Access to Information, Public Participation in Decision-making

[43]Palila v. Hawaii Dept. Of Land & Natural Res., 852 F.2d 1106, 1107 (9th Cir. 1991).

[44]61F Supp 2d 1001 (ND Cal 1999).

and Access in Environmental Matters (the Aarhus Treaty)[45] within the framework of the Economic Commission for Europe (UNECE), the so-called 'procedural environmental rights' granted to citizens and NGOs were further recognized and translated into more clear-cut international duties. Also, the EU has ratified the Aarhus Treaty back in 2005, which entails that this international convention even takes precedence over secondary EU legislation.[46] And although this international agreement remains notoriously silent on the topic of standing for natural objects and allows Member States to use certain standing criteria, it clearly obliges State Parties to reconsider traditional standing approaches, by virtue of which almost all environmental NGOs would be barred from challenging acts or omissions contravening environmental law (Schoukens 2015).

While a detailed analysis of the provisions on access to justice at EU level evidently falls outside of the scope of this chapter, I argue that three important stumbling blocks are to be overcome in order to operationalize wider standing for nature in the EU legal order.

First and foremost, there is the prevailing strict interpretation of the standing requirements enshrined in primary EU law, which renders direct access to justice before EU courts against EU acts largely illusionary in the context of environmental cases (Schoukens 2015). During the past decades several environmental NGOs have relentlessly tried to challenge the legality of EU acts in light of the environmental principles that are enshrined in the EU constitutional treaties. However, all of these actions have failed due to the prevailing interpretation that is given by the CJEU to the standing requirements set out in Article 263(4) of the TFEU (Schoukens 2019). In spite of wide-shared criticism and the subsequent adoption of the 2006 Aarhus Regulation,[47] which was precisely aimed at mitigating the relatively harsh jurisprudence with respect to standing in environmental cases, the plight of environmental NGOs remains. And thus potentially unsustainable EU decisions in the sphere of pesticides, genetically modified organisms, air quality as well as fisheries policy—the latter being an exclusive EU competence—are virtually immune from legal challenges by the environmental movement. Nature's interest cannot be defended in this context. Even a recent 'condemnation' by the Aarhus Compliance Committee for the lack of effective legal remedies available at EU level was unable to persuade the EU courts to revisit their conservative approach towards standing in environmental cases.[48]

---

[45]Convention on Access to Information, Public Participation in Decision-making and Access to Justice in Environmental Matters, done at Aarhus, Denmark, 25 July 1998 ('Aarhus Convention') (1999) 2161 UNTS 447; 38 ILM 517.

[46]Council Decision 2005/370/EC of 17 February 2005 on the conclusion on behalf of the European Community, of the Convention on access to information, public participation in decision-making and access to justice in environmental matters (2005) OJ L124/1.

[47]Regulation (EC) No 1367/2006 of the European Parliament and the Council of 6 September 2006 on the application of the provisions of the Aarhus Convention on Access to Information, Public Participation in Decision-making and Access to Justice in Environmental Matters to Community institutions and bodies (2006) OJ L 264/13.

[48]Communication ACCC/C/2008/32 (Part II) (European Union), adopted on 17 March 2017.

This lack of standing in environmental cases significantly hinders a further manifestation of rights of nature at EU level. For instance, in the recent *Melifera*-case, which revolved around a legal challenge by several environmental NGOs of the European Commission's decision relating to the approval period of the active substance glyphosate (a controversial weed killer), the General Court refused to allow environmental NGOs that promote the preservation of endangered bees to seek a legal review of the latter act.[49] Likewise, the General Court refused to reconsider its longstanding approach to standing for the environment in the first EU climate case. In its notable 2019 decision, it dismissed the legal action on procedural grounds. It was held that climate change affects all Europeans in many different ways and therefore the plaintiffs did not meet the strict standing requirement needed to challenge the contested acts.[50]

And thus, whereas Article 71 of Ecuador's Constitution grants liberal standing to all persons willing to enforce the rights of nature or, alternatively, other recent manifestations of rights of nature grant specifically appointed custodians the right to file legal claims on behalf of nature, the EU system of legal remedies does not appear to provide for much hope for rights of nature advocates. Even an indirect representation of endangered nature before an EU court via environmental NGOs seems to be illusionary. Nature remains voiceless at EU level.

Fortunately, this major deficiency of the EU legal order is somehow offset by the relatively progressive case law of the EU courts when it comes to access to justice for environmental NGOs at national level. A string of recent decisions of the CJEU underlines its willingness to grant more liberal standing in environmental cases at national level, where permits for construction works and industrial facilities can be challenged. The 2009 ruling of the CJEU in a Slovakian case in which decisions to hunt brown bears were challenged by Slovakian NGOs is a standout case for rights of nature. Here, the EU judges held that effective judicial protection is to be provided by national courts in cases where the fate of EU-protected species is at stake.[51] One might put forward that this helped the local NGO to indirectly guarantee the interests of protected brown bears in view of the purported hunting schemes. A similar rationale led the CJEU to dismiss the so-called German *Schutznorm*, which prevented German NGOs from filing lawsuits on behalf of EU-protected sites which are part of the afore-mentioned Natura 2000 Network.[52]

It is hard not to see in recent jurisprudence a beacon of hope for more rights-based approaches to environmental protection in the EU. Crucially, however, one needs to underline that—as of today—not a single EU or national court has held that EU-protected species should be granted *direct* standing before national courts. Moreover, it still remains to be seen whether the recent jurisprudential evolutions are apt to move in the direction of the liberal standing for the environment as a

---

[49]Case T-600/15 Pesticide Action Network Europe (PAN Europe) (2016) ECLI:EU:T:2016:601, para. 40–56.

[50]Case T-330/18, Armando Carvalho et al. (2019) ECLI:EU:T:2019:324, para. 41–54.

[51]Case C-240/09, Lesoochranárske zoskupenie VLK I (2011) ECLI:EU:C:2011:125, para. 50.

[52]Case C-115/09, Trianel (201)] ECR I-03673, par. 46.

whole, which seems to underpin many of the concrete manifestations of rights of nature (Schoukens 2018). If environmental NGOs lack funding or have strategic considerations not to file lawsuits on behalf of EU-protected nature, the latter remains at the discretion of national or regional competent agencies. It remains dubious at best whether private individuals can intervene on behalf of endangered ecosystems whenever they do not demonstrate a clear-cut interest in the preservation of the said biodiversity.

This being the case, reference is to be made to the European Commission, which can equally step in as guardian for EU-protected nature. However, one needs to bear in mind that Stone himself rejected an approach by which public authorities are to be framed as a performant guardian of nature or the environment (Stone 1972). Several authors have not refrained from criticizing the ambivalent role of the European Commission in enforcing EU nature conservation law (Krämer 2015). It is indeed a well-known fact of life that the Commission enjoys full discretion when contemplating the usage of its enforcement powers. That said, the recent infringement proceedings launched by the European Commission against Poland with respect to its destructive logging practices in the *Bialowieza* forest forcibly illustrate that the European Commission, if found willing and able, can nevertheless function as an effective guardian.[53] Of course, the latter case constitutes but one piece of anecdotical evidence: for every primeval forest that is saved, many other 'voiceless' nature disappears. And hence death by discretion is still lurking behind the corner.

Substantive Flaws and Deficiencies: Going Beyond Legal Formalism, Protecting Ecological Integrity in the EU?

Next to the procedural obstacles analysed above, the existing EU environmental legislation also contains important substantive loopholes when it comes to providing sufficient protection for our threatened nature. A first notable shortcoming of EU environmental law relates to the lack of an imperative of preserving ecological integrity in the EU treaties, alongside the notable absence of an overarching and binding No Net Loss instrument in secondary EU law. The lack of a clear-cut ban on activities liable to significantly impact EU's vulnerable ecosystems still marks the relative inaptness of the EU legal order at moving forward towards an ecosystem-based approach. Although the Habitats Directive is to be lauded for its focus on setting up an EU-wide ecological network of protected areas—the so-called Natura 2000 Network—the concept of 'ecosystem' is in itself not explicitly mentioned in the Habitats Directive. In fact, Natura 2000 leaves more than 80% of the EU territory 'unprotected'. Many endangered insect species, such as endangered bees, are not listed among the strictly protected species included in Annex IV to the Habitats Directives. As of today, the EU Nature Directives do not contain clear-cut restoration mandates for such 'ordinary nature', either. In close relation to these findings

[53]Case C-441/17R, Commission v Poland (2017) ECLI:EU:C:2017:877.

is the fact that the duty to carry out an Environmental Impact Assessment (EIA) is still being approached as a mere 'procedural' tool, which downplays its relevance for impacting the outcome of authorization procedures for unsustainable projects affecting not strictly protected nature in the EU.[54]

Second, even within the context of strictly protected areas and species, pieces of legislation such as the EU Habitats Directive contain some leeway and loopholes, which ultimately might result in the further decline of protected biodiversity. For one, the protected status of endangered species might always be lifted, which could result in the lifting of protected status for species like Gray wolves, whose presence remains controversial in many rural areas (Chapron et al. 2019). Even more so, the Habitats Directive contains a large number of derogation clauses, which entails that destruction of protected nature can be legalized if the right procedures are followed. For instance, pursuant to Article 6(4) of the Habitats Directive, plans and projects which can give rise to significant effects on an EU-protected site can nonetheless be carried out 'for imperative reasons of overriding public interest'. Although the substantive requirements to be met when applying the derogation clauses are relatively strict, administrative practice reveals that unsound economic motivation still manages to win the day for most of the project developments (McGillivray 2012). A similar provision is present in the context of strictly protected species—Article 16 of the Habitats Directive—and is increasingly relied upon by Member States to justify unsustainable hunting practices aimed at increasing public tolerance for the presence of large carnivores.

Third, the persisting lack of coherence and under-prioritization of the EU's environmental policy in relation to other policy areas remains another focal point. The lack of a set of overarching rights of nature at EU level hinders the prevalence of ecological considerations within sectoral policies, such as the EU's common agricultural and fisheries policies. To give but one concrete illustration, a 2018 ruling of the CJEU highlighted the under-prioritization of nature conservation interests in the context of the CFP.[55] Interpreting provisions regarding nature conservation that had been included in the 2013 Fisheries Regulation, the EU judges underscored that Member States are still precluded from adopting a ban on commercial fishing activities inside their own Natura 2000 sites when such measures affect fishing vessels of other Member States.

Interim Conclusion

The above analysis has unequivocally demonstrated that the operationalization of rights of nature within the current EU legal order would face substantial stumbling blocks. This gives rise to two distinctive preliminary conclusions: one, implementing rights of nature at the level of the EU constitutional treaties appears instrumental in

[54]Case C-420/11 Leth (2013) ECLI:EU:C:2013:166, para. 46.

[55]Case C-683/16 Deutscher Naturschutzring [2018] ECLI:EU:C:2018:433, para. 55–57.

repairing some of the persistent legal loopholes present within the existing EU legal framework regarding nature conservation; two, the operationalization of rights of nature at national level or through the adoption of secondary legislation (e.g. through an EU directive) might not be able to overcome all the existing obstacles to ecological governance.

Sparks of Hope: Promising Recent Jurisprudential Evolutions as New Pathways?

Before addressing potential new avenues to bolster rights of nature within the existing EU legal order, I aim to briefly bring under the spotlight some recent jurisprudential evolutions which—at least indirectly—might reveal some unlocked potential for a rights of nature rationale within the existing pieces of EU environmental legislation. And while the analysis of cases is inevitably selective—since it does not comprehensively address water management and public participation—it still points to an untapped reservoir of pieces of environmental legislation in terms of ecological governance.

The Legal Teeth of the Precautionary Principle as a Progressive Starting Point

In his seminal article on rights of nature, Stone (1972) correctly underlined that noting how 'early' irreparable damage to the environment might be occurring is often giving rise to many complexities, especially taking into account the multitude of uncertainties that come into play. And while authors like Heyvaert (2006) have concluded that the role of the precautionary principle in the EU environmental policy is to be regarded as 'marginal', a certain progressive shift is noticeable in recent jurisprudence . Reference must certainly be made to the relatively progressive interpretation of Article 6(3) of the Habitats Directive by the CJEU.[56] When reasonable scientific doubt remains as to the absence of significant effects, no permit can be given pursuant to the well-known *Waddenzee* rationale for plans and projects liable to cause damage to EU-protected sites. The relevance of the precautionary principle does not remain confined to the context of EU-protected nature, with the General Court relying upon the same principle when upholding an insecticides ban in 2018.[57]

The practical importance of the precautionary principle for safeguarding the substantive rights of nature cannot be underlined enough. In the above-mentioned *Bialowieza* case, one of the strategies of the Polish government to justify its logging activities was to claim that such intrusive actions were needed in order to contain the

[56]Case C-127/02, Waddenzee (2004) ECLI:EU:C:2004:482, para. 56–59.

[57]Case T-429/13 & T-541/13, Bayer CropScience AG (2018) ECLI:EU:T:2018:280, para. 109–111.

outbreak of the spruce bark beetle, which was allegedly compromising the forest's future. In their final ruling, however, the EU judges noted that there was still scientific controversy regarding the most appropriate methods to stop the spread of the spruce bark beetle and thus dismissed the Polish arguments with reference to the precautionary principle.[58] Most interestingly, though, the CJEU has recently also used the precautionary principle as an instrument to scrutinize recent biodiversity offsetting practices in the context of EU-protected nature. In a string of recent court cases, the EU judges were faced with project developments in which future habitat restoration measures were presented as genuine mitigation measures (Schoukens 2017).[59] The finding that the potential positive effects of the future creation of habitats are uncertain and highly difficult to forecast was cited here as the main line of reasoning to dismiss a more liberal understanding of the protection scheme.[60] This jurisprudence is to be hailed as an implicit recognition of the intrinsic value of EU-protected nature.

The Integrity of EU-Protected Nature Partly Preserved

Along similar lines, the CJEU has also consistently treated EU-protected nature and species as 'common heritage'. For instance, in its rulings regarding the destruction of the *Bialowieza* forest, the CJEU explicitly underlined that the purported forestry practices would cause irreparable damage to the EU's common heritage when reaching a substantive outcome.[61] Likewise, when interpreting the notion of 'integrity of a Natura 2000 site', the CJEU concluded that the latter is adversely affected if the project is liable to prevent the lasting preservation of the constitutive ecological characteristics of a site.[62] This appears to be in line with Stone's distinction between repairable damage to the environment—which might be balanced and weighed—and irreparable damage, which is to be precluded (Stone 1972). Interestingly, the EU judges also held that even small-scale interventions could be deemed unacceptable since, when accumulated, they might stand in the way of the much-needed recovery of a Natura 2000 site.[63] This is to be greeted as an important tool to combat the 'risk of a death by a thousand cuts'. The outcome of the *Weser* case, where further clarification had to be given to the concept 'deterioration of the ecological status of a waterbody' when applying Article 4 of the Water Framework Directive at concrete permit level, is also instructive in this respect. Here, the EU judges stated that the latter concept implies the drop in quality status of at least one quality element, for instance, fish populations, even if that fall does not affect other quality elements of the affected

[58]Case C-441/17, Commission v Poland (2018) ECLI:EU:C:2018:255, para. 179.

[59]Case C-164/17, Grace and Sweetman (2018) ECLI:EU:C:2018:593, para. 52–56; Cases C-387/26 and C-388/15, Orleans, ECLI:EU:C:2016:583, para. 50–56.

[60]Case C-521/12, Briels, ECLI:EU:C:2014:330, para. 32.

[61]Case C-441/17, Commission v Poland (2018) ECLI:EU:C:2018:255, para. 208.

[62]Case C-258/11, Sweetman (2013) ECLI:EU:C:2013:220, para. 46.

[63]Ibid.

waterbody.[64] However, decisions, such as the Weser ruling, are often criticized in legal literature for their alleged normative vigour, since this jurisprudence seems to reject a so-called balancing approach (e.g. Salm 2015). Such criticism further highlights the inherent difficulties of operationalizing a more eco-centric approach to nature within a primarily anthropocentric governance scheme.

A Restoration Imperative, with Legal Teeth?

Pursuant to Article 72 of the Ecuadorian Constitution, nature has the right to be restored, which constitutes but one illustration of the growing emphasis that is placed on the restoration narrative within the context of nature's rights and beyond. However, at first glance, most of the seminal legal instruments in the EU's environmental policy, such as the EU Nature Directives, do not seem to be explicitly preoccupied with putting forward ambitious restoration or enhancement plans. In a recent decision with respect to an attempt by the Dutch government to declassify a Natura 2000 site, though, the CJEU held that the restoration imperative which is present in Articles 2 and 6 of the Habitats Directive also requires Member States to include degraded areas with ecological potential in the Natura 2000 Network.[65]

A similar progressive attitude surfaced in a 2014 Italian case, where the CJEU repeated that Member States are principally tasked to prevent or, as the case may be, *remedy* further incremental degradation instead of simply abandoning an existing Natura 2000 site in exchange for the designation of other areas with similar characteristics.[66] Another interesting example of the restoration imperative is offered by the outcome of the French Hamster case, where France was ultimately condemned by the CJEU for its failure to come forward with a robust recovery policy for its plummeting populations of wild hamsters.[67] In sum, in light of the degraded state of many existing ecosystems, one might at least use the Habitats Directives as a lever to restore nature located in EU protected areas or, as the case may, biodiversity that is instrumental of the immediate survival of endangered species.

Nature Trumps Vested Interests

Operationalizing rights of nature inevitably requires a clear-cut mandate to interfere with unsustainable lock-in scenarios. The application of general principles of law, such as legal certainty, when combined with a classic understanding of property

[64]Case C-461/13, Bund für Umwelt und Naturschutz Deutschland eV (2015) ECLI:EU:C:2015:433, para. 66–67.

[65]Case C-281/16, Vereniging Hoekschewaards Landschap (2017) ECLI:EU:C:2017:774.

[66]Case C-301/12, Cascina Tre Pini Ss (2014) ECLI:EU:C:2014:214, para. 32.

[67]Case C-241/08, Commission v France (2010) ECLI:EU:C:2010:114, para. 34–39.

rights, could prevent national agencies and authorities from giving precedence to maintaining ecological integrity over short-term ecological interests. Even so, in a case concerning ongoing and authorized dredging works in a German protected site, the CJEU held that the principles of legitimate interests and legal certainty cannot be applied in such way that it would prevent new rules from being applicable to the future effects of previously authorized activities, which turn the conservation duties included in Article 6(2) and (3) of the Habitats Directive, for instance, into very powerful tools for advocating a paradigm shift.[68]

Most telling, though, was the 2016 decision of the CJEU in the *Grüne Liga Sachsen* case, where the CJEU was asked to shed light on the legal implications of the continuing adverse effects of an already constructed bridge. The CJEU clarified that 'so far as concerns the economic cost of the steps that may be considered in the review of alternatives, including the demolition of the works already completed, (…) it must be stated (…) that that is not of equal importance to the objective of conserving natural habitats and wild fauna and flora pursued by the Habitats Directive. Therefore, account being taken of the strict interpretation of Article 6(4) of that directive, (…), it cannot be accepted that the economic cost of such measures alone may be a determining factor in the choice of alternative solutions under that provision'.[69] The recent outcome of the Dutch case regarding the programmatic approach to nitrogen once again highlights the potency of the Habitats Directive, if rigorously applied, as an important lever to evolve towards more sustainable farming practices. In this respect, the CJEU reasoned that Member States cannot use generic conservation actions as a tool to mitigate additional nitrogen emissions linked to new project developments.[70] And as a result, the Dutch government is now required to rethink its industrial farming industry in view of the adverse effects it creates for many degraded EU-protected sites.

Interim Conclusion

Authors like Bétaille claim that EU environmental law as it stands offers enough levers for a substantive improvement of the environmental quality within the EU. The case law addressed in this subsection seems to partly confirm this strand of argumentation. Why do we need rights of nature when EU courts generally opt for a progressive interpretation of EU law? Would it not make more sense to focus all our efforts on a better enforcement of the existing environmental statutes? And could one not entertain that at least part of the substantive underpinnings of rights of nature are reflected in the existing EU environmental Directives? However, this approach only holds true when one negates the obvious difficulties to which a strict implementation

[68]Case C-226/08, Stadt Papenburg (2016) ECLI:EU:C:2010:10, para. 49.

[69]Case C-399/14, Grüne Liga Sachsen eV (2016) ECLI:EU:C:2016:10, paras. 40–46.

[70]Case C-293/17, Coöperatie Mobilisation for the Environment UA, ECLI:EU:C:2018:882, para. 121–132.

of such progressive jurisprudence is leading, especially within the current economic paradigms which are based upon the premise of infinite growth. For instance, in order to greatly reduce existing nitrogen deposition, a clear reduction of the dairy farming industry is required. Absent a legal translation of rights of nature in the constitutional framework of a legal order, well-trenched legal concepts such as property rights and legal certainty will be used as obstacles to prevent an effective implementation of the recent string of progressive rulings.

Avenues to the Future: A New Directive, Treaty Modification or National Legislation?

Having now assessed the leeway for operationalizing rights of nature within the existing legal EU order, it has become clear that new legal or regulatory initiatives might be needed to push the EU towards a rights-based approach to nature in the short term. But what options remain to operationalize rights of nature within the EU legal order? This is assessed in the penultimate section of this chapter.

A Rights of Nature Directive: A Tempting yet Unpersuasive Quick Legal Fix

At first glance, one of the most obvious pathways to operationalize rights of nature is to adopt an EU directive on this very topic. It is therefore no surprise to see the charitable organization called 'Nature's Rights' present a draft directive, which closely resembles the chapter on rights of nature that is included in the Ecuadorian's Constitution. Generally speaking, the draft seems to address the majority of the deficiencies of the existing EU legal order against the backdrop of rights of nature.[71]

According to Article 1, the proposed directive 'provides for the substantive and procedural rights of nature, the rights of people in relation to nature, establishing a duty of care, protection and enforcement and ecological governance'. When conflicts arise between rights of nature and the rights of other physical or legal persons, a certain hierarchy needs to be observed. The maintenance of 'the integrity, balance, health and equilibrium of nature as a whole' is to prevail. No less groundbreaking is Article 4(5), which stipulates that 'legal persons shall not enjoy any special rights or privileges that subordinates the rights of physical people and nature'. Likewise, the draft Directive also explicitly grants standing to nature, in terms that clearly go beyond anything that exists in existing EU environmental law. Breaches of the rights of nature are to give rise to compensation or restoration by the persons responsible for causing the breach. The proposed directive also obligates Member States to integrate

[71] This proposal can be consulted at https://www.citizensforeurope.eu/organisation/rights-of-nature-europe (accessed 15 September 2019).

and further protect nature's rights though what is dubbed 'ecological governance' aimed at restoring degraded landscapes.

While the directive is certainly to be lauded as an interesting effort to operationalize rights of nature within the EU legal order, it seems very unlikely that it will be adopted any time soon (Schoukens 2019). Yet even if adopted, I contend that this directive cannot be regarded as a meaningful step forward for rights of nature in the EU. In view of its grand ambition, though, the more fundamental point is whether the choice for the instrument of an EU directive is the right one. Implementing this directive would entail the amendment and/or outright revoking of many existing environmental and other norms. At several points, the Directive openly goes against the hierarchy of norms present within the EU legal order: environmental directives are to be interpreted in view of the provisions of primary law, such as the TFEU, and not vice versa.

Amending the EU Treaties: Balancing Between Theory and Pragmatism

In view of the preceding analysis, I submit that the only sustainable manner to achieve an effective operationalization of rights of nature in the EU consists in an inclusion of certain rights in the EU constitutional treaties. As of today, however, the prospect of initiating a lengthy and complex treaty review for introducing rights of nature within the EU's constitutional order seems to significantly downplay the success rate of this scenario. Even when encompassed in the context of a more general treaty revision, the question remains whether enough political support will be found for such amendments.

National Legislation and Litigation: The Ultimate Lever

A more hands-down way to implement rights of nature within the EU legal framework is to engage national parliaments and/or courts. For instance, in the Netherlands, several NGOs have already advocated the granting of legal personhood to the Wadden Sea, which has led to the adoption of a local ordinance advocating the establishment of an impartial management committee for the Wadden Sea.[72] As such, conferring legal rights to ecosystems would not necessarily be hindered by the existing EU legal order. However, operationalizing rights of nature at municipality level will unavoidably lead to questions regarding the apparent (?) lack of competences of local authorities in this particular domain. This being the case, granting legal personhood might equally be done through the passing of national legislation and/or

[72]See https://www.omropfryslan.nl/nieuws/895902-noardeast-fryslan-stemt-met-motie-voor-bijzondere-rechten-waddengebied (accessed 15 September 2019).

strategic litigation. As environmental protection is a shared competence within the EU legal framework, sufficient leeway is offered to the Member States to consider the conferral of legal personhood to nature. Moreover, such approach might also be in line with the subsidiarity principle, another fundamental principle of EU law. Pursuant to this principle, which is further defined in Article 5 TEU, the EU does not take action (except in the areas that fall within its exclusive competence), unless it is more effective than action taken at national, regional or local level. Of course, national legislation granting legal personhood to nature could still be challenged before national courts. When faced with new legislation passed for granting legal rights to nature, national courts could be tempted to refer legal questions relating to the alignment with EU law to the EU courts in Luxemburg in view of possible interferences with the EU legal order (Schoukens 2019). However, this is not necessarily a bad thing: it might indeed offer the CJEU the ultimate opportunity to shed its view on the thorny legal questions to be faced when aligning rights of nature with primary EU law as well as EU nature conservation law. Moreover, ligation strategies focused on the judicial recognition of standing of EU protected species could provide for another interesting playground of rights of nature activists.

Conclusion and Outlook

This chapter had no ambition to provide the reader with definitive answers to all questions at stake regarding the complex interplay between rights of nature and the existing EU legal order. It merely wanted to initiate a conversation about a rights-based approach to nature by presenting a thorough analysis of how rights of nature might interact within a legal order which is still dominated by an anthropocentric approach to ecological governance. In the first section, I argued that introducing rights of nature in existing legal frameworks is to be approached as a crucial push to a more eco-centric and holistic approach to nature. Granted, the simple conferral of certain rights to nature itself will certainly not suffice to completely end all destruction of nature. However, by allowing nature to take a central place in our constitutional framework, also within the EU, a further impetus is given to a novel type of ecological governance, which corrects the prevailing dichotomy between Man and Nature. The second section of this paper, which focused on the EU legal order, aptly demonstrated that the concrete operationalization of rights of nature within a primarily anthropocentric legal framework is liable to give rise to a multitude of legal issues. In this respect, this chapter's main conclusion is that the emerging case law of the CJEU, especially in the field of precaution, EIA, water management and habitat restoration might offer new, more gradual pathways towards ecological governance. Conversely, the doors of the EU courts remain firmly closed for nature in the context of direct standing before the EU courts. Likewise, EU environmental law lacks a clear 'ecological integrity' imperative in order to mitigate and/or correct its anthropocentric underpinnings. Whereas implementing rights of nature through the adoption of an EU directive sounds tempting it would inevitably give rise to countless legal objections and questions. As a treaty reform appears unrealistic for the time being, a more realistic short-term option would consist in implementing rights of nature through national legislation or litigation, which might ultimately present itself as the more appropriate bottom-up solution in this context.

References

Bétaille, J. (2019). Rights of nature: Why it might (not) save the world. *Journal for European Environmental and Planning Law, 16*(1), 36–64.

Bosselman, K. (2008). The way forward: Governance for ecological integrity. In L. Westra et al. (eds.), *Reconciling Human Existence with Ecological Integrity* (pp. 319–332). Earthscan.

Bosselman, K. (2016). A normative approach to environmental governance: Sustainability at the apex of environmental law. In D. Fisher (Ed.), *Research Handbook on Fundamental Concepts of Environmental Law* (pp. 22–50) Edgar Elgar.

Boyd, D. (2017). *The rights of nature, a legal revolution that could save the world*. ECW Press.

Burdon, P., & Williams, C., Rights of nature: A constructive analysis In D. Fisher (ed.), *Research Handbook on Fundamental Concepts of Environmental Law* (pp. 196–221). Edgar Elgar.

Chapron, G., Epstein, Y., & Lopez-Bao, J. V. (2019). A rights revolution for nature. *Nature, 363*(99), 1392–1395.

Cullinan, C. (2011). Wild law. *A manifesto for earth justice*. Green books.

Darwin, C. (1874). *The descent of man and selection in relation to sex*. United Kingdom: Plume.

de Waal, F. (2013). *The bonobo and the atheist: in search of humanism among the primates*. W. W. Norton & Company.

Driessen, B. (2017). Fundamental animal rights in European law. *European Public Law 23*(3), 574–586.

Elder, P. S. (1984). Legal rights for nature—the wrong answer to the Right(s) question. *Osgoode Hall Law Journal 22*(2), 285–295.

Emmenegger, S., & Tschentscher, A. (1994). Taking nature's rights seriously: The long way to biocentrism in environmental law. *Georgetown International Environmental Law Review 6*(3), 545–592.

Gaines, S. E. (2014). Reimagining Environmental Law for the 21st Century. *Environmental Law Reporter 44*, 215–101188.

Giagnocavo, C., & Goldstein, H. (1990). Law reform or world re-form: The problem of environmental rights, *Mc.Gill. L.J.*, 346–386.

Harari, Y. N. (2018). 'The myth of freedom', *The Guardian*, Retrieved September 14, 2018 from https://www.theguardian.com/books/2018/sep/14/yuval-noah-harari-the-new-threat-to-liberal-democracy.

Hart, H. L. A. (1982). *Essays on Bentham: Studies in jurisprudence and political theory*. Oxford: Clarendon Press.

Heyvaert, V. (2006). Facing the consequences of the precautionary principle in European Community law. *European Law Review 31*(2), 185–206.

Hovden, K. (2018). The best is not good enough: Ecological (Il)literacy and the rights of nature in the European union. *Journal for European Environmental and Planning Law 15*(3–4), 281–309.

Hsiao, E. C. (2013). Whanganui River Agreement—Indigenous Rights and Rights of Nature. *Environmental Policy and Law 42*(6), 371–375.

Iorns Magallanes, C. (2013). Maori cultural rights in Aoteroa new zealand: Protecting the cosmology that protects the environment. *Widener Law Review 21*(2), 273–327.

Kauffman, C. M., & Martin, P. M. (2017). Can Rights of Nature Make Development More Sustainable? Why some Ecuadorian Lawsuits Succeed and Others Fail. *World Development 22*, 130–142.

Kauffman, C. M., & Martin, P. M. (2018). Constructing Rights of Nature Norms in the US, Ecuador and New Zealand. *Global Environmental Politics 18*(4), 43–62.

Kotzé, L., & Calzadilla, P. V. (2017). Somewhere between Rethoric and Reality: Environmental constitutionalism and the rights of nature in Ecuador. *Transnational Environmental Law 6*(3), 1–33.

Krämer, L. (2015). Implementation and enforcement of the habitats directive. In C. H. Born et al. (Eds.), *The EU Habitats Directive in its Environmental Context* (pp. 229–244) Routledge.

La Follette, C. Maser, C. (2017). Sustainability and the rights of nature. CRC Press.

Leopold, A. (1949). A Sand County Almanac. Oxford University Press.

McGillivray, D. (2012). Compensating biodiversity loss: The EU Commission's approach to compensation under Article 6(4) of the Habitats Directive. *Journal of Environmental Law 24*(3), 417–450.

Nash, J. (1993), The case for Biotic Rights, *Yale Journal of International Law 18*, 235–250.

O'Donnell, E. L., & Talbot-Jones, J. (2018). Creating legal rights for rivers: Lessons from Australia, New Zealand, and India. *Ecology and Society, 23*(1), 7.

Pietrzykowski, T. (2018). *Personhood beyond humanism: Animals, Chimeras, Autonomous Agents and the Law.* Springer.

Raz, J. (1986). *Morality of freedom.* Oxford University Press.

Sagoff, M. (1974). On preserving the natural environment. *Yale Law Journal 84*, 205–267.

Salm, D., The case for smart governance in European Water law. In: B. Vanheusden & L. Squitani (Eds.), *EU Environmental and Planning Law Aspects of Large-Scale Project*, (pp. 215–235) Intersentia.

Salt, H. S. (1894). *Animals' rights considered in relation to social progress.* Macmillan.

Schilmoller, A., & Pelizzon, A. (2013). Mapping the terrain of earth jurisprudence: Landscape, thresholds and horizons. *Environmental and Earth Law Journal 3*(1), 1–32.

Schoukens, H. (2015). Access to Justice in Environmental Cases after the Rulings of the Court of Justice of 13 January 2015: Kafka revisited? *Utrecht Journal of International and European Law 31*(8), 46–67.

Schoukens, H. (2017). Proactive habitat restoration and the avoidance of adverse effects on protected areas: Development project review in Europe after Orleans. *Journal of International Wildlife Law & Policy 20*(2), 125–154.

Schoukens, H. (2018). Granting legal personhood to nature in the European Union: Contemplating a Legal 'R)evolution to avoid an ecological collapse? *Journal for European Environmental and Planning Law 15*(4), 309–332.

Schoukens, H. (2019). Granting legal personhood to nature in the European Union: Contemplating a Legal 'R)evolution to avoid an ecological collapse? *Journal for European Environmental and Planning Law 15*(3), 65–90.

Shelton, D. (2015). 'Nature as a legal person', Vertigo Hors-Série, Retrieved September 2015 from https://journals.openedition.org/vertigo/16188.

Singer, P. (1975). *Animal liberation: A new ethics for our treatment of animals.* Random House.

Somsen, H. (2017). The End of European union environmental law: An environmental programme for the Anthropocene. In L. Kotzé (ed.), *Environmental Law and Governanance for the Anthropocene*, (pp. 353–372) Hart.

Staker, A. (2017). Should chimpanzees have standing? The case for pursuing legal personhood for non-human animals. *Transnational Environmental Law 6*(3), 485–507.

Stone, C. D. (1972). Should trees have standing? Towards legal rights for natural objects. *Southern Californian Law Review 9*(4), 451–501.

Stone, C. D. (2010). *Should trees have standing? Law, morality and the environment.* Oxford University Press.

Stone, C. D. (2012). Response to commentators. *Journal of Human Rights and the Environment, 3*(Special Issue), 100–120.

Tanasescu, M. (2016). The rights of nature in Ecuador. *Environment, Political Representation, and the Challenge of Rights,* 85–106.

Taylor, P. E. (1998). From environmental to ecological human rights: A new dynamic in international law? *Georgetown International Environmental Law Review*, 309–397.

Vedder, H. (2010). The treaty of lisbon and european environmental law and policy. *Journal of Environmental Law 22*(2), 285–299.

Verschuuren, J. (2016). The growing significance of the principle of sustainable development as a legal norm. In: D. Fisher (ed.), Research handbook on Fundamental Concepts of Environmental Law, (pp. 276–306) Edward Elgar.

Villalba, U. (2013). 'Buen Vivir vs. development: A paradigm shift in the andes. *Third World Quarterly 34*, 1427–14442.

Warnock, M. (2012). Should trees have standing? *Journal of Human Rights and the Environment, 3*(Special Issue), 56–68.

Whittemore, E. (2011). The problem of enforcing nature's rights under Ecuador's constitution: Why the 2008 environmental amendments have no bite. *Pacific Rim Law & Policy Journal 20*(3), 659–691.

Wood, C. (2014). *Nature's trust. Environmental Law for a New Ecological Age.* Cambridge University Press.

The European Union's Diplomacy: Protecting Non-Human Nature?

Simon Schunz, Bram De Botselier, and Sofía López Piqueres

Abstract The European Union (EU) is widely considered as a 'leader' in global environmental politics. Over the past forty years, it has gradually adopted a corpus of primary/secondary environmental law allowing it to become the strongest regional environmental protection regime in the world. As a global actor, the EU attempts to export this *acquis*, following a mainstreaming approach that in principle demands attention to the protection of non-human nature in all of its external action. Employing four different approaches to the human-NHN relationship derived from the debate about global environmental justice, this chapter provides a critical discussion of EU external activities related to the protection of non-human nature and their effects. It focuses on three domains that are central to tackling the causes of the Anthropocene: EU external action aimed at mitigating climate change, halting biodiversity loss and governing the Arctic as a major theatre of environmental degradation processes. The chapter finds that the EU's diplomacy primarily pursues a 'reformist' justice agenda, aimed at better 'managing' non-human nature. It fares rather well with regard to the conclusion of international environmental agreements (institutional effectiveness), but less so when it comes to the actual protection of non-human nature (ecological effectiveness).

S. Schunz (✉) · B. De Botselier · S. López Piqueres
College of Europe, Brugge, Belgium
e-mail: simon.schunz@coleurope.eu

B. De Botselier
e-mail: bram.debotselier@coleurope.eu

S. López Piqueres
e-mail: sofia.lopez@coleurope.eu

J. C. Pereira and A. Saramago (eds.), *Non-Human Nature in World Politics*,
Frontiers in International Relations, https://doi.org/10.1007/978-3-030-49496-4_12

Introduction

Earth System scientists contend that our planet has entered the Anthropocene, a geological epoch in which the human species has come to profoundly alter atmospheric, biospheric, geologic, hydrologic and other Earth System processes (Rockström et al. 2015). This observation at the same time confirms and challenges the long-standing view, best expressed by neoclassical economics, that humans (should) dominate Non-Human Nature (NHN), that is, 'the phenomena of the physical world collectively, including plants, animals, the landscape, and other features and products of the earth, as opposed to humans or human creations' (Oxford Dictionary 2019).

On the one hand, the Anthropocene is an expression of the effects of human practice resulting from the neoclassical view that NHN is simply a resource to be exploited. On the other hand, it also fundamentally challenges the assumption of boundless human dominance of NHN: by highlighting how human activities exceed 'planetary boundaries' in areas such as biological diversity, biogeochemical flows and climate change, classifying this geological epoch as the Anthropocene underscores that humanity is in the process of moving beyond the realm of manageability of its biophysical bases ('safe space') (Rockström et al. 2015).

If it is thus evident that 'the ecological underpinnings of the Earth are under monumental assault' by the 'extraordinary rhythms of human life, consumption and industrialisation', the relationship between humanity and NHN, and the way it is politically treated, moves to the very heart of debates about global politics (Harris 2016, p. 1). And with it the question: 'can we match the planet with our politics?' (Burke et al. 2016, p. 5–6).

'Global politics', in this context, does not merely refer to the international political decision-making processes on the 'issue area' typically associated with NHN, 'the environment' (Meyer 2008, p. 773). Rather than being about this one 'issue area', the recognition of the scale of the challenge posed by the Anthropocene implies a need to think holistically in terms of 'planet politics' (Burke et al. 2016). This in turn makes it necessary to investigate the human-NHN relationship at the more fundamental level of global ethics, that is, the 'exploration of the … moral values, norms, and responsibilities that we acknowledge in regard to the relations between states and *the relations individuals have with one another and the natural world*' globally (Dower, cited in Harris 2016, p. 2, emphasis added).

Thinking in terms of 'planet politics' thus invites a reflection on questions of justice in the Anthropocene: what rights and duties do human beings have not only vis-à-vis each other, but also and especially towards 'the natural world'? In this chapter, we are primarily interested in this latter aspect, asking: '*what does humanity owe to the planet*'? Posing the question of our duty vis-à-vis NHN in the Anthropocene directly implies asking what human beings owe to each other, as damages to NHN caused by some will inevitably threaten the livelihood of others—and ultimately of all of humanity if planetary boundaries are durably exceeded.[1]

[1] The literature on 'planetary boundaries' does not advocate NHN protection for the sake of NHN, but is fundamentally anthropocentric, as the boundaries it identifies delimit the safe operating space

Different takes on the human-NHN relationship co-exist in the contemporary debate about global (environmental) justice. Ehresman and Stevis (2018) identify four positions, which are also employed here: *status quo* environmental justice (presupposing human dominance of NHN: 'humanity owes nothing to NHN'), *reformist* environmental justice (which upholds but critically engages with human dominance of NHN and advocates institutional reforms to better 'manage' the environment: 'humanity owes NHN its sustainable use'), *transformational* environmental justice (which commends profound socio-economic systemic transformations to (re-)establish a more balanced human-NHN relationship: 'humanity owes NHN strong consideration and protection') and *ecological* justice (arguing that humans should live in harmony with NHN: 'humanity owes NHN equal treatment').

Rather than in making a mere theoretical argument, we are interested in analysing the human agency to protect NHN of one collective actor, the European Union (EU). More precisely, we focus on what the EU does as a diplomatic actor to protect NHN. The EU has long been considered as a global frontrunner, if not leader, when it comes to the issue area 'environment' (see, e.g. Kelemen and Vogel 2010). Over the past forty years, it has gradually adopted a corpus of environmental law to become the strongest regional environmental protection regime in the world (see Chap. 11 in this volume). Simultaneously, it has pursued the ambition to influence the global politics that touch upon NHN, as aptly illustrated by the Treaty on the Functioning of the EU's (TFEU) Article 191.2, which calls for 'promoting measures at international level to deal with regional or worldwide environmental problems, and in particular combating climate change'. Central to the EU's positions on NHN are a high level of attachment to 'preserving, protecting and improving the quality of the environment' (Art. 191.1 TFEU) in all its actions (Art. 11 TFEU), as well as the notion of precaution, the rule of law and the pursuit of global multilateral solutions (Art. 21 TEU) (see also Delreux 2013). *Prima facie*, one may thus assume that the EU has the intention to engage in the protection of NHN via its diplomatic action.

The chapter critically scrutinises this assumption by analysing those of the EU's external policies which can a priori be expected to have the biggest impact on NHN protection globally. They comprise efforts that are central to tackling the causes of the Anthropocene and attaining the objective of remaining within planetary boundaries, and are taken from the broader issue area 'environment': EU external action aimed at mitigating climate change, halting biodiversity loss and governing the Arctic as a major theatre of environmental degradation processes. We ask the following research question: *to what extent does the EU's diplomacy*—via *its discourse (external policy positions) and practice (external policy behaviour) and its (institutional and ecological) effects on environmental issues at the global level—contribute to the protection of non-human nature?* To determine the extent of the EU's efforts, we will establish if they correspond to a status quo, reformist, transformational or ecological justice perspective.

for humanity on this planet. This chapter is agnostic about the motivation for pursuing the protection of NHN, whether they are anthropo- or NHN-centric. What it is interested in is whether protection measures are pursued and effective.

By answering this question, we intend to add a novel perspective to the literature on the EU's environmental diplomacy, which has produced conceptually and empirically very valuable insights (see, e.g. Adelle et al. 2018a), but has tended to focus on institutionalist accounts of EU external environmental action, consequently emphasising also its institutional effectiveness. The scale of the challenge posed by the Anthropocene makes it necessary to complement existing studies by engaging in analyses of how the EU's action fares in terms of justice, as defined above, as well as ecological effectiveness.

We first develop an analytical framework that defines and interlinks the key concepts of EU external action, justice and effectiveness. Subsequently, we briefly discuss the EU's internal environmental regime as a prerequisite for its external action. After applying the framework to the three cases of climate change, biodiversity and the Arctic, special attention is given to considering the EU's effectiveness when it comes to NHN protection. To conclude, we argue that the EU mostly acts in line with a reformist understanding of global environmental justice, and reflect on its potential for protecting NHN.

Analytical Framework

This section discusses the EU as a diplomat before operationalising the extent to which the Union can contribute to NHN protection by linking different global justice positions to the scale of change needed to remain within 'planetary boundaries' in the Anthropocene. It then introduces various understandings of EU external effectiveness.

First, while the EU acts as a diplomat on issues ranging from trade to development and foreign and security policy, and though these activities can and do have (in)direct effects on NHN, this analysis focuses on policies that are primarily concerned with the *protection* of NHN. We take these to be the activities undertaken in the pursuit of EU treaty objectives that are a priori aimed at high levels of environmental protection and precaution (Art. 191 TFEU). As the environment is a shared competence between the EU and its members (Art. 4 TFEU), EU external environmental action is a collective effort of the EU institutions and member states (Delreux 2013). Its external policy positions can be proposed by the Commission, but are typically adopted via conclusions of the (Environment) Council of the EU (ibid.). Relevant Commission and Council documents therefore also form the key sources for analysing EU discourse on NHN protection. It is on the basis of the Council conclusions that the EU's external representatives act in bilateral and, especially, multilateral contexts—including negotiations of Multilateral Environmental Agreements (MEAs)—to pursue objectives related to NHN protection. As diplomacy is about 'the conduct of relations between states and other entities with standing in world politics by official agents and by peaceful means' (Bull 2012, p. 156), EU diplomatic practice is scrutinised by examining how its 'agents' act (means), vis-à-vis whom and in what fora (targets).

To what extent the EU's diplomacy contributes to NHN protection can be captured by linking the four global justice positions evoked in the introduction to the action needed to remain within 'planetary boundaries' (Rockström et al. 2015). These four positions differ in their view of the human-NHN relationship (Ehresman and Stevis 2018, p. 108):

- *Status quo environmental justice* presupposes human dominance of nature, whose resources are perceived as continuously available to meet human needs.[2] It embodies a 'Promethean … view of the human-nature relationship' (Ehresman and Stevis 2018, p. 111): 'The world is made for man, not man for the world' (Bacon, cited in Hopwood et al. 2005, p. 38–39). Adherents of this position would typically either not accept the notion of 'planetary boundaries' *at all* or not consider it as reason for concern, but primarily place faith in human ingenuity and technology[3] to guarantee sustained human well-being.
- *Reformist environmental justice* advocates are in favour of reforming the current socio-economic system to render it more 'environmentally just'. They recognise that 'our lives, activities and society are nested within the environment', but 'do not consider that a collapse in ecological or social systems is likely or that fundamental change is necessary' (ibid.: 43).
- *Transformational environmental justice* perspectives see 'mounting problems in the environment and/or society as rooted in fundamental features of society today and how humans interrelate and relate with the environment' (ibid.: 45). Its advocates therefore argue for a profound change of socio-economic structures. While many variants of this perspective, including eco-feminism and eco-socialism, co-exist, on the human-NHN relationship the eco-socialist perspective is emblematic: 'We by no means rule over nature like a conqueror over a foreign people …—but … we … belong to nature, and exist in its midst' (Engels 1968, quoted in Hopwood et al. 2005, p. 46).
- *Ecological justice* adherents most unequivocally embrace the idea that 'nature is not only instrumentally, but intrinsically and objectively, valuable' (Ehresman and Stevis 2018, p. 115). Given this inherent value of nature, the 'moral community' entitled to rights must be extended to include NHN (Leopold 1968/1948). Advocates of this position promote the internalisation of the value of NHN into institutions and policies at all levels of governance, including in and through international law. This can go as far as seeking agreements on collective (self-)coercion to control NHN exploitation (Clapp and Dauvergne 2011, p. 17).

The status quo position holds an unequivocal and the reformist position a predominantly instrumental view of NHN, whereas the transformational and ecological

[2]The status quo position largely corresponds to an understanding of justice also held by 'Ecomodernists', who desire to 'use humanity's extraordinary powers in service of creating a good Anthropocene' (Ecomodernism 2019).

[3]The central reliance on technology as the solution to any potential challenge arising from NHN is characteristic of the status quo justice position (and ecomodernism). At the same time, adherents of the reformist and transformational justice perspectives also seek to develop and manage technologies (e.g. renewable energy technologies) in pursuit of (environmental) justice-related goals.

justice positions see value in protecting NHN per se, imbuing it with certain rights. However, the dividing line between reformist and transformational positions can be rather thin. The reformist position does pay attention to NHN, and activities corresponding to this view can have—intended and unintended—effects contributing to its protection.

If placed on a continuum, the four justice positions allow for heuristically determining the extent to which EU diplomacy contributes to NHN protection (Table 1). Located at one end of the spectrum is the status quo position, which exhibits a low level of attention to NHN clearly *not* 'matching the reality of our changing planet' in the Anthropocene. By contrast, this match might *possibly* be given, and the EU might contribute to humanity remaining within critical boundaries when it acts in line with a reformist justice perspective if this leads to ecologically effective results. It is *probably* given, meaning that EU diplomacy has a higher likelihood of contributing to remaining within such boundaries when the EU's discourse and practice correspond

Table 1 EU diplomacy and approaches to global justice and their positions on human-NHN relations

Analytical units	Status quo env. justice	Reformist env. justice	Transformational env. justice	Ecological justice
EU Discourse With its positions, …	…, the EU privileges growth and technological innovation, which are expected to contribute to solving NHN-related challenges, should they arise	…, the EU identifies ways in which the current global system can be reformed to sustainably use NHN	…, the EU expresses the view that socio-economic structures need to be altered in order to effectively protect NHN	…, the EU places the protection of NHN at the heart of its policies
EU Practice With its actions, …	…, the EU privileges growth and technological innovation, which are expected to contribute to solving NHN-related challenges, should they arise	…, the EU engages with others to reform global institutions to manage and sustainably use NHN	…, the EU works for a change of the socio-economic structures at the global level allowing for the effective protection of NHN	…, the EU pursues the recognition—at the global level (in international law)—of the view that NHN has an intrinsic value equal to that of human beings
What human beings owe to NHN	Nothing	Sustainable use	Protection	Equal treatment
Staying within planetary boundaries	No	Possibly (if effective)	Probably	Certainly

to transformational justice, and it is *certainly* given when its diplomacy corresponds to ecological justice perspectives located at the other end of the spectrum.

To classify the EU's diplomacy, it is necessary to determine to which of the four justice positions its discourse and practice correspond. To this end, we engage in a 'pattern-matching' (Yin 1984) exercise between the definitions of the forms of justice developed in Table 1 and the empirical evidence gathered for the analysed cases. They will be investigated primarily on the basis of document analysis (e.g. Council conclusions) and secondary literature. Emphasis is placed on the EU's most recent attempts to engage externally on these issues.

A final step in the design of the analytical framework, and in the analysis per se, is that of defining and analysing the effectiveness of the EU's diplomacy, that is, its impact on NHN protection. Effectiveness is regularly defined as 'goal attainment', notably in the context of EU external environmental action (e.g. Oberthür and Groen 2018). This goal attainment can be assessed in terms of both *output* and *outcome.*

When it comes to output, assessing goal attainment leads to establishing 'institutional effectiveness' (Gysen et al. 2006, p. 99). The basic method for establishing whether the EU has been institutionally effective is to compare its policy objectives (e.g. to adopt a legally binding agreement on CO_2 emissions reductions) to the output of an international negotiation process (e.g. an international treaty enshrining emissions reduction targets), so as to determine the degree of 'match' (ibid.). If there is a match and no other factor is deemed decisive in bringing about this output, a certain degree of EU institutional effectiveness can be considered established. In terms of outcome, measuring goal attainment must yield an evaluation of 'the degree to which the change of the policy issue, caused by the policy, is in line with the policy goals' (ibid.). The question here is whether the ultimate policy goal the EU has managed to get into an international agreement (e.g. to reduce emissions) is de facto attained. This 'impact effectiveness' (ibid.) can, in environmental contexts, be usefully referred to as EU (external) 'ecological effectiveness', defined as the impact of EU diplomacy on the actual protection of NHN.

These understandings of effectiveness can be linked to the above global justice perspectives. Whereas a status quo position is not interested in institutional nor ecological effectiveness, which it sees at best as by-products of activities stimulating technological innovation and growth, transformational environmental and ecological justice positions are only interested in ecological effectiveness, as they aim for systemic change with direct ecological rather than mere institutional effects. The reformist position, by contrast, is interested primarily in institutional effectiveness, which it sees as a precondition for ecological effectiveness: 'getting necessary reforms right' to pave the way for attaining desired outcomes.

The analysis concentrates on what sort of effectiveness EU diplomacy attains by comparing its objectives to both the outputs and outcomes of global political processes in the three domains of climate change, biodiversity and Arctic environmental governance.

The EU's Internal Environmental Regime

The (societies in the) EU impact(s) on NHN in ways that contribute to the exceeding of planetary boundaries. At the same time, the Union has also long been engaged in efforts aimed at protecting NHN, through internal and external action, notably in the area of the environment. This section considers EU internal environmental action to the extent necessary to understand its diplomatic activities aimed at NHN protection.

Since the 1970s, the EU has gradually developed strong treaty bases for the environment, culminating in the 2009 Lisbon Treaty's Article 191 TFEU, alongside the build-up of secondary law that comprises hundreds of legal acts on issue areas ranging from air pollution to water quality (Delreux and Happaerts 2016, p. 19). All this has resulted in a regional regime that is comprehensive in scope, operates with a mix of regulatory, planning, market based and information tools (ibid.), and follows a vision of European societies that is embodied in the notion of 'ecological modernization', best expressed by the 2000 Lisbon and 'Europe 2020' Strategies' goal of pursuing 'smart and sustainable growth' (Machin 2019, p. 213). This vision, which aims to 'reconcile … economic growth and ecological targets to offer a "win-win" situation' (ibid.: 223), and the EU's corresponding internal practice demonstrates that its domestic policies with relevance for NHN protection primarily express a reformist justice position.

When it comes to its effectiveness, the EU's regime has a mixed record. Its institutional effectiveness, which can be understood here as the extent to which treaty objectives have been transposed into secondary legislation and policies, is—given the very existence of the above-mentioned *acquis*—undoubtedly high. However, policies suffer from a non-negligible implementation deficit, which even the EU's (in theory) solid compliance procedures have been unable to overcome (Jordan and Tosun 2013). While this deficit may jeopardise the EU's internal ecological effectiveness, its regime has positively impacted NHN protection. The European Environment Agency's (EEA) periodic reports on the 'State of the Environment in Europe' (SOER) testify to advances in terms of reduced air and water pollution, resource efficiency and climate change, thanks to EU measures (EEA 2015). On climate change, for instance, the EU was and is well underway to meet its 2020 and 2030 targets, although it is still 'far from 1.5 °C compatible action' demanded by the Paris Agreement (CAT 2019a, b). In other domains, such as land degradation, species and habitat conservation, the EU is 'not on track', implying that its ecological effectiveness remains more limited (ibid.).

The EU's internal environmental regime and its effects are in three ways relevant for the discussion of its diplomacy and external effectiveness related to NHN protection. First, through its internal action, the EU can and, as seen, does—at a regional level—to a certain extent contribute to NHN protection. Second, being a (successful) regional environmental regime has important repercussions for the EU's role as a diplomatic actor, as it provides the credibility and bases—in terms of priorities, targets and policy instruments—for engaging externally. Third, intra-EU policies can have unintended effects on third parties, leading others to adopt similar levels of environmental regulation to access the Union's single market (Damro 2012). The EU relies heavily on this 'market power' diffusion effect when it engages with parties in its neighbourhood, albeit with varying degrees of success when it comes to the actual (limited) institutional and (very limited) ecological effectiveness (Buzogány 2018).

More significant than its unintended external effects is its diplomacy in areas of importance for NHN, which the chapter turns to now.

EU Diplomacy and the Protection of Non-Human Nature

With its diplomatic action, the EU could protect NHN in manifold ways, including, for instance, by integrating environmental considerations into its trade and development diplomacies. Most profoundly, however, the EU can contribute to NHN protection by engaging in *environmental* diplomacy in areas central to tackling the causes of the exceeding of planetary boundaries. This section therefore focuses on its diplomatic positions (discourse) and behaviour (practice) regarding global political efforts to deal with climate change, biodiversity loss and the governance of the Arctic environment.

Climate Change

At the heart of the debate about the Anthropocene, climate change represents a major challenge when it comes to the protection of NHN. The combat against it has primarily been confined to the United Nations climate regime, created around the 1992 UN Framework Convention on Climate Change (UNFCCC). Within and beyond this multilateral regime—for instance through bilateral cooperation with key partners—the EU has been a lead player (Wurzel et al. 2017).

The global climate regime has gone through different phases: until the end of the 2000s, parties had the aspiration to adopt a legally binding international agreement that enshrines top-down, strong and enforceable emissions reduction targets for major emitters—initially industrialised countries, and as of the mid-2000s also 'emerging economies'. Following the 'failed' 2009 Copenhagen Conference of the Parties (COP) 15, this logic was called into question notably by the United States and the emerging economies from the 'BASIC' group (Brazil, China, India, South Africa). Since then, and particularly with the 2015 Paris Agreement, the climate regime has transformed in nature, combining the identification of an indicative, top-down global mitigation target of keeping greenhouse gas emissions 'well below 2 °C above pre-industrial levels and … pursue efforts to limit the temperature increase to 1.5 °C' (UNFCCC 2015: Article 2) with bottom-up 'Nationally Determined Contributions' (NDCs), anchored in national law, and a reporting and review system that is binding at the global level.

This section discusses the EU's contribution to NHN protection through its climate diplomacy during the pre-2009 and post-2009 periods, focussing on the negotiation and implementation of the Paris Agreement.

EU Discourse

EU discourse on climate change in general and on its climate diplomacy in particular has been rather stable across time. Since 1996, it has revolved around the objective of limiting global temperature increase to 2 °C. Besides this commitment to a science-guided response to the challenge, the EU's position has included a strong attachment to the precautionary principle (Art. 191.2 TFEU) and to finding multilateral, legally binding solutions.

Illustrative of this discourse was the EU's position for the 2009 Copenhagen COP. A 2007 Commission communication called for 'urgent action' to limit climate change to 2 °C and proposed that developed countries, including the EU, should reduce their emissions by 30% from 1990 levels by 2020 (European Commission 2007b, p. 2). Endorsed by the Council, the EU's position was to offer a unilateral 20% reduction to its negotiation partners, paired to the promise of a 30% reduction conditional on other actors' ambition, 'to demonstrate international leadership on climate issues' (ibid.). The communication continued to state that '[b]y 2050 global emissions must be reduced by up to 50% compared to 1990, implying reductions in developed countries of 60–80% by 2050. Market-based instruments such as the EU ETS [Emissions Trading System] will be a key tool to … reach … targets at least cost' (ibid.).

In the post-2009 context, the EU continued to rely on a similar discourse. The Lisbon Treaty's call for 'promoting measures at international level to deal with regional or worldwide environmental problems, and in particular combating climate change' (Art. 191.1 TFEU) promoted NHN protection to a quasi-constitutional principle. The September 2015 Environment Council conclusions outlined the position that 'to stay below 2 °C, global greenhouse gas emissions need to peak by 2020 at the latest, be reduced by at least 50% by 2050 compared to 1990 and be near zero or below by 2100' and promoted 'an ambitious and durable legally-binding agreement under the UNFCCC … applicable to all Parties and addressing in a balanced and cost-effective manner mitigation' (Council of the EU 2015).

With these positions, the EU clearly acknowledged the need to transcend the status quo in global climate politics. The major means to attain this was seen in a reform of international legal structures to pursue targets at the 2020, 2030 and 2050 time horizons. While this points to a belief in the 'manageability' of climate change—*if* the right reform measures are taken and implemented—the EU's attachment to 'cost-effectiveness', and notably market-based measures that are strongly anchored in the predominant 'sustainable growth' rationale embodied in the Europe 2020 strategy, further suggests that its positions are primarily an expression of a reformist understanding of environmental justice.

Interestingly, the EU's post-Paris climate discourse seems to be going a step further, underscoring that the dividing line between a reformist and a transformational justice position is indeed thin. Following the science-guided logic with which the EU has approached the climate topic for more than 20 years, which led to its promotion of the 2 °C target, it would be politically difficult for it to ignore the Paris Agreement's more ambitious 1.5 °C target. The Commission's November 2018 'Clean Planet for All' communication therefore takes the October 2018 IPCC Special Report on this temperature goal into account, which argues that 'Pathways limiting global warming to 1.5 °C … require rapid and far-reaching transitions in energy, land, urban and infrastructure … and industrial systems. These systems transitions are unprecedented in terms of scale' (IPCC 2018). Sketching the EU's 2050 climate strategy, the 'Clean Planet' communication contends: 'Immediate and decisive climate action is essential' (European Commission 2018, p. 1). This observation in itself, which contrasts with the prior use of the term 'urgent action', evokes notions of a transformational justice understanding. Key claims of the communication then combine elements of past discourse with novel aspects: 'The aim of this long-term strategy is to confirm Europe's commitment to lead in global climate action and to present a vision that can lead to achieving net-zero greenhouse gas emissions by 2050 through a socially fair transition in a cost-efficient manner. It underlines the opportunities that this transformation offers to European citizens and its economy, whilst identifying challenges ahead' (ibid.: 3). While the leadership ambition and cost-effectiveness are reminiscent of earlier documents, the target itself is more ambitious. The reference to 'transformation' is also less equivocal than the ubiquitous use of the term 'low-carbon transition'. At the same time, 'the proposed Strategy does not intend to launch new policies', which in any case would need to be negotiated between the member states and the Parliament (ibid.: 3). This and the list of measures proposed to reach 'decisive climate action' again point to a rather reformist stance. The Commission's position was reiterated by the Foreign Affairs Council in early 2019 (Council of the EU 2019). Its 'conclusions on climate diplomacy' depict climate change as a 'direct and existential threat' of 'extreme urgency', which 'requires rapid and far-reaching societal transformation, environment protection and economic transition' (ibid.: 1, 4).

Altogether, from a long-term perspective, the EU has discursively solidly committed to the protection of NHN via ambitious climate diplomacy that must be considered as reformist, although more recently it has straddled the dividing line with a transformational justice perspective.

EU Practice

The EU's climate diplomacy practice has arguably evolved more than its key positions across time. Prior to the 2009 Copenhagen summit, relevant EU external action was primarily about 'leadership by example' (Schunz 2014). Especially as of the late 1990s, the EU designed internal policies on climate change that strongly informed its external action. Central to this position was the message that the EU can reform, ecologically modernise and decouple growth from emissions (Machin 2019). The way the EU engaged in its multilateral and bilateral climate diplomacy was thus primarily about explaining its internal policies to third countries (Schunz 2014). At its heart was, as of the mid-2000s, the ETS.

In the post-COP 15 context, this practice has visibly evolved (Schunz 2019). No longer a key player given its declining emissions relative to other major emitters, the EU has had to adapt to the changing global context. It has done so especially by diversifying its outreach strategy in terms of both target actors and forms of interaction, transcending the mere showcasing of its 'model' and addressing other parties' interests. Concretely, the EU's post-COP 15 outreach strategies pursued two intertwined objectives: they were aimed at committing other parties to progress in the multilateral regime context *while* seeking practical advancements through hands-on cooperation with non-EU partners.

In this vein, the EU has reached out to the developing countries before COP 21, and later for implementing the Paris Agreement. This has helped create the 'High-Ambition Coalition' that shaped the target included in the Agreement (Oberthür and Groen 2018). To enable this and make additional progress, notably with regard to countries' NDCs, the EU has intensified its cooperation with especially small-island states, African and Latin American countries—linking climate and development policies—through dialogues, capacity-building measures and support to climate mitigation and adaptation processes (for overviews, see, for Latin America: Dominguez 2018; for Africa: Adelle and Lightfoot 2018). It has also begun to systematically demand the Paris Agreement ratification of countries with which it concludes trade agreements. Moreover, the EU has intensified ties with the major emerging economies China and India (Belis et al. 2018). Based on practical cooperation around emissions trading (in the Chinese case) and energy technologies (in both cases), it has been able to stabilise climate-specific relations and create a high level of trust that was instrumental to the adoption of the Paris Agreement and its implementation (ibid.).

Assessing the EU's climate diplomacy across time from a justice perspective, its behaviour comes closest to that of a reformist justice understanding. Initially, it was rooted in a strong conviction that its own model (based on a reformist justice understanding) would provide the basis for persuading others to reform in similar ways and to jointly agree on international law that would enshrine the commitment to global reform in line with the 2 °C target. The EU's more recent strategy, though broader, remains reformist in both its objectives (pursuing a reform of the climate regime) and means (promoting, e.g., emissions trading).

Biodiversity

The loss of biological diversity is one of the main challenges of the Anthropocene: when it comes to genetic diversity it is said to have already crossed critical boundaries (Steffen et al. 2015). Referred to as the 'sixth extinction', the human-driven loss of species that are now 'critically endangered' would result in a mass extinction on a scale that has only happened five times in the last 540 million years (Barnosky 2011,

p. 56). However, this extinction is bound to be aggravated by ecological stressors such as rapid climate change (ibid.).

The global biodiversity regime, created for the purpose of halting that loss, is centred around the 1992 Convention on Biological Diversity (CBD) and its two protocols, the 2000 Cartagena Protocol on Biosafety and the 2010 Nagoya Protocol on Access and Benefit Sharing (ABS) of genetic resources. Contrary to its changing role in the climate regime, the EU has continuously been a key player in the CBD framework because several major powers have not ratified key treaties in the biodiversity regime, notably the United States.

Other, more issue-specific conventions equally play a role in the biodiversity regime. For instance, the EU is also a member of the Convention on the Conservation of Migratory Species and, since 2015, the Convention on International Trade in Endangered Species of Wild Fauna and Flora (CITES) (Kettunen 2018).

EU Discourse

The EU adopted Biodiversity Strategies in 1998, 2006 and 2011, all of which had an explicit external dimension (European Commission 1998, 2006, 2011). Considering that the first EU strategy was adopted only six years after the CBD, it is clear that the international biodiversity framework drove EU internal biodiversity policy rather than vice versa. The latest strategy acknowledges the benefits of biodiversity use while recognising the negative impact of EU economic activities on biodiversity loss around the world.

Concretely, the 2011 strategy sets out targets for 2020, which correspond to the international Aichi targets aimed at addressing causes, reducing pressures and enhancing benefits from biodiversity, alongside a 2050 'vision' (Kettunen 2018). Many of these targets heavily focus on the implementation of existing initiatives. For example, with target one, the EU aims to fully implement its Birds and Habitat Directives (European Commission 2011). Target six—'step[ping] up action to tackle the global biodiversity crisis'—touches most clearly on the EU's diplomatic action. Under this target, the Union aimed to ratify the Nagoya Protocol—which it also did in 2014. Furthermore, it stipulates improving the integration of biodiversity in trade and development cooperation policies (European Commission 2011).

In its approach to the negotiations on the post-2020 international biodiversity framework, the EU calls for a better implementation of the CBD and its protocols, aiming to include targets that are 'ambitious' but 'realistic' (Council of the EU 2018, para. 4). It continues to see mainstreaming of biodiversity into and across sectors as 'pivotal' for achieving these targets and stresses that 'transformational change is urgently required' (ibid.: para. 10).

Over time, the EU's discourse thus appears to correspond to a reformist justice perspective. While it does recognise the challenges related to biodiversity and even that urgent change is needed, the measures it proposes are primarily about an improved implementation of already existing policies.

EU Practice

The EU was a very active participant in the negotiations of the Convention on Biological Diversity and the Cartagena and Nagoya Protocols, dating back to at least a decade (Delreux 2012; Oberthür and Rabitz 2014). More recently, it has used a wide variety of tools to promote these biodiversity MEAs—most notably via issue-linkages and mainstreaming in its trade and development policies.

To begin with, the EU requires that the CBD, CITES and the Cartagena Protocol are ratified by beneficiaries of its Generalised System of Preferences + scheme (Morgera 2012). Moreover, it has attempted to mainstream biodiversity policy in Free Trade Agreements (FTA), with limited success. Few FTAs contain explicit and detailed articles on biodiversity, the ones with Colombia/Peru, Georgia and Moldova being the clearest examples (Kettunen 2018). In most other FTAs, the language used is vague, without references to increasing ambition either internally or externally. The EU-Japan Economic Partnership Agreement, which entered into force in February 2019, can serve as an example. The article on biodiversity covers about one page, while the relevant chapter in the EU-Colombia/Peru FTA takes more than three. The parties commit to respecting the CBD, its protocols and CITES, they are to encourage 'the use of products which were obtained through sustainable use of natural resources' and to 'exchange information' (European Commission 2019, p. 437–438). There are nevertheless no concrete commitments to increase the protection of natural resources.

Several EU budget lines are specifically dedicated to the protection of natural resources. Specific parts of the geographical programmes under the Development Cooperation Instrument (DCI) are aimed at 'natural resources management, including land, forestry and water' (European Parliament and Council of the EU 2014: art. 5(3)). As to the thematic programmes of the DCI, the Global Public Goods and Challenges (GPGC) programme is particularly relevant. In 2014–2020, this included a budget of €398–504 million aimed at protecting ecosystems, among others, through Biodiversity for Life (B4LIFE) and the Forest Law Enforcement, Governance and Trade (FLEGT) programmes (Adelle et al. 2018b). Also the European Development Fund, directly financed by the EU member states outside of the Union's budget and targeting former colonies, has a set of thematic objectives on the environment and natural resources. However, even if biodiversity protection has thus been increasingly connected to development, the EU's development policy continues to privilege poverty reduction (ibid.).

Finally, the EU also strives to play a leading role in the negotiations on the post-2020 international biodiversity targets. For example, it has joined the 'Coalition of the Willing on Pollinators', which aims to increase international cooperation and sharing of experience and forms part of a broader effort called the 'EU Pollinators Initiative' (European Commission 2018). Pollinators, particularly bees, have become a symbol of the 'sixth extinction' and internal EU measures (such as the banning of harmful neonicotinoids pesticides) have been put in place (Livingstone et al. 2018). Nevertheless, alternative pesticides may be equally harmful and the EU has been less active on those. Moreover, as of May 2020, seven EU member states—including Italy and Poland—have not yet ratified the Nagoya Protocol, meaning that the EU's credibility for the negotiations of the post-Aichi biodiversity targets is also compromised.

Altogether, the EU's practical engagement on biodiversity exhibits a desire to reform global governance in line with a reformist justice understanding, but more limited efforts when it comes to target-setting and mainstreaming.

Arctic Environmental Governance

Given its importance as a key regulator of the Earth's climate and home to significant populations of non-human species found nowhere else on the planet (IUCN 2013), but also the fragility of its ecosystems, the Arctic region is at the core of the discussions about the Anthropocene. The IPCC's 2018 Special Report on the 1.5 °C target (IPCC 2018) refers to the Arctic as a 'unique and threatened system' and a region at 'disproportionately high risk' (IPCC 2018, p. 13,11) that is currently suffering from a warming rate two to three times higher than the global annual average (IPCC 2018, p. 6).

Concrete attempts to protect the Arctic such as the European Parliament's call for 'international negotiations for the adoption of an international treaty … along the lines of the existing Antarctic Treaty' (European Parliament 2009), which guarantees that Antarctica remains a 'natural reserve, devoted to peace and security' (Article 2, Protocol on Environmental Protection to the Antarctic Treaty), were flatly rejected and later abandoned. Arctic (non-EU and EU) states felt the region should instead be governed by the less-constricting UN Convention on the Law of the Sea (UNCLOS). Consequently, EU efforts to protect NHN through a composite Arctic policy were scaled down on paper and in practice.

EU Discourse

The first key document on the Arctic published by EU institutions was the 2008 Communication 'The European Union and the Arctic region', drafted by DG Mare and DG Relex. It contained three priorities: the protection and preservation of the Arctic, the promotion of a sustainable use of resources subject to the adherence to 'strict environmental standards' and the fostering of multilateral governance in the region (European Commission 2008). On this last point, two concrete proposals were made: the 'establishment of a network of marine protected areas' and the discussion of 'possibilities at international level to promote measures for protecting marine Biodiversity in areas Beyond National Jurisdiction' (BBNJ) (ibid.:11).

The 2008 efforts to advance multilateral governance hit the wall of Arctic states' reluctance to see third parties involved in the region, and were not incorporated in the 2012 Joint Communication by the European Commission and the High Representative. This document opted instead for focusing on research, sustainable development and international cooperation 'based on existing international law (notably UNCLOS)' (European Commission and High Representative 2012, p. 17).

In 2016, the idea of creating 'marine protected areas' under the aegis of UNCLOS was resuscitated (European Commission and High Representative 2016). However, the expression 'sustainable use' of such biodiversity was included ibid. Moreover, despite the stipulation of respect for the 'precautionary principle', the Communication also embraced the goal of economic development (ibid.; Stępién and Raspotnik 2016).

In a report on the Arctic environment, the European Environment Agency (EEA) pointed out the inadequate addressing of this dilemma as a hurdle to the realisation of the 2050 European Commission's 'vision of a Europe that live[s] well, within the planet's ecological limits' (EEA 2017; European Commission 2014). Pointing out the EU's 'short-termism currently dominating political and economic thinking' with regard to the Arctic, it called for a 'reconfigur[ation] (of) systems of production and consumption so that they operate within planetary limits and thereby ensure the wellbeing of current and future generations' (EEA 2017, p. 22).

As long as the EEA's transformational environmental justice call for action remains an unheard minority position, the EU's discourse on the 'sustainable use' of the Arctic corresponds most closely to a reformist understanding of justice.

EU Practice

Despite the fact that the Arctic region was first acknowledged as a priority in the pre-Lisbon era (European Commission 2007a), the treaty reform kept the EU's hands tied in areas crucial for NHN protection, such as the choice of energy sources and the structure of the energy supply (Article 194 TFEU). A sign of the relevance of this issue is that a large share of the hydrocarbons consumed by EU member states originates in the Arctic, notably in Russia and Norway (Eurostat 2017, p. 15). At aggregate level, EU energy-related activities thus continue to exploit Arctic fossil fuels in ways that contradict its aforementioned climate diplomacy efforts and are detrimental for NHN in the region and globally. Additionally, economic growth in logistics, cruise tourism, fisheries and mineral resources, sectors that depend on shipping, is to continue, and EU member states are expectant, as evidenced in their official documents on the Arctic (De Botselier et al. 2018). Of these sectors, the EU only has an exclusive competence in the conservation of marine biological resources under its common fisheries policy (Art. 3(d) TFEU), which is why its room for active intervention is severely limited.

When it comes to addressing other threats to the Arctic environment, the EU has been more active. The Arctic Council (AC), the central regional forum established in 1996, has been documenting the impact on Arctic ecosystems of pollutants produced and used in distantly located places. Most of what the Arctic Monitoring and Assessment Programme (AMAP) considers as 'chemicals of emerging Arctic concern', that is, chemicals 'with characteristics similar to legacy pollutants, including a potential to reach the Arctic', are not 'subject to international (global) regulation' (AMAP 2017, p. 3). There are, however, two MEAs that cover, albeit to a limited extent, pollution from chemicals: the 2001 Stockholm Convention on Persistent Organic Pollutants and the 2013 Minamata Convention on Mercury, resulting from negotiations in which the EU acted as a leader (Biedenkopf 2018, p. 194–196). Although it is too early to evaluate the EU's role in the actual implementation of the Minamata Convention (De Botselier et al. 2018), its role in the nomination of chemicals such as Perfluorooctanoic Acid (PFOA) to be limited or banned under the Stockholm Convention has been successful (Chemicalwatch 2015).

Finally, Arctic ecosystems are also very vulnerable to Heavy Fuel Oil (HFO), the most consumed marine fuel in the region. Not only could a spill be pernicious, but the very combustion of HFO is the source of harmful emissions of pollutants such as black carbon. This component of particulate matter lessens the reflecting capacity of ice-covered areas and exacerbates climate change. To mitigate these risks, EU Arctic and non-Arctic members have been lobbying in the International Maritime Organisation for an HFO ban similar to the 2011 ban in the waters around Antarctica (De Botselier et al. 2018, p. 27). Negotiations are ongoing, and should be finalised by 2021. Similarly, the EU has been able to play a role in the preparatory committee for the equally ongoing negotiations of a legally binding instrument on the conservation and sustainable use of marine Biodiversity in areas Beyond National Jurisdiction, a treaty with potentially important repercussions for the Arctic (Young and Friedman 2018, p. 125–126).

Overall, EU practice in the Arctic oscillates between status quo-oriented activities (mostly by its member states) aimed at exploiting Arctic economic opportunities and genuinely EU-level efforts, driven by the EU institutions and some member states, aimed at reforming global institutions to better manage the environment in the region.

Comparative Summary

Across the three cases discussed here, a rather unequivocal picture of EU diplomacy's contribution to NHN protection emerges, pointing to a strong orientation towards reformist environmental justice positions.

In the case of climate change, the EU's diplomatic approach is decidedly reformist: the EU acknowledges the problem, and has long argued that it is manageable with the 'right' institutional reforms allowing for policies that help spur necessary transitions. While this understanding permeates its discourse and practice, most recently, elements of transformational justice positions are detectable in its discourse.

On biodiversity, while the EU has been relatively successful in shaping the global regime, uploading its internal policies to the international level via the existing MEAs (Groen 2019), it has not been as active in mainstreaming biodiversity objectives across other policies, and its practical commitment to strengthening international efforts to halt biodiversity loss appears also limited. By consequence, the EU's discourse and practice seem to be located in the reformist environmental justice category.

Finally, in the Arctic context, and in addition to what has been argued on climate change, contrasting approaches to Arctic policy are visible in both the discourse and the practice of the EU and its members states (see also De Botselier et al. 2018). They correspond to global justice understandings in the range from status quo to reformist.

EU Diplomacy and Non-Human Nature: Effects

This section discusses the effects of EU diplomacy on NHN, distinguishing between institutional and ecological effectiveness.

EU Institutional Effectiveness

In the climate context, the assessment of EU institutional effectiveness makes it necessary to consider the impact of its activities on the global climate regime. Three efforts were undertaken to reform the multilateral regime, resulting in the 1997 Kyoto Protocol, the 2009 Copenhagen Accord and the 2015 Paris Agreement.

The Kyoto Protocol was negotiated at a time when parties collectively believed in top-down legally binding solutions to 'manage' the climate challenge. With the adoption of this treaty and its targets for industrialised countries, the EU was able to attain some of its key positions, and was hence institutionally rather effective (Schunz 2014, p. 67–112). Given the nature of the outcome—a regime reform aimed at incremental steps towards targets that were not overly ambitious, especially if cost-effective 'flexible mechanisms' such as emissions trading could be used to meet them—EU institutional effectiveness embodied a reformist justice position. The Copenhagen Accord represented a weak, status quo-oriented outcome of a two-year regime reform negotiation process: a political declaration that testified more to the failure of these negotiations than to their effectiveness. In this context, the EU was clearly ineffective in the pursuit of its reformist goals, as aptly remarked by then European Commission President Barroso: 'in ambition, we were always leading, but … others were much more influential when it was about reducing the ambitions' (cited in Schunz 2014, p. 221). Finally, the Paris Agreement changed the nature of the regime, starting an open-ended process in which parties agreed to enhance ambition over time, based on a transparency framework and guided by the 2 °C/1.5 °C targets, but with NDCs anchored in national laws. In this process, the EU was institutionally effective: many of its positions—regarding the global target, the need for a legally binding transparency framework—were taken up in the Agreement (Oberthür and Groen 2018). The regime itself, and the elements of it that the EU had co-determined, clearly transcend COP 15's status quo outcome: targets are not only stated, but also ideas are presented on how they could realistically be attained, assuming all parties' continuous commitment and ambition. This institutional effectiveness corresponds to a reformist justice understanding, recognising the need for action that is not radical but incremental.

Further institutional effectiveness of the EU's climate diplomacy can be detected in the ways it has cooperated with other key players following the 2017 US announcement to withdraw from the Paris Agreement. In co-leadership with China and Canada, through institutionalised dialogue (European Commission 2017), additional commitments were made to the Paris framework, which importantly allowed for the operationalization of the Paris Agreement via a 'rulebook' adopted at the 2018 Katowice COP 24 (UNFCCC 2018).

When it comes to biodiversity, the EU's impact on the CBD and especially its two Protocols is comprehensively demonstrated in the literature: the Cartagena Protocol is widely credited as carrying the EU's hallmark, whereas the Union was also a key player in the talks on the CBD and the Nagoya Protocol (Groen 2019). However, in all cases, this was mainly the result of the EU's intermediate position between developed and developing countries and the absence of the US at the negotiation table, which left the EU as the lead industrialised player in the negotiations (ibid.). In the Nagoya Protocol negotiations in particular, its relatively conservative policy objectives were easier to defend than more reformist aims like the ones it had previously defended in the climate regime (Oberthür and Rabitz 2014). As the EU did thus not have a particularly progressive stance in either of the negotiations, and this contributed successfully to the final outputs in the form of international treaties, its non-negligible institutional effectiveness was the expression of an at best reformist justice position.

Regarding the governance of the Arctic, the EU's reformist environmental justice orientation has proven its very limited institutional effectiveness, leading to status quo outputs. The European Parliament's initial goal of adopting a treaty similar to the Antarctica Treaty was never seriously considered. Similarly, the idea of establishing a network of marine protected areas never translated into concrete policies. When it comes to BBNJ protection, treaty negotiations—supported by the EU—are underway, but their output and its implications for the Arctic remain to be seen.

Across the three cases, with the partial exception of the Arctic, the EU has thus demonstrated that it can be institutionally effective, uploading its policy objectives into international agreements. This institutional effectiveness is an expression of the EU's overarching adherence to a reformist justice position, which embodies the belief that institutional reform will ultimately lead to NHN protection.

EU Ecological Effectiveness

Does the EU's diplomacy make a difference when it comes to NHN protection globally? Does it contribute to stabilising 'greenhouse gas concentrations in the atmosphere at a level that would prevent dangerous anthropogenic interference with the climate system' (Art. 2 UNFCCC), a level which has been suggested to lie at the equivalent of 2 °C or better 1.5 °C? Does it contribute to the protection of biodiversity and the fragile Arctic environment?

On climate change, the notion of 'leadership by example' as a diplomatic strategy makes it first necessary to consider if the EU is achieving its internal objectives to be credibly able to 'show the way'. As discussed above, the EU should meet its 2020 and 2030 targets, but its policies currently fall short of internationally agreed targets. Globally, emissions have been steadily growing. The Kyoto Protocol, hailed as an EU 'diplomatic success', did not deliver due to the non-participation of certain Annex I countries, notably the United States, and a formal compliance of the participating parties that heavily relied on the use of flexible mechanisms (Shishlov et al. 2016). The Paris Agreement may inspire trust for better results: it could be a long-term signal for decarbonization, desired and supported by the EU. Based on currently pledged targets, however, global temperature increase is projected to lie between 2.7 and 3 °C (CAT 2019b). Far from the Paris targets, this cannot be considered ecologically effective.

In the context of biodiversity, in its internal policies, the EU has adopted progressive measures, such as eradicating neonicotinoids that have proven to pose serious danger to the life of bees and designating marine protected areas. However, its biodiversity strategy for 2020 falls short on almost all of its self-set targets (Commission 2015). In the light of this record, the EU has not (yet) been able or willing to promote very reformist policies globally. While, contrary to other major economies, such as the US, it is at the forefront of international biodiversity negotiations, its positions have veered to the moderately conservative rather than progressive side, which does not enable its diplomacy to positively contribute to enhancing the ecological effectiveness of global biodiversity policies.

The EU's ecological effectiveness in the Arctic context is closely linked to progress in the combat against climate change. Given that little progress has been made on this, and that on other key matters—BBNJ, chemicals, heavy fuel oil—progress has yet been limited, it remains overall elusive. In sum, and given its internal contradictions, the EU's efforts point to a status quo outcome when it comes to the environmental governance of the Arctic.

Across cases, the ecological effectiveness of the EU's activities corresponds to a reformist environmental justice position: progress regarding NHN protection in the medium term cannot be ruled out, but depends on the effective implementation of the institutional advances, which exist primarily in the area of climate change (with repercussions for biodiversity and the Arctic). As they only insufficiently deliver for the time being, EU ecological effectiveness—and ultimately thus its contribution to NHN protection—remains limited.

Conclusion

This chapter examined to what extent the EU's diplomacy—through its discourse and practice on major environmental issues and their global effects—contributes to NHN protection. To determine this extent we reasoned in justice terms, identifying four ways in which humans (in this case the EU as an ensemble of citizens) conceptualise their obligations vis-à-vis NHN.

We found that the EU's diplomacy is predominantly in line with a reformist understanding of global environmental justice (for an overview of the main insights, see Fig. 1). As its treaties, secondary law and policies mandate the Union to 'preserve, protect and improve the quality of the environment' (Art. 191.1 TFEU), its discourse often exhibits tenets of reformist, and at times—recently on climate change—even transformational justice positions. At the same time, its diplomatic practice tends to be reformist at best, embodying an instrumental view of NHN, and a commitment to its 'sustainable use' rather than its unwavering protection. Particularly conservative positions on aspects of biodiversity and Arctic governance even point to the conclusion that EU practice at times corresponds to status quo notions of environmental justice.

The EU's environmental diplomacy has displayed a rather high level of institutional effectiveness: it has been able to 'upload' important policy objectives into international treaties reforming the global climate and biodiversity regimes, as well as aspects of Arctic governance. These outputs of EU diplomacy correspond to a reformist justice perspective. However, they have to date not translated into ecological effectiveness: the outcomes of EU diplomacy when it comes to NHN protection are limited, corresponding de facto to a status quo justice position. They do not preclude NHN protection in the medium term, as especially the Paris Agreement architecture *may* allow for attaining its underlying objectives, but this heavily depends on the effective implementation of the institutional advances.

Overall, our attempt to pinpoint the extent to which EU diplomacy matches the demands of NHN in the Anthropocene locates its efforts in the sphere of the

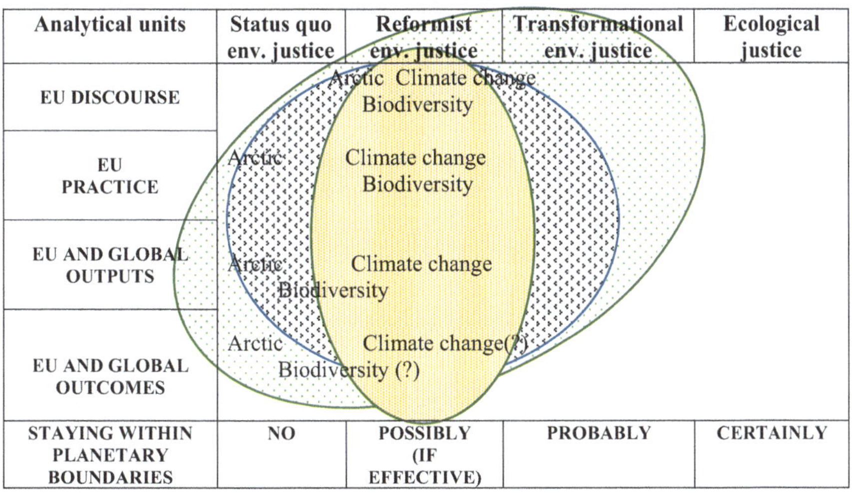

Fig. 1 EU diplomacy and global justice positions on the human-NHN relationship

'possible': it *might* contribute to the objective of remaining within planetary boundaries. However, and although its treaties would allow for it, the EU has so far not adopted positions—and has not empowered itself to the extent needed to defend NHN-related objectives vis-à-vis other (e.g. trade-related) objectives—to be able to act in ways that would enhance the likelihood that NHN is protected (bearing in mind that the EU operates in complex global contexts).

These findings have implications for the academic study and practice of EU (environmental) diplomacy. First, our approach reveals the shortcomings of current research, which—by focusing on what the EU does and how it fares as environmental diplomat—naturally tends to focus on analysing the EU as a reformist justice player, judging its action against self-set targets without sufficiently considering whether its undeniably resource-intense efforts are engaged in doing 'the right thing' and enhancing the chances for greater ecological effectiveness. More studies should investigate what the EU does once an institution has been created or reformed at the global level. A reform is never an end in itself, but only a means to attaining outcomes in line with the need to protect NHN.

Second, the findings clearly expose that EU internal policies and diplomacy are currently to a certain extent geared towards NHN protection, but this objective is not considered as important as other, notably (a specific type of) socio-economic aims related to the promotion of jobs and growth. With the European Commission's December 2019 'European Green Deal' proposal and its ambitious roadmap of (legislative) actions, including the proposals of a novel 'European Climate Law' (March 2020) and Biodiversity Strategy 2030 (May 2020), the Union has—at least on paper—sketched out a more holistic and transformational perspective of the human-NHN relationship, which resonates with the transformational justice-oriented discourse of the Commission's November 2018 'Clean Planet for All' communication, on which it also builds. However, at the time of writing, the broader take-up, adoption by the co-legislators and implementation of all these initiatives remains uncertain. It may be expected that, if successful, these initiatives are likely to also impact the EU's future 'green deal diplomacy' in ways that point to a stronger transformational justice orientation. Permeating its diplomacy to date, however, is the strong conviction that the current socio-economic system can effectively be reconciled with 'sufficient' levels of NHN protection, and that institutional reforms—in the EU and globally—suffice to prevent humanity from exceeding planetary boundaries. The limited ecological effectiveness of this approach makes it necessary to reflect on the conditions under which policies that would adequately address the challenge of planetary boundaries while promoting also social justice could emerge. On this point, the EU's elaborate treaty bases on matters pertaining to NHN protection and its political system allowing for transcending short-term electoral cycles contrast with the constraint that major initiatives require the consensus of all member states. Whether the EU continues to adhere to reformism in the current socio-economic context, which can only produce effective outcomes if reforms are successfully implemented, or opts for a different set of more ambitious policies is thus ultimately a question of political choice—a choice that represents the major contemporary challenge facing European societies desirous of 'matching the reality of the planet'.

References

Adelle, C., Biedenkopf, K., & Torney, D. (Eds.). (2018a). *European Union external environmental policy*. Cham: Palgrave Macmillan.

Adelle, C., Delputte, S., De Roeck, F., & Nichelson, S. (2018b). Environmental instruments in development cooperation: promoting better development and environmental outcomes? In C. Adelle, Biedenkopf, K., & Torney, D. (Eds.). *European Union External Environmental Policy* (pp. 81–101). Cham: Palgrave Macmillan.

Adelle, C., & Lightfoot, S. (2018). Africa: Searching for shared issues and overcoming asymmetries. In C. Adelle, Biedenkopf, K., & Torney, D. (Eds.), *European Union External Environmental Policy* (pp. 253–273). Cham: Palgrave Macmillan.

AMAP (2017). *Chemicals of emerging arctic concern*. Oslo: Summary for Policy-makers.

Barnosky, A. D., et al. (2011). Has the Earth's sixth mass extinction already arrived? *Nature, 471*(7336), 51.

Belis, D., Schunz, S., Jayaram, D., & Wang, T. (2018). Climate diplomacy and the rise of 'multiple bilateralism' between China, India and the EU. *Carbon and Climate Law Review, 12*(2), 85–97.

Biedenkopf, K. (2018). Chemicals: Pioneering ambitions with external effects. In C. Adelle, Biedenkopf, K., & Torney, D. (Eds.). *European Union External Environmental Policy* (pp. 189–206). Cham: Palgrave Macmillan.

Bull, H. (2012). *Anarchical society: A study of order in world politics*. London: Palgrave Macmillan.

Burke, A., Fishel, S., Mitchell, A., Dalby, S., & Levine, D.J. (2016). Planet politics: A manifesto from the end of IR. *Millennium: Journal of International Studies, 44*(3), 499–523.

Buzogány, A. (2018). Neighbourhood countries: Promoting environmental protection close to home. In C. Adelle, Biedenkopf, K., & Torney, D. (Eds.), *European Union External Environmental Policy* (pp. 233–252). Cham: Palgrave Macmillan.

CAT (2019a). Climate Action Tracker—EU. January 2019. Retrieved 25, Feb 2019, from https://climateactiontracker.org/countries/eu/.

CAT (2019b). Climate Action Tracker—Global Warming Projections. January 2019. Retrieved 25, Feb 2019, from https://climateactiontracker.org/global/temperatures/.

Chemicalwatch (2015). EU tells Stockholm Convention meeting of PFOA ban proposal. 7 May.

Clapp, J., & Dauvergne, P. (2011). Peril or prosperity?. In J. Clapp & P. Dauvergne (Eds.). *Paths to a Green World—The Political Economy of the Global Environment* (pp. 1–18). Cambridge, MA: MIT Press. 2nd ed.

Council of the EU (2015). Conclusions on the UN Climate Change Conference in Paris 2015. Retrieved 25, Feb 2019, from https://www.consilium.europa.eu/en/press/press-releases/2015/09/18/counclusions-un-climate-change-conference-paris-2015/.

Council of the EU (2018). Council conclusions on CBD meetings. 12948/18. Brussels. 9 October.

Council of the EU (2019). Climate Diplomacy—Council conclusions. Brussels. 18 February.

Damro, C. (2012). Market power Europe. *Journal of European Public Policy, 19*(5), 682–699.

De Botselier, B., López Piqueres, S., & Schunz, S. (2018). Addressing the 'Arctic Paradox': Environmental policy integration in the European Union's emerging arctic policy. *EU Diplomacy Paper* 03/2018. Bruges: College of Europe.

Delreux, T. (2012). The EU in negotiations on the Cartagena Protocol on Biosafety. In J. Wouters, Bruyninckx, H., Basu, S., & Schunz, S. (Eds.) *The European Union and multilateral governance: Assessing EU participation in United Nations human rights and environmental fora* (pp. 214–231). London: Palgrave Macmillan.

Delreux, T. (2013). The EU as an actor in global environmental politics. In A. Jordan & Adelle, C. (Eds.), *Environmental Policy in the EU—Actors, Institutions and Processes* (pp. 287–306). Oxon: Earthscan from Routledge.

Delreux, T., & Happaerts, S. (2016). *Environmental policy and politics in the European Union*. Basingstoke: Palgrave Macmillan.

Dominguez, R. (2018). Latin America: A Pragmatic approach and a modest contribution. In C. Adelle, Biedenkopf, K., & Torney, D., *European Union External Environmental Policy* (pp. 211–231). Cham: Palgrave Macmillan.
Ecomodernism (2019). *An Ecomodernist Manifesto*. Website. Retrieved 28, June 2019, from http://www.ecomodernism.org/.
EEA (2015). *SOER—Executive Summary*. Retrieved 25, Feb 2019, from https://www.eea.europa.eu/soer-2015/synthesis/report/0c-executivesummary.
EEA (2017). *The Arctic environment*. Copenhagen: European perspectives on a changing Arctic.
Ehresman, T., & Stevis, D. (2018). International environmental and ecological justice. In G. Kütting, & Herman, K. (Eds.). *Global Environmental Politics. Concepts, Theories and Case Studies* (pp. 103–120). London and New York: Routledge.
European Commission (1998). A European Community Biodiversity Strategy. COM(1998) 42. Brussels. 4 February.
European Commission (2006). Halting the Loss of Biodiversity by 2010 and Beyond—Sustaining Ecosystem Services for Human Well-Being. COM(2006) 216. Brussels. 22 May.
European Commission (2007a). An integrated Maritime Policy for the European Union. COM (2007) 575 final. Brussels. 10 October.
European Commission (2007b). Communication: Limiting Global Climate Change to 2 °C: The way ahead for 2020 and beyond. Brussels. 10 January.
European Commission (2008). The European Union and the Arctic Region. COM(2008) 763. Brussels. 20 November.
European Commission (2011). Our Life Insurance, Our Natural Capital: An EU Biodiversity Strategy for 2020. COM(2011) 244. Brussels.
European Commission (2014). 'General Union environment action programme to 2020. Living well, within the limits of our planet'. Directorate-General for Environment.
European Commission (2015). Mid-term review of the EU biodiversity strategy to 2020. COM(2015) 478. Brussels. 2 October.
European Commission (2017). EU co-hosts major international climate meeting with Canada and China. Brussels. 15 September.
European Commission (2018). EU Pollinators Initiative COM(2018) 395 final. Brussels.
European Commission (2019). EU-Japan Economic Partnership Agreement: Texts of the agreement. Retrieved 25, Feb 2019, from http://trade.ec.europa.eu/doclib/press/index.cfm?id=1684.
European Commission & High Representative (2012). Developing a European Union Policy Towards the Arctic Region: Progress Since 2008 and Next Steps. JOIN(2012) 19 final. Brussels. 26 June.
European Commission & High Representative (2016). An Integrated European Union Policy for the Arctic. JOIN(2016) 21 final. Brussels. 27 April.
European Parliament (2009). Resolution on the international treaty for the protection of the Arctic. Brussels. 26 March.
European Parliament and Council of the EU (2014). Regulation No 233/2014 on establishing a financing instrument for development cooperation for the period 2014–2020. Brussels. 11 March.
Eurostat (2017). EU energy in figures. Luxembourg: Publications Office of the EU.
Groen, L. (2019). Explaining European Union effectiveness (goal achievement) in the Convention on Biological Diversity: The importance of diplomatic engagement. *International Environmental Agreements: Politics, Law and Economics, 19*(1), 69–87.
Gysen, J., Bruyninckx, H., & Bachus, K. (2006). The modus narrandi. A methodology for evaluating effects of environmental policy. *Evaluation, 12*(1), 95–118.
Harris, P. G. (2016). *Global Ethics and climate change* (2nd ed.). Edinburgh: Edinburgh University Press.
Hopwood, B., Mellor, M., & O'Brien, G. (2005). Sustainable development: Mapping different approaches. *Sustainable Development, 13*(1), 38–52.

IPCC. (2018). Global Warming of 1.5 °C—an IPCC special report on the impacts of global warming of 1.5 °C above pre-industrial levels and related global greenhouse gas emission pathways—Summary for Policymakers. 6 October.
IUCN (2013). The Arctic Biodiversity Assessment. Retrieved 3, March 2019, from https://www.iucn.org/content/arctic-biodiversity-assessment.
Jordan, A., & Tosun, J. (2013). Policy implementation. In A. Jordan & Adelle, C. (Eds.), *Environmental Policy in the EU—Actors, Institutions and Processes* (pp. 247–266). Oxon: Earthscan from Routledge.
Kettunen, M. (2018). Biodiversity: Strong policy objectives challenged by sectoral integration. In C. Adelle, Biedenkopf, K. & Torney, D. (Eds.), *European Union External Environmental Policy* (pp. 147–166). Berlin: Springer.
Kelemen, D., & Vogel, D. (2010). Trading Places: The role of the United States and the European Union in international environmental politics. *Comparative Political Studies, 43*(4), 427–456.
Leopold, A. (1968/1948). *A sand county almanac*. Oxford: Oxford University Press.
Livingstone, E., Paravicini, G., & Marks, S. (2018). Commission a step closer to Juncker's pledge to save the bees. *Politico*, 27 April. Retrieved 25, Feb 2019, from https://www.politico.eu/article/european-commission-a-step-closer-to-junckers-pledge-to-save-the-bees-neonicotinoid-ban-pesticides.
Machin, A. (2019). Changing the story? The discourse of ecological modernisation in the European Union. *Environmental Politics, 28*(2), 208–227.
Meyer, J. M. (2008). Political theory and the environment. In J. S. Dryzek, Honig, B., & Phillips, A. (Eds.), *The Oxford Handbook of Political Theory* (pp. 773–791). Oxford: Oxford University Press.
Morgera, E. (2012). The trajectory of EU biodiversity cooperation. In E. Morgera (Ed.), *The external environmental policy of the European union: EU and international law perspectives* (pp. 235–259). Cambridge: Cambridge University Press.
Oberthür, S., & Groen, L. (2018). Explaining goal achievement in international negotiations: the EU and the Paris agreement on climate change. *Journal of European Public Policy, 25*(5), 708–727.
Oberthür, S., & Rabitz, F. (2014). On the EU's performance and leadership in global environmental governance: the case of the Nagoya Protocol. *Journal of European Public Policy, 21*(1), 39–57.
Oxford Dictionary (2019). *Nature*. Retrieved 25, Feb 2019, from https://en.oxforddictionaries.com/definition/nature.
Rockström, J., & Klum, M., with Miller, P. (2015). *Big world, small planet: Abundance within planetary boundaries*. New Haven: Yale University Press.
Schunz, S. (2014). *European Union foreign policy and the global climate regime*. Brussels: Peter Lang.
Schunz, S. (2019). The European Union's environmental foreign policy: From planning to a strategy? *International Politics, 56*(3), 339–358.
Shishlov, I., Morel, R., & Bellassen, V. (2016). Compliance of the Parties to the Kyoto Protocol in the first commitment period. *Climate Policy, 16*(6), 768–782.
Steffen, W., Richardson, K., Rockström, J., Cornell, S. E., Fetzer, I., Bennett, E. M., et al. (2015). Planetary boundaries: Guiding human development on a changing planet. *Science, 347*(6223), 1259855.
Stępień, A., & Raspotnik, A. (2016). *The EU's new Arctic Communication: not-so-integrated, not-so-disappointing? ArCticles*. Rovaniemi: University of Lapland.
UNFCCC (2015). Adoption of the Paris Agreement. FCCC/CP/2015/L.9/Rev.1. 12 December.
UNFCCC (2018). Preparations for the implementation of the Paris Agreement and the first session of the Conference of the Parties serving as the meeting of the Parties to the Paris Agreement. Decision -/CP.24. 19 December.
Wurzel, R. K. W., Connelly, J., & Liefferink, D. (Eds.). (2017). *The European Union in international climate change politics. Still taking a lead?*. Abingdon: Routledge.
Yin, R. K. (1984). *Case study research. Design and methods*. Thousand Oaks, CA: Sage.
Young, M., & Friedman, A. (2018). Biodiversity beyond national jurisdiction: Regimes and their interaction. *AJIL Unbound, 112,* 123–128.

Trust and the WWF in the Arctic

Danita Catherine Burke

Abstract The perception of International Non-Governmental Organisations (INGO), particularly environmentally focused ones, is mixed throughout the North and the Arctic. In the North American North and Arctic, the legacy of the anti-sealing campaign has left an indelible mark that reinforces local views that protection of non-human nature is the environmental INGO priority. This chapter explores the questions: how do you develop a trusted reputation as an INGO operating in the North and how does perceived trustworthiness influence organisational capacity? It argues that balancing the human and non-human aspects of the Arctic and North is central to INGO ability to be trusted and seen as trustworthy in those parts of the world. The ability of an INGO to demonstrate that it accounts for both dimensions is reinforced by its brand. Lying at the heart of the brand, and trust in it, are organisational legacy and networks. Using the case study of the World Wide Fund for Nature/World Wildlife Fund (WWF), this chapter demonstrates that legacy and networks, with the access they can provide, are essential to INGOs ability to demonstrate to different audiences in the North and Arctic that they can work with them. The primary audience for INGO work used in this piece is the Arctic states and their representatives to the Arctic Council, the Arctic region's pre-eminent forum for environmental protection and sustainable development discourse.

Introduction

The Arctic region has suffered from many misunderstandings about it. A central theme in representations of it is the idea of the pristine. The idea of the pristine focuses on the non-human nature of the region. Instead of seeing the Arctic as a lived-in space and homeland, it is portrayed as an empty space, a *terra nullius* (Steinberg et al. 2015, p. 12). The pristine narrative has become immortalised in art, film and poetry over centuries and an unintended, or perhaps unconsidered, repercussion of the popular narrative is that it has marginalised, and in some cases, removed, the

D. C. Burke (✉)
University of Southern Denmark, Odense, Denmark
e-mail: burke@sam.sdu.dk

J. C. Pereira and A. Saramago (eds.), *Non-Human Nature in World Politics*,
Frontiers in International Relations, https://doi.org/10.1007/978-3-030-49496-4_13

people that live there from depictions, discussions and decision-making about their homelands, feeding into their disenfranchisement (Grant 1989, 1998, 2001; Burke 2018). While not the only actors to make such depictions popular, International Non-Governmental Organisations (INGOs), particularly environmentally focused INGOs, perpetuate the idea of a pristine Arctic. It is an easy mistake to make, especially as it is a commonly held connotation of the Arctic by those not in or from the region and a very marketable idea as epitomised by imagery of a lone polar bear meandering through an open, tree-less landscape or marooned on a shrinking piece of ice adrift with the flowing ocean surrounding it for miles.

In particular, past INGO campaigns and movements, such as the anti-sealing movement in Canada in the 1970s and 1980s, reinforced a perception amongst northerners that INGOs see nature as more important than people. The anti-sealing movement destroyed a core part of traditional land-based economies in many Northern and Indigenous communities in the North American North, sometimes intentionally (e.g. protests against Newfoundland and Labrador commercial hunting) and sometimes not (e.g. Inuit subsistence hunting economies) (Phelps Bondaroff and Burke 2014). Regardless of the initial intent and objectives, the implementation of the campaign took on a deep ecology bent that saw the presence of humans in the non-human world as interference causing damage since 'the future health of the eco-system is of primary importance' (Lee 1995, p. 35) and 'all sealing is wrong' (Phelps Bondaroff and Burke 2014, p. 174). Nobel as the ideal of seal and animal protection may have come across at the time with the images of blood-soaked white ice-sheets making for stark imagery that played into the idea of the pristine being destroyed, what environmentalist did not foresee, or think ahead about, was the fallout of the movement. It caused the loss of traditional social, economic and cultural structures with major long-term consequences for local peoples, such as international cultural shaming of traditionally well-reputed local peoples, major job loss in rural and isolated areas with limited economic diversification options, and social structure decline due to unemployment and outmigration (Phelps Bondaroff and Burke 2014).

The legacy of the anti-sealing campaign is so strong in parts of the Arctic and North that the view that non-human nature is the INGO priority exists. This perception has been reinforced by past experience. The perception, however, fails to account for the non-monolithic nature of INGOs and the spectrum of organisations that try to work on finding a balance for human and non-human nature co-existence, thereby challenging the pristine narrative, while recognising a need for change in the human/non-human relationship for sustainable and symbolic interaction (Phelps Bondaroff and Burke 2014).

For an INGO to positively stand out in the Arctic region, their brand and reputation must distinguish itself by providing a clear balance between the human and non-human dimensions of the region, acknowledging the peoples as well as the environment of their homeland as important. The ability of an INGO to signal the human and non-human dimensions of its work and cultivate trust in the duality of their focus is reinforced by the INGO brand with branding being a central component to their organisational capacity building and stakeholder trust in their work. Lying at the heart of the brand, and trust in it, are organisational legacy and networks. Legacy and networks, with the access they can provide, are essential to INGOs and perhaps one of the most high-profile examples of an INGO that has mastered and guarded their brand is the World Wide Fund for Nature/World Wildlife Fund (WWF). For this chapter, I will explore one part of the WWF, its Global Arctic Programme.[1]

The Global Arctic Programme is the WWF's circumpolar programme started in 1992. According to their website: 'WWF's Arctic Programme is the only circumpolar Environmental NGO present at the Arctic Council, where we hold observer status'; (WWF n.d.) a position it has held since 1998 (Arctic Council 2015a). This status, as the only environment INGO, has changed somewhat with the 2017 admission of Oceana and the National Geographic Society (Arctic Council 2015a), with the latter, for example, also being a 'global nonprofit organisation that pushes the boundaries of exploration' resulting in the WWF now sharing the Arctic Council spotlight with other high-profile global non-profits active in the Arctic (National Geographic n.d.). Regardless, the fact remains that the WWF has a very strong brand recognition in Arctic affairs. The WWF's Arctic Programme stands out for its work, especially for its work in the Arctic Council, which is the Arctic region's pre-eminent regional forum for environmental protection and sustainable development discourse (Arctic Council 2015b).

In the Arctic region, the WWF and its Arctic Programme have a solid reputation with Arctic states. The WWF is known as a key contributor to sustainable development and environmental protection discussions, for having a permanent presence in the region through offices and personnel, for demonstrating recognition for the need to consult with Indigenous peoples in its work and plans, and for strong working relationships with states and many corporate actors. These attributes have contributed to the WWF Global Arctic Programme's generally high status with Arctic states as a positive example of how NGO observers to the forum should participate (Burke and Phelps Bondaroff 2019). To really understand why the WWF Global Arctic Programme is received well in the Arctic and the Arctic Council, an exploration is needed of the questions: how do you develop a trusted reputation as an INGO operating in the North and how does perceived trustworthiness influence organisational capacity?

There are many facets to trust building and trustworthiness, but in this chapter, I focus on exploring two major components—legacy and networks. I argue organisational legacy is irrevocably tied to reputation, and as future behaviour cannot be predicted, trust in an organisation is largely situated within the context of the past. As such, the past is defining in organisational trust building and the legitimacy that an organisation may obtain as a result. Networks feed into organisational trust and perceived trustworthiness by aligning organisations with access to people, resources and opportunities to prove themselves, while also playing the role of audience to

[1] I acknowledge that the role of Arctic Indigenous peoples is very important in Arctic governance and the Arctic Council. A fuller exploration of WWF, trust and the Arctic region is in progress in the form of a monograph with Manchester University Press which will provide more insight into this dimension of Arctic cooperation and trust. At this time, however, this piece focuses on the Arctic states as the primary audience (trustors) in this analysis of the WWFs (trustee) work.

present actions which are tomorrow's past. As such, networks serve the dual purposes of cultivating potential access and audience.

The Project and Data Collection

This chapter is part of my project titled, 'Breaking the Ice: INGOs as Arctic Council Observer Status Applicants (ACOSA)' funded by the European Union's Horizon 2020 research and invocation programme under the Marie Skłodowska-Curie grant agreement No. 746312. The project explores how international non-governmental organisations balance their organisational pursuits with expectations of conduct as pre-conditions for their involvement in institutions and forums, specifically within the Arctic Council. This project was inspired by my previous project (2016–2018), supported by the Carlsberg Foundation, on diplomacy in the Arctic Council.

For this project, semi-structured interviews were conducted with 37 people in between April 2018 and March 2019. The purpose of these interviews includes: to give insight into the types of people attracted to NGO work and how NGO work has been carried out; what it is like to work with, or for, an NGO; perceptions of NGOs and their role in global governance and the Arctic; and Arctic Council observer evaluations procedures and factors that influence application assessments. Additionally, interview data collected in 2016 and 2017 for the Arctic Council diplomacy project influences this project. These interviews were conducted throughout the Arctic region, with state and Indigenous representatives, as well as academics, researchers and NGO input. In total 66 interviews were conducted, 7 of which were research design and advisory conversations. A consolidation of this research can be found in the book, *Diplomacy and the Arctic Council* (Burke 2019).

In 2018–2019, interviews were conducted with: Arctic Council officials that have, or had, been involved in observer application assessments; representatives from the World Wide Fund for Nature/World Wildlife Fund (WWF); representatives of Greenpeace with a focus on people in Arctic state offices and/or who have, or had, worked on the 'Save the Arctic' campaign; Permanent Participant and Indigenous community representatives; and other individuals such as a consultants to the Arctic Council with years of Arctic advocacy experience; and a couple of academics, including a retired academic with over 20 years of experience researching, and who had consulted for the Arctic Council.

The conversations and interviews were conducted in person in various locations throughout the Arctic states and a select number of other locations, as well as by telephone and over Skype. As part of the fieldwork, which was primarily between September 2018 and January 2019, I travelled to seven of the eight Arctic states, Switzerland and the Netherlands. I did not travel to Russia for this project. After considering the potential risks and gains for myself and for my project participants, I decided to do interviews with Russian representatives by telephone and Skype. In total 11 people from WWF, from the Global Arctic Programme and from Arctic state offices were spoken to as part of this project. I also spoke to nine Arctic state officials who were closely involved, or formerly involved, in their state's assessment of observers and representation at the Arctic Council. In the chapter all interviews are referenced using generic pseudonyms. In terms of direct citation in this chapter, mainly interview data from Arctic states officials collected in 2018 and 2019 is used.

In the interest of transparency, it must be noted that the majority of the WWF personnel had originally consented to the use of their names in the project and those that contributed an on-the-record conversation or interview were transparent and professional in their reflections and manner. Three individuals, however, requested pseudonyms but were comfortable being identified as associated with WWF and four requested to approve quotes before publication. To best respect and accommodate their participation requests and to protect identities, particularly as the Global Arctic Programme is a finite and identifiable group (see Table 1), I have concealed the identities of all WWF and Arctic state project participants, minus having their permissions to acknowledge their broad Arctic state and WWF affiliations in the attributions. The purpose is to help protect the participation of those who did not want their identity disclosed in publications associated with the ACOSA project, but to also acknowledge that the interviews were with people directly involved in aspects of the organisation's Arctic work. Additionally, to further protect identities, some references to interviews have minor modifications to remove potentially identifying information from inclusion.

The Interconnectedness of Trust and Networks

For INGOs seeking to develop their networks and have access to people and places involved in decision-making, trust is a major component to the foundation of their brand recognition and ability to build relationships. This is especially true as 'trust only becomes relevant in strategic settings where cooperation cannot be taken for granted' (Brugger et al. 2017, p. 408). Trust, however, is not static and 'increases collaboration between participants when they perceive the relationship to be long-term and one in which all members will enjoy a benefit' (Porras et al. 2004, p. 347).

Russell Hardin (2006, p. 17) argues that trust is a cognitive notion in the same category as knowledge and belief. Therefore, you do not choose to trust; you either trust or you do not. As Marek Kohn states:

> Like love, trust is involuntary. You can't help whether you love or trust someone; you either do or you don't. Although situations may arise where one says, 'I'll just have to trust you then', that is not really trust. Rather it means 'I'll just have to take the chance that you will act in my interests,' or 'I will have to act in the same way that I would if I really trusted you' (Kohn 2008, p. 9)

The literature on 'trust', however, is a vast area of study with authors such as Hardin (2006), Sztompka (1999), Keating and Thrandardottir (2017), Laura Considine (2015), Brugger et al. 2017 and Kohn (2008) exploring different dimensions of how trust impacts all aspects of interaction. These dimensions include the theoretical dimensions of trust, donor-NGO relationships; public belief and accountability

Table 1 WWF Arctic Programme team

Name	Level of education	Country office	Name	Level of education	Country office
David Aplin	Masters	US	Monique Newton	Masters + trained mediator	Canada[a]
Elena Agarkova Belov	J.D. (Juris Doctor)	US	Colleen Parker	Masters	Canada
Elisabeth Kruger	*No data*	US	Kaare Winther Hansen	*No data*	Denmark (Greenland)
Sybille Klenzendorf	Ph.D.	US	Mette Frost	*No data*	Denmark
Margaret Williams	*No data*	US	Martin Sommerkorn	*No data*	Norway[a]
Peter Winsor	Ph.D.	US[a]	Nils Harley Boisen	Masters	Norway
Mark Brooks	*No data*	Canada	Tom Arnbom	*No data*	Sweden
Doug Chiasson	*No data*	Canada	Liisa Rohweder	Ph.D.	Finland
Leanne Clare	Masters	Canada[a]	Lotta Manninen	Ph.D.	Finland[a]
Paul Crowley	*No data*	Canada	Alexey Knizhnikov	*No data*	Russia
Andrew Dumbrille	*No data*	Canada	Ekaterina Khmeleva	Ph.D.	Russia
Martine Giangioppi	Masters	Canada	Alexey Kokorin	Ph.D.	Russia
Erin Keenan	Masters	Canada	Sergey Rafanov	*No data*	Russia
Brandon Laforest	Ph.D.	Canada	Mikhail Stishov	*No data*	Russia
Melanie Lancaster	Ph.D.	Canada[a]	Oleg Sutkaitis	*No data*	Russia
Martha Lenio	Ph.D.	Canada	Gert Polet	*No data*	Netherlands
Melissa Nacke	Masters	Canada	Femke Hilderink	*No data*	Netherlands
Rob Downie	*No data*	UK			

[a]WWF Arctic Coordinating Team; The Director of the WWF Arctic Programme is Peter Winsor
Information gathered from profiles found on the WWF Global Arctic Programme website—https://arcticwwf.org/work/team/ (WWF, *How we work*)
This chart represents the officially disclosed team make-up of the Global Arctic Programme as of 13 March 2019

of NGOs; bi-lateral state and organisational alliances and agreements; international organisation-state relations; nuclear deterrence.

For INGOs, the perception of trustworthiness is vitally important for their capacity to do their work, though perceptions of trustworthiness are often audience-specific. Hardin states that '[t]he best device for creating trust is to establish and support trustworthiness…without the latter, there is no value in trust' (Hardin 1996, p. 29). Since 'actors are unable to reliably predict the future behavior of others' (Brugger et al. 2017, p. 407), trustworthiness 'revolves around whether one party has positive expectations that another party will fulfil agreements where there is the possibility of a loss to the first party if the second party defects' (Keating and Thrandardottir 2017, p. 136). Therefore, in the absence of certainty, expectations and predictions based on past performance inform perceptions of an actor or entity and its relative trustworthiness.

Brugger et al. (2017, p. 408) argue that there are four key features of trust: (1) 'trust only becomes relevant in strategic settings where cooperation cannot be taken for granted'; (2) 'cooperative response by the trustee will produce certain gains for the trustor, whereas uncooperative behavior will cause damage and costs'; (3) 'due to both, uncertainty and the potential damage … the trustor's decision whether or not to cooperate becomes a risky decision'; and (4) 'a trustful expectation enables the trustor to bridge this risk'. The risk of misplaced trust can be massive, potentially undermining both the trustor and trustee, depending on the situation and on the audiences involved in supporting the work of both sides and those targeted with their narratives and counter-narratives in the wake of a violation of trust.

The risks involved in trust and the lack of future knowledge results in actors having their past ideas and actions play an important role in perception generation about the actor's level of trustworthiness, though that past is not an absolute guarantee of future trust. This process of looking to the past as a predictor of future behaviour is generally referred to as the organisation's reputation. Rightly or wrongly, the past can both bolster and hinder an actor's efforts to evolve and to engender trust with various audiences, whether they are possible members, alliances and partners, or donors.

Trust is the cornerstone of INGO legitimacy and efforts to implicitly or explicitly protect or cultivate that trust involve the organisation as a whole and its representatives, who effectively operate as diplomats or brand ambassadors for the organisations as part of the daily operations for many INGOs. What is more malleable in discussions about trust and legitimacy of INGOs is by whom do they want to be seen as trustworthy? While the category of INGO can imply homogeneity, is it comprised of a wide range of different organisations with various sizes, focus areas, objectives, membership types and repertories for pursuing their strategic objectives (Phelps Bondaroff and Burke 2014). As such, the audiences and targets of their work vary and to whom they wish to be seen as a trustworthy actor and to whom they wish to be seen as an adversary does as well. Therefore, the processes of cultivating and protecting their legitimacy through being perceived as a trustworthy actor depends very much on their networks and their access to networking opportunities with their key target audiences and stakeholders.

According to Keck and Sikkink (1999, p. 90): 'Networks are communicative structures. To influence discourse, procedures and policy, transnational advocacy networks may become part of larger policy communities that group actors from a variety of institutional and value positions'. When part of the larger policy communities, the process of networking transpires. Networking is the

> the development, maintenance, or use of social or professional contacts for the purpose of exchanging information, resources, or services. A professional network can be thought of as a web or series of interconnected webs—whereby links or ties exist between focal individuals and the individuals or entities with whom they share a connection or relationship. Networking typically occurs between two individuals but can be examined as an interaction between groups, companies, or institutions (Britannica Academic 2019).

Networking enables INGO members to have discourse with and facilitate the possible exchange of information, personnel, ideas and resources.

Networks help to 'promote norm convergence or harmonisation at the regional and international levels' and they also assist in 'norm implementation, by pressuring target actors to adopt new policies, and by monitoring compliance with regional and international standards' (Keck and Sikkink 1999, p. 90). In addition to helping to empower and legitimise an INGO to pursue its agendas, networks can also attribute and/or reinforce status through a process of mutual recognition; status means 'an effective claim to social esteem in terms of positive or negative privileges' (Weber 1978, p. 305). This process of mutual recognition of the status is important in trust building and perceptions of trustworthiness of INGOs, their representatives and their actions and outputs because, as Weber (1978, p. 306) explains, using the example of money and entrepreneurial position: 'Money and entrepreneurial position are not in themselves status qualifications, although they may lead to them, and the lack of property is not in itself a status disqualification, although this may be a reason for it'.

Actor and stakeholder engagement are needed for operations, and what defines local actors and stakeholders can vary depending on objectives and working arenas. This in turn can cause conflicts within an INGO more broadly and potentially impact trust in their work and negatively reflect on their status with certain audiences. As such, the bigger an INGO gets and the more complex its work and strategic objectives are as an organisation, the more at risk a holistic perception is of organisational trust. Consequently, the bigger the INGO, the more policing is required by the organisation, both internally and externally, to protect that trust.

The WWF and Its Global Arctic Programme

The WWF was founded on the notion of finding a more sustainable human/non-human nature relationship balance, and this notion has carried through the WWF actions and discourse thereby fostering a legacy, institutional mindset and approach towards problem identification and solution proposition. Attempting to maintain this balance means that WWF will not satisfy all, with some seeing it as not going far enough to protect non-human nature, with partnership and problem selection playing key roles in conveying that negative impression to some but also acting as positive signals to other actors that it is an organisation they can work with rather than feel like they are being lectured. Using the Arctic states as the primary audience in this case analysis, I demonstrate that the WWF is generally trusted by Arctic state representatives. But to understand why this is the case, you have to understand the WWF's legacy, the role that human and non-human nature play in the legacy, and how the organisation operates in order to conceptualise why this would appeal to the Arctic states.

WWF: An Overview

The WWF has always sought to have a science-driven and diplomatic approach towards improving global awareness about nature and encourage the sustainable co-existence of humans with non-human nature. The WWF was formed in 1961 as a conservation organisation part of the first wave of environmentalism in the 1950s and 1960s (Carter 2001, p. 132). From its inception, the WWF has been associated with elite figures in society, government and academia. It was founded by the *Morges Manifesto* which was signed 'by 16 of the world's leading conservationists, including biologist and African wildlife enthusiast Sir Julian Huxley, IUCN vice president Sir Peter Scott and director-general of the British Nature Conservancy E. M. Nicholson'. It also includes elite individuals such as His Royal Highness Prince Philip, the Duke of Edinburgh as an emeritus president. Other notable individuals associated with the WWF include former US President Dwight Eisenhower and former US Ambassador Philip K. Crowe (WWF, *History*).

The WWF is quite an institutionalised organisation. Carter (2001, p. 136) argues that '[c]onservation groups have become more institutionalised, in so far as they are now mass-membership organisations which have acquired greater legitimacy and better access to policymakers'. Its mass-membership includes doing work 'in 100 countries and is supported by more than one million members in the United States and close to five million globally' (WWF, *About us*). It also has primary and associated offices in over 40 countries (WWF, *How is the WWF run*; also see Raustiala 1997, p. 728).[2]

WWF International is the WWF's Secretariat and it has a Director General and they are based out of Gland, Switzerland. The secretariat's role is to 'lead and coordinate the WWF Network of offices around the world, through developing policies and priorities, fostering global partnerships, coordinating international campaigns and providing supportive measures in order to help make the global operation run as smoothly as it can' (WWF, *How is WWF run*). The WWF offices fall into two categories: national organisations and programme offices. National organisations

[2]Kal Raustiala states that WWF 'has major affiliates in twenty-eight key states'. While this is a number from 1997, it does raise the question that to what extent the offices in the 40 countries are considered 'major' operations is a point open to interpretation.

'can raise funds and carry out work autonomously' and programme offices 'must work under the direction of one of the independent WWF offices' (WWF, *How is WWF run*).[3]

The Global Arctic Programme was set up in 1992 amidst the structural environment and legacy of the WWF. It involves participation from the national offices in the US, Canada, Russia, Norway, Denmark/Greenland, Sweden, Finland, UK and the Netherlands (there is no WWF office in Iceland), and it is based out of the Canadian national office (previously it was based in Norway) (WWF, *How we work*). Though it is not uncommon to hear people refer to the WWF as an observer within the Arctic Council, technically it is the Global Arctic Programme and not WWF International, though coordination with the international office does occur. Coordination occurs to varying degrees between the Programme and WWF International and it is hard to pinpoint it as an actor external to the organisation but, from my interviews with WWF representatives, it has been noted that some coordination occurs to ensure that the work of the Global Arctic Programme continues to adhere to the broader WWF ideals, objectives and standards.

According to Carter (2001, p. 136): 'Most conservation groups are wedded to conventional forms of pressure. Their political campaigning focuses on the dissemination of information, lobbying and using the legal system to protect the environment'. The WWF very much follows this model of conventional pressure tactics and partnerships with a wide range of state, corporate and third-sector actors. As their website states on its 'about us' page: 'Together, in partnership with foundations, governments, businesses, communities, individuals and our more than six million members, we can conserve many of the world's most ecologically important regions' (WWF, *About us*).[4] It has worked with a range of stakeholders including the Colombian Foundation for Higher Education, the Conservation Foundation,[5] the Government of Malawi, Unilever, the Global Environment Facility, Walmart, the Government of the Russian Federation, and Coca Cola (WWF, *History*; also see Adderley and Mellor 2013).

Some see the insider approach of INGOs like WWF as 'selling out'. There were those that pushed against this insider approach towards advocacy, and this belief in a need for a different direction for advocacy contributed to the second wave of environmental NGOs in the 1970s, which included the rise of organisations like Greenpeace (Carter 2001). Greenpeace sees this traditional insider approach to advocacy as undermining organisational legitimacy. Taking money from governments and corporations is seen as a way for these actors to try and exert control over INGOs; neutralising them through dependency on funding and conflicts of interest.[6] If an INGO takes

[3]A specialist office is located in Brussels to work on EU related politics and activities and another office focused on international economic issues (e.g. the World Bank) is located in Washington, D.C.

[4]There are conflicting numbers provided by WWF about its membership, but it appears to range between 5 and 6 million mass members.

[5]The WWF and the Conservation Foundation merged in 1990.

[6]Many of the people I spoke to from Greenpeace pointed out their belief that their organisation is more free to pursue their work by not taking corporate or government money.

the money, some see it as hindering the INGOs ability to effectively protest the benefactor if they behave in a manner that is counter to the INGOs mandate.[7]

A great example of this kind of conflict between donor and INGO is the Sierra Club and The Clorox Co. In 2007, the Sierra Club partnered with The Clorox Co. 'to promote a new line of eco-friendly Clorox products in exchange for a share of the profits' (Associated Press 2008). Clorox built its brand on chlorine bleach (Barringer 2008) and the Sierra Club had fought them for decades (Associated Press 2008). The decision to partner, however, resulted in a bit of a mutiny within a segment of the Sierra Club and reputational damage for the INGO.

It was reported that the deal 'angered and embittered Club members all across the country [US]'. The result of the fallout was that the Sierra Club's US national board felt that it had to 'remove the leaders of the Club's 35,000-member Florida chapter, and to suspend the chapter for four years' because the 'leadership of the Florida chapter had been highly critical of the national board's decision in mid-December 2007 to allow The Clorox Company to use the Sierra Club's name and logo to market a new line of non-chlorinated cleaning products called 'Green Works' (Grist 2008). Supporters of the Sierra Club are publicly divided and vocal with some feeling that the credibility of the organisation has been corroded by the association with Clorox, which it actively sought out, and the fact that it was financially incentivised to lend its brand for money (Kamenetz 2008). As such, organisations like Greenpeace are active and overt about distancing themselves from governments and corporate actors by not accepting their money to ensure their capacity to criticise these actors and hold them to account (Comi et al. 2015).

WWF is not the Sierra Club, but it too, like all INGOs, must navigate partnerships and the boundaries between what it can do with others and its mandate, reputation and legacy. Presently, WWF International and some national offices in Asia and Africa are facing serious accusations of misconduct. In 2016, the humanitarian NGO Survival International launched a formal complaint against WWF to the Organization for Economic Cooperation and Development (OECD) (Barkham 2017; OECD Watch 2017) on behalf of the Baka Indigenous people of Cameroon (Survival International 2016). The case centred on WWF's leading role in the creation of several national parks and other protected areas in Cameroon on the land of Indigenous peoples, which included lands used by the Baka. Survival's complaint argues that WWF violated its own policy that 'any such projects must have the free, prior and informed consent of those affected'. Instead, Survival argues that WWF helped create conservation zones on Baka land without their consent and WWF repeatedly failed to 'take action over serious human rights abuses by wildlife guards it trains and equips' (Survival International 2017).[8]

More accusations emerged again in March 2019 with WWF being accused of funding guards who torture and kill in national parks in Asia and Africa. Among the allegations is that 'Indigenous people and villagers have been shot, beaten unconscious, sexually assaulted, and whipped by armed guards in parks in places like

[7] A deeper exploration of WWF, trust and its insider approach in its work is in progress.

[8] Work towards a resolution of Survival's complaint against WWF is ongoing.

Nepal and Cameroon' and that the organisation acts as 'global spymaster' (BBC News 2019; also see McVeigh 2019). WWF has hired an international lawyer and has begun an independent review to investigate and address the accusations (WWF, *Statement*).[9]

In contrast with problems in Asia and Africa, within the Arctic region and the Arctic Council, the WWF is largely seen as a legitimate and trusted actor. According to Mitchell and Stroup (2017, p. 398) '[a]n NGO's reputation is a critical determinant of its authority and ability to act independently or collaboratively to influence global politics'. WWF's willingness to consider and incorporate both the human and non-human dimensions of the Arctic in its approach help it to cultivate a reputation as a legitimate and trusted actor with the peoples and governments there who generally see the deep ecology approach towards the region as untenable but environmental and ecological protection and preservation as important. WWF, however, cannot take its reputation as a mid-road voice between human and non-human nature for granted and it is aware that trust is something that the organisation must nurture to maintain. It nurtures this trust through consultation processes with key actors and stakeholders, coupled with a local presence in the region through satellite offices, as well as through partnerships.

Working with WWF: Reflections from Arctic State Officials

In the Arctic Council, the Arctic states and Permanent Participants recognise the risk of admitting observers (Burke and Phelps Bondaroff 2019). At its most basic level, there is a risk of letting the wolf through the door if they misjudge an actor and admit it into their space. What one considers a 'wolf' depends on perspective, and for the Arctic states, the concern is that INGO actors that are too far on the non-human/deep ecology side of advocacy will not be open to hearing the types of interests and concerns being voiced by Arctic states and peoples. As a result, such organisations might disrupt attempts at workable solutions that the majority there would consider, preferring instead hard-line approaches towards what they consider acceptable outcomes in research and negotiations.

Prudently the Arctic states, with input from the region's Indigenous peoples, have formalised rules for admission of observers, conducted upon admission and have retained the power to expel observers (also see Arctic Council 2013). The legacy of actors, like the WWF, help to bridge the perceived risk of the admission of INGOs as observers by fostering expectations of them based on past work. The cooperative legacy and actions of WWF in the Arctic Council, for instance, has added to the forum's work by contributing expertise and scientific evidence to reports, Working

[9]WWF has launched an investigation into the allegations as of March 2019 and the investigation is ongoing as of September 2019. At no point in time during the research or writing of this book was there any accusation of misconduct by WWF, the WWF Global Arctic Programme, their national offices in the Arctic states, their representations or partners in the Arctic or the North raised to me by interviewees.

Groups and Task Forces. At the same time, WWF representatives push actors at the Council, using scientific evidence, to more strongly consider the long-term effects of different forms of economic development on regional and global ecosystems (Burke and Phelps Bondaroff 2019).

Generally, observations and experiences of Arctic state representatives support the view that WWF is seen as trustworthy and therefore warrant continued access to the Arctic Council and the officials there.[10] For starters, the WWF brand is very strong and entrenched in public awareness around the world. As one state official remember: 'When I grew up I remember that their brand was so visible. It sort of created, from childhood, a sort of bond with the organisation—the panda bear' (Arctic state representative 3).

A key factor that underpins the receptiveness of the Arctic states and their officials to the WWF Global Arctic Programme is the belief that the organisation is open to genuine dialogue; which is an experience that contrasts with some other experiences that the states have had with other INGOs.

> With WWF we are sort of used to having them around and part of the discussion. We always have, at least it has been my experience, very informative and enlightening discussions with them. You can have these exchanges of views and listen to them and they will listen to you. That is not the case with [some] other [INGO] applicants. They act like they know everything best and their opinion is the right one (Arctic state representative 2).

As another representative similarly reflected on their working relationship with WWF; 'You can disagree but have a rational discussion [with WWF] about your disagreements' (Arctic state representative 3).

The belief you can enter into dialogue with WWF representatives stems, in part, from the long legacy of its approach towards conservation. It also comes from WWF's history of having elite members of society, such as former royalty and politicians, as high-ranking members of the organisation (Arctic state representative 3). It comes down to the feeling that like is talking to like, to a certain extent; this feeling is in the sense that WWF, like many government officials and department representatives, tends to work with facts and figures and interact in certain behind-closed-doors ways typical in diplomatic environments, whereas some other more radical organisations may communicate in ways that government officials may be less likely to respond to positively.

One state official reflected that: 'I think in general [the perception of the WWF by the Arctic states] is quite good…WWF don't lead with advocacy. They lead with science. They lead with policy contributions' (Arctic state representative 4). The comment that WWF leads with science rather than advocacy came up repeatedly with interviewees, both from WWF and from the Arctic state representatives. What is generally meant by this statement is that WWF, through its Arctic Programme experts

[10] I acknowledge that the opinions of Indigenous Peoples representatives are equally valuable to that of Arctic state representatives in a review of trust in WWF and INGOs. This chapter focuses on the perspectives of Arctic state representatives. Examples of opinions expressed by some Permanent Participant representatives are forthcoming in another piece I am writing on WWF, trust and its role in the Arctic region.

and their partners, aims to have all their decisions informed by scientific experts and research and use this data to try and open dialogue through meetings, reports and consultation in order to connect decision-makers with the scientific data that underpins their suggestions and recommendations. They do not picket or interrupt meetings, stage media stunts, interfere in business practices (e.g. illegally board an oil rig or interfere in traditional hunting practices) or other such tactics to try and convey their ideas to relevant issue stakeholders. This approach towards engaging regional players has been relatively successful for the organisation and its Arctic Programme.

The experience with WWF's participation in the Working Groups and Task Forces of the Arctic Council, where much of the daily work of the forum unfolds, has also been very well received. The common words used to describe WWF in the state official interviews were 'professional' and 'prepared'. The impressions of their contributions were quite high. One official noted that they and their state delegation had 'a very good experience with WWF'. In particular:

> They have been very, I must say, important for the Arctic Council work and they have been very active, they have participated and they have something to say and contribute into the work. They participate also in the Working Groups and also in Task Forces. They are always prepared when they come to the meetings…they put a lot of effort in it and a lot of manpower into their participation in the Arctic Council. They are willing and contributing in a very positive and meaningful way (Arctic state representative 2).

This kind of experience was a common reflection. While it was not uncommon to have disagreements with WWF, in large part because of their approach towards cooperation, these disagreements could lead to a healthy and fruitful discussion or debate which pushed Arctic states and officials in their thinking about certain topics.

At the same time, the ability of WWF to lead with science and policy contributions is directly linked in the minds of some state officials to the networks WWF, and its representatives, have amassed over the decades. On one level this is in part because 'WWF wants to cooperate with governments and corporations' (Arctic state representative 1). On another level, state representatives are acutely aware of how far-reaching WWF's networks extend and the potential that access to these networks may hold.

> It's a global and recognized organization which has clear goals which is very connected to governments, other NGOs, scientific and business communities. They have a network of their own which is very influential. If you want to be connected to a NGO that has anything to do with environmental protection or climate issues, it's the WWF (Arctic state representative 1).

The sheer size of the WWF network, in the Arctic and beyond, means that state officials are aware that WWF has a degree of access, awareness and knowledge that can make it a very useful partner and the states are interested in tapping into that network to get things done.

At the individual representative level in the Arctic, the experiences with the representatives from the Global Arctic Programme also factor into the trust Arctic states and their representatives are willing to place in the organisation. Reflecting on their own working relationships, one state official commented that:

> I certainly have [long] standing relationships with a lot of WWF people who have been there for many years. Yes, they do have some turn over, every organisation does, but they have quite a few people who have been at this for a while and often have been at it longer than the individual Senior Arctic Officials and other government officials who rotate on a more regular basis among jobs in their respective governments. And so the WWF has an impressive institutional memory about how things have gone in the past and can be a source of advice, informally, to people like me in that sense (Arctic state representative 5).

Another official summarised the experience working with the Global Arctic Programme and its representatives: 'I just expect that when WWF says something it is probably their honest opinion and probably based on some fact or science' (Arctic state representative 3).

Conclusion

At the heart of developing a trusted reputation in the North and the Arctic is identifying who your target audiences are and using your organisational brand, as supported by its legacy and networks, to foster the INGOs ability to deliver on its work and in a way that resonates with those they are pitching that work towards. Trying to please all audiences is not a sustainable approach in INGO work. Even mainstream INGOs, like WWF, do not appeal to all. The organisation has found its niche and is fostering it by distinguishing itself in the eyes of Arctic states and peoples as separate from the negative environmental INGO legacy in the North and Arctic, while fully aware that distancing itself from the anti-sealing movement will not be a popular move with those audiences that see the outcome of the movement as a positive thing. WWF and its supporters, past and present, never petitioned for the outright ban on sealing products. Rather they sought industry reform and sustainable management of the hunt. Therefore, the organisation's approach to distancing itself from the more radical fridges of the anti-sealing movement that prioritised non-human nature and pushed for and got the industry to collapse is unlikely to alienate core WWF supporters.

WWF is finding a place for itself within the discourse of the Arctic and the North by striking a balance that works within its mandate and organisational focus between pursuing avenues to protect non-human nature (e.g. setting up the polar bear protection programme) and trying to acknowledge the history, culture, traditional knowledge and needs of the people that live and interact with that nature (e.g. Inuit and other Indigenous peoples). Maintaining this balance is not easy and this organisational direction is not for all environmental INGOs, with those with more deep ecology leanings, for example, being much less inclined to see merit in making allowances for the human experience, priorities and aspirations in the region.

Furthermore, in the pursuit of its work, WWF, like many other INGOs, has entered into partnerships with foundations, various levels of government, and corporations. This approach towards its work can come at a cost. Depending on your perspective on advocacy and the prioritisation of human versus non-human nature, the cost may result in dissatisfaction with the extent to which an organisation like WWF can pursue certain agendas given its links to other parties with their own interests, limitations and objectives. The fact remains, however, that at least with the representatives of the Arctic states, the ways in which WWF approaches its Arctic and Northern work and its attempts to balance human and non-human nature within it, are generally well received. While states, specifically the eight Arctic states, are just one broad audience cluster, in the matter of Arctic Council observer membership, they weighed the risk of admitted WWF and have chosen to trust the organisation. So far, the decision has only served to further integrate the WWF's high standing in the eyes of many actors in the forum.

A range of other audiences exists in Arctic governance and diplomacy: other observers (e.g. other INGOs, intergovernmental organisations, non-Arctic states who are interested in the region); Indigenous peoples throughout the Arctic region and the Permanent Participants that represent many of them at the Arctic Council and other WWF national offices in other parts of the world. As such, this chapter is not a definitive take on trust in the WWF internationally, rather a snapshot of trust in it with a specific audience at a given point in time. Ultimately trust is not static and WWF, like all INGOs appealing to particular audiences, they must take care to not undermine that trust as it is hard to gain and easy to damage.

References

Adderley, S., & Mellor, D. (2013). Who's influencing whom? Developing sustainable business partnerships. *EuroMed Journal of Business, 9*(1), 60–74.

Associated Press (2008, July 16). Some in Sierra Club feel sullied by Clorox deal. Retrieved October 10, 2019, from http://www.nbcnews.com/id/25708115/ns/us_news-environment/t/some-sierra-club-feel-sullied-clorox-deal/#.XN1bmI4zaUl.

Arctic Council (2013). Arctic Council Observer Manual for Subsidiary Bodies. Retrieved October 10, 2019, from http://hdl.handle.net/11374/939.

Arctic Council (2015a). Observers. Retrieved October 10, 2019, from https://arctic-council.org/index.php/en/about-us/arctic-council/observers.

Arctic Council (2015b). The Arctic Council: A backgrounder. Retrieved October 10, 2019, from https://arctic-council.org/index.php/en/about-us.

Barkham, P. (2017). Human rights abuses complaint against WWF to be examined by OECD. *The Guardian*. Retrieved October 10, 2019, from https://www.theguardian.com/environment/2017/jan/05/oecd-to-examine-complaint-against-wwf-over-human-rights-abuses-in-cameroon.

Barringer, F. (2008). Clorox courts sierra club, and a product is endorsed. *The New York Times*. Retrieved October 10, 2019, from https://www.nytimes.com/2008/03/26/business/businessspecial2/26cleanser.html.

BBC News (2019). WWF accursed of funding guards who torture and kill in poaching war. Retrieved October 10, 2019, from https://www.bbc.com/news/world-47444297.

Britannica Academic (2019). Networking. Retrieved October 10, 2019, from https://academic-eb-com.proxy1-bib.sdu.dk/levels/collegiate/article/networking/610311.

Brugger, P., Hasenclever, A., & Kasten, L. (2017). Trust among international organizations. In J. A. Koops & R. Biermann (Ed.), *Palgrave handbook of inter-organizational relations in world politics* (pp. 407–426). London: Palgrave Macmillan.

Burke, D. C. (2018). *International disputes and cultural ideas in the Canadian Arctic: Arctic sovereignty in the national consciousness*. London: Palgrave Macmillan.

Burke, D. C. (2019). *Diplomacy and the Arctic council*. Montreal and Kingston: McGill-Queens University Press.
Burke, D. C., & Phelps Bondaroff, T. (2019). Becoming an Arctic council NGO observer. *Polar Record, 54*(5–6), 349–359.
Carter, N. (2001). *The politics of the environment*. Cambridge: Cambridge University Press.
Comi, A., Lurati, F., & Zamparini, A. (2015). Green alliances: How does ecophilosophy shape the strategies of environmental organizations? *VOLUNTAS: International Journal of Voluntary and Nonprofit Organizations, 26*(4), 1288–1313.
Considine, L. (2015). 'Back to the Rough Ground!' A grammatical approach to trust and international relations. *Millennium Journal of International Studies, 44*(1), 109–127.
Grant, S. D. (1989). Myths of the North in the Canadian ethos. *The Northern Review, 3/4*(Summer/Winter), 15–41.
Grant, S. D. (1998). Arctic wilderness–and other mythologies. *Journal of Canadian Studies, 32*(2), 27–42.
Grace, S. E. (2001). *Canada and the idea of North*. Montreal/Kingston: McGill-Queen's University Press.
Grist. (2008, March 29). Sierra Club removes leadership of its Florida chapter. Retrieved October 10, 2019, from https://grist.org/article/the-clorox-debacle-continues/.
Hardin, R. (1996). Trustworthiness (Vol. 107, No. 1, pp. 26–42). The University of Chicago Press.
Hardin, R. (2006). *Trust*. Cambridge: Polity Press.
Kamenetz, A. (2008, January 9). Clorox goes green. *Fast Company*. Retrieved October 10, 2019, from https://www.fastcompany.com/958579/clorox-goes-green.
Keating, V. C., & Thrandardottir, E. (2017). NGOs, trust, and the accountability agenda. *The British Journal of Politics and International Relations, 19*(1), 134–151.
Keck, M. E., & Sikkink, K. (1999). Transnational advocacy networks in international and regional politics. *International Social Science Journal, 51*(159), 89–101.
Kohn, M. (2008). *Trust: Self-interest and the common good*. Oxford: Oxford University Press.
Lee, M. (1995). *Earth first! Environmental apocalypse*. New York: Syracuse University Press.
McVeigh, K. (2019). WWF accursed of funding guards who 'tortured and killed scores of people.' *The Guardian*. Retrieved October 10, 2019, from https://www.theguardian.com/global-development/2019/mar/04/wwf-accused-of-funding-guards-who-allegedly-tortured-killed-scores-of-people.
Mitchell, G. E., & Stroup, S. S. (2017). The reputations of NGOs: Peer evaluations of effectiveness. *Review of International Organisations, 12*(3), 397–419.
National Geographic. (n.d.). About us. Retrieved October 10, 2019, from https://www.nationalgeographic.org/about-us/.
OECD Watch. (2017). Survival International vs. WWF. Retrieved October 10, 2019, from https://complaints.oecdwatch.org/cases/Case_457.
Phelps Bondaroff, T., & Burke, D. C. (2014). Bridging troubled waters: History as political opportunity structure. *Journal of Civil Society, 10*(2), 165–183.
Porras, S. T., Clegg, S., & Crawford, J. (2004). Trust as networking knowledge precedents from Australia. *Asia Pacific Journal of Management, 21*(2004), 344–363.
Raustiala, K. (1997). States, NGOs, and international environmental institutions. *International Studies Quarterly, 41*(4), 719–740.
Steinberg, P. E., Tasch, J., & Gerhardt, H. (2015). *Contesting the Arctic: Politics and imaginaries in the circumpolar North*. London: IB Tauris.
Survival International (2016). Survival International accurses WWF of involvement in violence and abuse. Retrieved October 10, 2019, from https://www.survivalinternational.org/news/11107.
Survival International (2017, September 5). Survival-WWF OECD talks break down over tribal consent. Retrieved October 10, 2019, from https://www.survivalinternational.org/news/11800.
Sztompka, P. (1999). *Trust: A sociological theory*. Cambridge: Cambridge University Press.
Weber, M. (1978). Economy and society: An outline of interpretive sociology. Berkeley, Los Angeles, and London: University of California Press.

WWF (n.d.). About Us: Who we are. Retrieved March 13, 2019, from https://www.worldwildlife.org/about.

WWF (n.d.). History. Retrieved March 13, 2019, from https://www.worldwildlife.org/about/history.

WWF (n.d.). How is WWF run? Retrieved March 13, 2019, from https://wwf.panda.org/organization/.

WWF (n.d.). How we work: The WWF Arctic Programme. Retrieved March 12, 2019, from https://arcticwwf.org/work/team/.

WWF (2019). Statement on safeguarding human rights in conservation, and addressing allegations of human rights abuse. Retrieved September 15, 2019, from https://wwf.panda.org/wwf_news/wwf_independent_review_/?349754/Statement-on-safeguarding-human-rights-in-conservation-and-addressing-allegations-of-human-rights-abuse.

A Typology of Direct Action at Sea

Teale N. Phelps Bondaroff

Abstract Direct Action (DA) is a sometimes confrontational strategy employed by those seeking to bring about the change they would like to see in the world. It is typically applied within the boundaries of states, however, the high seas provide an ideal theatre for the use of DA at the international level. The most prolific practitioners of DA at the international level are marine conservation organizations. In recent years, there has been increased diversity of DA strategies. This chapter develops a typology of DA at sea, using the categories of traditional DA, service provision, monitoring/surveillance, deterrence and compellence. The latter three categories of the typology eschew the moral framing traditionally invoked by activists, in favour of legalistic framing. These forms of 'Direct Enforcement' (DE) frame ecologically harmful activity as criminal in nature, and propose DA as a form of law enforcement. In so doing, practitioners expand the role of non-state actors into a territory traditionally reserved for the state. While this expansion has the potential to put activists and governments into conflict, this is not exclusively the case. Several activist-state relationships are identified: acquiescence, antipathy, concurrence and collaboration.

Introduction

Transnational activists have typically been perceived as depending primarily on the soft power of 'advocacy' in order to achieve their desired ends. Advocacy typically relies on 'information politics' and 'moral persuasion', strategies aimed at indirectly changing behaviour by raising awareness of an issue among the general public (Keck and Sikkink 1998, p. 16; Epstein and Barclay 2013, p. 99). Often activists will use

I am grateful for the ideas, editorial assistance, and advice of Dr. Mette Eilstrup-Sangiovanni, Stan Phelps, Bryce Casavant, and Danita Catherine Burke. A portion of this research was conducted with the support of a Visiting Researcher Fellowship at the Centre for Global Studies, University of Victoria, and I am thankful for this assistance. I am especially indebted to Stephanie Ferguson for all of her support.

T. N. Phelps Bondaroff (✉)
The Idea Tree Consulting, Victoria, BC V8X 1M2, Canada
e-mail: hello@teale.ca

J. C. Pereira and A. Saramago (eds.), *Non-Human Nature in World Politics*, Frontiers in International Relations, https://doi.org/10.1007/978-3-030-49496-4_14

their positions as moral authorities to make persuasive appeals and to 'mobilize shame' in order to urge other actors to change their behaviour (Keck and Sikkink 1998, p. 24). Material leverage can also be brought to bear by organizing boycott and divestment campaigns, where activists will typically use persuasive techniques to mobilize other actors, typically members of the public, to harm the economic interests of a target (see for example Trentmann 2019; Sasser et al. 2006). When monitoring compliance, activists typically rely on 'naming and shaming', using information to draw attention to non-compliance with norms, standards, laws or agreements (see Keck and Sikkink 1999, p. 97). In adopting this approach, activists are seldom seen as enforcing norms or coercing compliance, rather they play the role of information brokers, increasing compliance indirectly through the strategic use of information.

In contrast to these strategies, an increasing number of activists have turned to Direct Action (DA), an approach that uses direct, sometimes confrontational, actions that actively seek to bring about the change they desire. DA has long been used by activists operating in domestic settings, but the study of its application on transnational issues and campaigns has been limited. Given that the efforts of most activists take place on land, and therefore within the jurisdictions of states, this is unsurprising. However, DA has been used by transnational marine conservation organizations at sea, and most notably on the high seas.

In recent years, the Sea Shepherd Conservation Society (SSCS) has risen in prominence as a practitioner of DA, using this approach as part of its many marine conservation campaigns. Founded by Paul Watson in 1977, the SSCS employs what it describes as 'innovative direct-action tactics to investigate, document, and take action when necessary to expose and confront illegal activities on the high seas' (SSCS n.d. 'Mission Statement'). Many of its tactics are confrontational, and have included ramming, boarding and sinking of vessels, and the use of prop-foulers, stink bombs and smoke flares (see Scarce 2006; Cameron 2009, p. 20; Darby 2007). The organization presents itself as an international conservation law enforcement organization, and claims to use 'aggressive non-violent direct action' in order to enforce international conservation law (Watson quoted in Stuart et al. 2013, p. 9). This type of framing is contrary to how most activists operating in the (international) arena frame themselves and their actions.

In examining the strategies employed by the SSCS in its Southern Ocean anti-whaling and Mediterranean bluefin tuna campaigns, Eilstrup-Sangiovanni and I posited that SSCS uses 'Direct Enforcement' (DE), a form of DA and a strategy whereby activists seek 'to compel compliance with international agreements through denial and coercion' (Eilstrup-Sangiovanni and Phelps Bondaroff 2014, p. 356). The contention is that in the international arena, groups employing DA often frame their actions as enforcement. We observed that the strategy did not appear to be unique to the SSCS, and noted a number of breakaways and offshoots from the SSCS, as well as groups independent of the SSCS, using versions of DE.

The use of DA as part of transnational marine conservation campaigns has been adapted to accommodate various political opportunity structures, issues and theatres. Organizations, such as The Black Fish/FishAct (TBF), have conducted undercover investigations, gathering information which is then shared with authorities and used to prosecute poachers. Community groups, such as the Mendocino Abalone Watch (MAW), conduct high visibility patrols of California beaches, deterring would-be poachers and confronting those found to be violating harvesting regulations. Even the SSCS, which often pitted itself against states in its earlier campaigns, is now working with West African states, offering its vessels as platforms for Santomean and Gabonese naval personnel to conduct fisheries monitoring and enforcement operations.

This chapter seeks to further our understanding of DA and its application by activists on transnational issues by surveying a wide range of marine conservation organizations which appear to be employing versions of DA, and by categorizing these various strategies into a typology: traditional DA, service provision, monitoring/surveillance, deterrence and compellence. While the use of DA is often confrontational, and has the potential to put activists and governments into conflict, this need not be the case. Recent years have seen collaboration between DE practitioners and states. The typology is further refined through the examination of several activist-state relationships, which are characterized by acquiescence, antipathy, concurrence, and collaboration.

The typology was developed through the monitoring of the operations of marine conservation organizations engaged in DA at sea over the past decade. Initial examination of the strategy of the SSCS, and the theory undergirding DE was developed through the use of participant observation conducted onboard a SSCS vessel as it participated in *Operation No Compromise*, the SSCS's 2010–2011 Antarctic anti-whaling campaign. Subsequently, the operations of organizations engaged in DA at sea were monitored both as a passive external observer and through more direct association on the part of the author, who has worked as a researcher, consultant and grant writer for a number of the organizations herein examined. The activities of these organizations were triangulated through press releases, news stories, and where possible, through informal interviews with participants.

This chapter begins with a brief overview of DA, offering a summary of its history, basic strategy and use in campaigns. DA is characterized as either constructive or obstructive depending on whether activists are seeking to prevent an action from occurring, or to enact a specific outcome. In presenting a definition of DA, this section explores the nuanced interplay between DA and image events. The subsequent section delineates the adaptations and variations that are made to both DA and its practitioners in order for this strategy to be effectively employed at sea. The chapter then outlines the typology of DA at sea, beginning with traditional DA, which is encapsulated by the conventional obstructive 'image events' of Greenpeace. The constructive counterpart to this type, service provision, is then detailed. Service provision entails activists working to clean and restore marine ecosystems.

The strategy of DE is then introduced, and the basic components of this subset of DA, as developed by Eilstrup-Sangiovanni and myself, are explained. In this iteration, DE is seen as relying principally on denial and coercion. Since this formulation however, practitioners have made numerous modifications to the strategy, and from these various forms, three sub-types of DE are identified and added to the overall typology of DA at sea. Examples from a range of marine conservation organizations are used to explore the overlapping sub-types of DE:

- Monitoring/surveillance, whereby activists gather information on illegal activities and then give this information to the authorities to support arrest and prosecution, or use the information to justify further types of DE.
- Deterrence, whereby activists' perceived threat capacity and presence serve to induce would-be criminals to refrain from undertaking a specific action or actions.
- Compellence, whereby activists undertake to stop illegal activities which are already underway.

The chapter then turns to the question of activist-state relations. DA has traditionally been seen as a strategy which puts activists in conflict with states due to its confrontational nature. While this is often the case, variations in activist-government relationships result in a number of different applications of DE. These applications of DE may vary in form, but each has been successfully used to achieve desirable outcomes on the part of activists:

- Acquiescence—the reluctant acceptance of activists' use of DE by states without overt objections, typically resulting when a state is disinterested in the issue at hand, or when a state is unwilling to engage on an issue but equally reluctant to actively frustrate activist operations.
- Antipathy—typified by conflict, escalating hostilities and efforts on the part of the state and activists to frustrate the actions of the other.
- Concurrence—where activists and governments share an equal concern for an issue, and conduct parallel operations. In these situations, there may be some unidirectional intelligence sharing, either overt or through signalling, with information typically flowing from activists to governments and not in the opposite direction.
- Collaboration—where activists and states actively cooperate in joint operations to achieve desirable outcomes, on issues of shared concern. These relationships typically emerge when states have a diminished capacity to act relative to the capabilities of activists.

The use of DA at sea is not restricted to organizations concerned with marine conservation issues. The chapter concludes with a call for additional research into these organizations as a means of testing the applicability of the typology to issues beyond marine conservation. Other areas ripe for future research are also identified.

Figure 1 is useful in conceptualizing the typology of DA at sea.

Direct Action

Direct action has its origins in the repertoires of the early radical labour movement (Tracy 1996, p. 150). Drawing on both activist and academic sources, a clear and useful definition of DA can be presented as follows: a form of action whereby

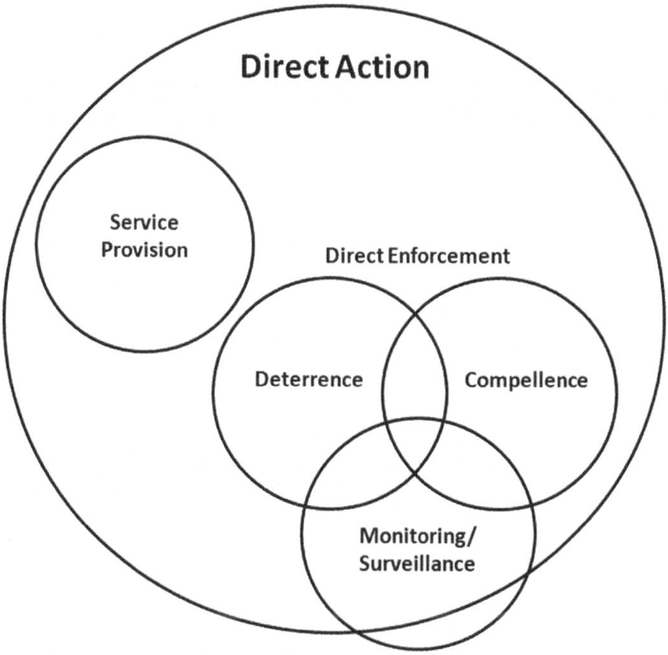

Fig. 1 A typology of direct action at sea

activists seek to directly enact the changes they desire or to stop actions to which they are opposed (see inter alia Graeber 2009, p. 207; Jordan 2002, p. 62). As Doherty, Paterson and Seel elaborate, '[n]ot content to try to influence politicians or institutions, protesters have seen direct action as disruption seeking to delay ecologically damaging projects and to escalate their costs' (2000, p. 2). In this way DA is not a form of protest, whereby activists entreat other actors to intervene on their behalf; instead activists take matters into their own hands, entirely bypassing mediators. As such, practitioners cannot be accurately described as 'protesters', and hence will be referred to as 'activists' in this chapter.

The goal of DA in the context of environmental activism is to directly protect non-human nature rather than appealing to some other actor to intervene on ones behalf. The basic strategy of DA entails activists employing a range of constructive or obstructive tactics to bring about change. The strategy behind constructive DA is straightforward; activists create or enact the thing that they want to see in the world. For example, those desiring increased food security and lower carbon meals may engage in 'illicit cultivation of someone else's land', otherwise known as guerilla gardening (Reynolds 2008, p. 3).

Obstructive variations of DA typically rely on two mechanisms: (1) stopping or preventing ecologically harmful activity directly at the source, and (2) increasing the costs of ecologically destructive activity to the point that it becomes economically unviable. The first component of DA takes the form of denial—the act of disrupting or delaying some activity. Forestry/anti-logging campaigns, for example, in the domestic setting, may use blockades to prevent access to a cut block. Other forms of denial in this context may include the damage or destruction of extractive equipment, also known as 'ecotage' (Phelps Bondaroff 2008). Such actions can also inflict direct and indirect economic impact on targets by increasing the cost of operation or discouraging investors (see Vanderheiden 2005, p. 444).

Efforts to establish a useful definition of DA that effectively differentiates it from other strategies are frustrated by those using the term to describe anything which appears militant or confrontational, and by activists who used the term in order to appear more militant or 'edgy'. However, as with constructive DA, not all DA is militant, and not all militant activism is DA. For example, tearing down barricades at a protest might be militant but not constitute DA (Graeber 2009, p. 207–208). Others have attempted to brand any form of dramatic symbolic politics as DA, where the drama and confrontational nature of DA is used to draw attention to an action or issue, or to demonstrate the depth of commitment felt by activists (Wapner 1995, p. 321). One might consider this to be the case, when organizations, such as Greenpeace, create powerful image events using DA, where the goal of the action is the creation of the image itself, and the direct goals of the DA secondary. This is exemplified by one Greenpeace activist who noted that 'Greenpeace believes that an image is an all important thing. The direct actions call attention to the issues we're involved in' (Steve Loper, Action Director for Greenpeace USA cited by Durland 1987). Some image events are also DA, but not all DA is an image event. As Graeber notes, 'it would be possible to have a secret direct action', because regardless of the attention it receives, an act of DA still accomplishes something direct (Graeber 2009, p. 204).

Ultimately, DA can be used to add drama to a conventional image event. However, the problem with defining DA as protest with an 'edgy' modifier is that this robs us of an effective term to describe actions whose purpose is not to appeal to an intervener and urge them to act, but to proactively seeking to bring about change, to 'build the future now' (Graeber 2009, p. 202). The discussion concerning the extent to which an action must prioritize achieving direct impacts as compared to serving as an image event, and the role this plays in classifying an action as either DA, or an image event, is a discussion for another place. For the purposes of this chapter, the focus will remain principally the DA aspect of actions, and upon actions with a greater emphasis on DA.

Conducting Direct Action at Sea

When considering the application of DA in a maritime theatre, variations and adaptations of repertoires used in land-based operations can be expected, as operating in this theatre has certain requirements and limitations, and presents activists with some unique opportunities. Operating at sea is expensive; ships, fuel, victuals, communication and navigation equipment, registration, certification are all requisites. Even the most sophisticated land-based DA cannot be compared, in terms of the cost, to purchasing and maintaining a vessel at sea. For instance, the *Bob Barker*, a patrol vessel operated by the SSCS, was purchased at a cost of $5 million USD (Murphy 2010), the custom-built Ocean Warrior cost the SSCS over €8.3 million (SSCS Australia 2016), and Greenpeace launched the purpose-built *Rainbow Warrior III* at a reported cost of £16 million (Alter 2011; Bonnie 2011). The cost of maintaining and operating a vessel often exceeds its purchase cost. For its 2009–2010 Antarctic anti-whaling campaign, SSCS deployed two vessels for over three months, at a cost of $5.5 million USD, whereas the 2008–2009 campaign during which only one vessel was deployed cost $2.3 million USD (SSCS 2008 "990"; SSCS 2009 "990"). This is not to say that smaller groups with concomitant budgets cannot conduct DA at sea, but only that their operations may well be restricted to shore and near-shore actions.

One consequence of considerable operating costs is the need to attract donations, which comes with additional organizational imperatives, such as a need for professional fundraising efforts and visibility. Greenpeace is well known for its creative and ever-evolving fundraising techniques, such as direct mail and face-to-face fundraising (Shaiko 1993, p. 93). The SSCS was able to dramatically expand its fleet after increased visibility resulting from the airing of the reality TV show *Whale Wars* in 2008, on the *Animal Planet* network. In 2008, the tax year prior to the premier of *Whale Wars*, the SSCS received $3.4 million USD in contributions; the following year contributions reached $9.4 million USD (SSCS 2008 "990"; and SSCS 2009 "990"). The SSCS, whose fleet comprises 11 seafaring vessels along with smaller boats, is now able to claims to operate the world's largest private navy (SSCS n.d. "Our Flagship"; Rawley and Berube n.d.).

Given the high visibility of vessels, the use of covert tactics is severely limited on the high seas. While it is relatively simple for covert saboteurs to escape prosecution by blending into an urban environment following an action, it is challenging for activists operating a vessel to 'go underground'. Maintaining a public profile is non-optional; ships are difficult to hide and must eventually dock to re-supply. As such, the extent to which activists engaged in DA at sea can overtly violate the law may be restricted as compared with land-based counterparts. Expensive vessels are vulnerable to seizure,[1] and as a result, activists must devise various measures to avoid losing these valuable assets. This may involve avoiding engaging in certain actions, or, shielding oneself through the use of various legal tools, as will be explored below.

Whereas horizontal organizational structures are associated with land-based DA practitioners (Lee 1995, p. 67; Phelps Bondaroff 2008, p. 68–69), operating a vessel requires 'a well-defined command structure as found in any quasi-military operation' (Burnett 2003, p. 248). Furthermore, while the barriers to entry for personnel operating on land may be low, vessels require trained, skilled and potentially certified, officers and crew. Organizations are not able to crew vessels on passion alone, but must recruit and retain at least some trained personnel.

[1]See for example the 2008 Royal Canadian Mounted Police seizure of the SSCS's flagship *Farley Mowat* (Hopper 2015); or the Russian 2013 seizure of the Greenpeace vessel *Arctic Sunrise* (BBC News 2013).

As a result of the many demands of operating a larger offshore vessel at sea, a considerable amount of DA conducted by marine conservation organizations and activists is coastal in nature, occurring on, or near the shore. This presents a challenge with regard to classifying various strategies and tactics as either land- or sea-based. The application of DA which is of most interest to international relations scholars is clearly that which occurs on the high seas, beyond the direct jurisdiction of states 200 nautical mile Exclusive Economic Zones (EEZs), as these areas present a unique theatre with complex overlapping, and sometimes ambiguous or non-existent jurisdictions.

Examples of these kinds of operations are rare, largely as a result of the constraints elaborated above. Likewise, many maritime operations will occur in multiple jurisdictions, beginning and ending on shore, traversing various national jurisdictions, and sometimes entering the high seas. Focusing on only those components of an operation that occur in a narrowly defined jurisdiction would produce an overly myopic typology. Likewise, looking at DA that occurs closer to shore provides insight into strategies which could be applied further offshore. As a result, in developing the typology of DA at sea, this chapter will consider examples of both offshore, nearshore and shore-based operations.

Applications of Direct Action at Sea

Traditional Direct Action

One of the earliest examples of DA at sea was the 'Don't Make a Wave' campaign, which was an attempt to halt nuclear tests in the Pacific by sailing vessels into the waters near test sites by a group that would later organize itself into Greenpeace. Here the purpose of the action was to prevent tests from occurring by placing people on a vessel within harm's way. These early actions around Amchitka, United States of America (USA) were launched with a strong moral, anti-nuclear and anti-war message (Hunter 2004; Weyler 2004, p. 55), while later actions against French testing around Mururoa maintained these goals, but also incorporated claims concerning legality, specifically the legality of the exclusion zone established by the French around the test sites (Weyler 2004).

During the early years of the anti-whaling movement, Greenpeace and Friends of the Earth (FoE) sent vessels onto the high seas to clash with Japanese and Soviet whalers. These early actions were often a combination of 'bearing witness' style campaigns and DA (Weyler 2004, p. 28, 481). For example, in 1975, the Greenpeace vessel the *Phyllis Cormack* encountered the Soviet whaling fleet in the Pacific Ocean. In addition to taking many evocative 'David and Goliath' images of activists in small Zodiac contrasted with hulking rusted giant whalers, activists also placed themselves between the whalers and their prey (see for example Zelko 2013; Weyler 2004).

This type of action using DA to produce powerful image events, or 'mind bombs', is the hallmark of Greenpeace's use of DA, and Greenpeace in general (Hunter 2004). We can see it replicated across numerous Greenpeace campaigns. For example, in July 1978, off the coast of Spain in the Bay of Biscay, Greenpeace activists attempted to prevent the dumping of toxic chemicals by the British vessel *Gem* by interposing themselves between the dumping platforms and the sea. The action resulted in a Greenpeace Zodiac being struck by a barrel, which produced iconic images, helped galvanized public opinion around the issue of waste disposal at sea, and ultimately played an important role in the development of the Convention on the Prevention of Marine Pollution by Dumping of Wastes and Other Matter (the London Dumping Convention) (Parmentier 1999).

In such cases, Greenpeace used DA to draw attention to an issue, with the aim of changing laws and norms. These actions can still be considered DA, despite being predominantly focused on the creation of image events. The DA elements of these actions follow the strategy of traditional obstructive land-based DA, seeking to prevent an ecologically harmful action from occurring and raising the costs of the target. These campaigns can be considered the first application of DA at sea in the typology: the straightforward application of obstructive DA at sea.

Service Provision

Efforts to cleanup and restore marine ecosystems typify the application of traditional constructive DA at sea. Here, activists directly improve the marine environment by removing pollutants or by restoring previously degraded habitat. A seminal example of this kind of DA is the removal of abandoned lost or otherwise discarded fishing gear (ghost gear) from marine ecosystems. Ghost gear has a devastating impact on marine animals and habitat; it entangles millions of animals annually, introduces synthetic material into the marine food web, damages habitat, smothers reefs, transports invasive species, impedes navigation and negatively impacts various industries (see inter alia FAO 2009; World Animal Protection 2014). Efforts on the part of non-state actors to remove ghost gear from the marine environment are plentiful and vary in scale.

Generally, those cleanup efforts that involve DA can be categorized into shore and beach cleanups, reef and wreck removals, and coastal and open oceanic retrieval operations. The former category is the most accessible, and beach cleanups can be ad hoc, such as people picking up trash while working out—'plogging' (Poon 2018), or coordinated by an organization or network, such as events like the International Coastal Cleanup Day (Ocean Conservancy 2014). However, remote beach cleanups are less accessible for volunteers and resource intensive (Surfrider Pacific Rim n.d.).

The safe removal of ghost nets from wrecks and reefs can be a little more involved, requiring the use of certified SCUBA divers. The California Lost Fishing Gear Recovery Project encourages the public to report the presence of lost gear, and hires experienced and certified SCUBA divers to remove gear from near-shore waters in a safe and ecologically sensitive manner. This process is very resource intensive and as such is only capable of removing small amounts of gear. Despite these limitations over 60 tonnes of ghost nets have been retrieved by this organization between 2006 and 2012 (SeaDoc Society n.d.; and see World Animal Protection 2014). The Netherlands-based Ghost Fishing Foundation similarly uses volunteer divers to recover nets from wrecks in the North Sea and elsewhere (Ghost Fishing n.d. 'Introduction').

With regard to open ocean and coastal retrieval operations one organization, GhostNets Australia has been very active in removing ghost nets from the beach and coastal waters of the Gulf of Carpentaria (see Gunn et al. 2010). Since 2004, this organization has worked in cooperation with rangers from 31 indigenous communities in Northern Australia to remove over 13,000 nets from close to 3000 kms of coastline (GhostNets Australia 2012, n.d. 'Indigenous Rangers').

In addition to removing pollutants from marine habitat, activists can also actively work to restore damaged or degraded ecosystems. In 2016, Sea Shepherd Dive, a subsidiary of the SSCS, partnered with the Malaysian company Ocean Quest, to offer a coral propagation course. The course trains divers in an ecologically sensitive method to rebuild damaged reefs. It is now offered by a number of Sea Shepherd Dive partner dive centres, and those who complete the programme receive a qualification and actively participate in rehabilitating coral reefs (SSCS 2016c).

Direct Enforcement

An examination of the strategy of the SSCS, a prominent practitioner of DA at sea, Eilstrup-Sangiovanni and I found that this organization was engaging in a particular variation of DA (2014). On its Antarctic anti-whaling and Mediterranean bluefin tuna campaigns in particular, the SSCS appeared to be using DA to enforce international law and regulations, a strategy we described as 'direct enforcement'.

In order to effectively carry out DE, the SSCS framed itself as an environmental law enforcement organization and its actions as law enforcement. Watson claims that 'Sea Shepherd's empowered to intervene by virtue of the United Nations World Charter for Nature, which allows for non-governmental organizations (NGOs) and individuals to uphold international conservation law' (specifically Sects. 21 and 24, see Watson in *Whale Wars* 2012 at 14:40; Watson quoted in Cameron 2009, p. 5). While there is considerable debate within the legal community around the nature of these claims, whether or not the SSCS can technically enforce (international) law is beside the point (see inter alia Roeschke 2009; Caprari 2010; Moffa 2012; Khatchadourian 2007; Anton 2010a, b; Jabour and Iliff 2009; Honniball 2015; Nucci 2013; Kanehara 2009). The organization makes these claims and acts upon them through the use of DA, and by so doing, enforces the law.

The rhetoric of enforcement is also bolstered by practitioners using the imagery and symbolism of enforcement. On the campaigns examined, the SSCS also adopted the language and iconography of an enforcement organization. Individual actions were often described as 'enforcement actions', and targets characterized as 'poachers', 'criminals' engaged in 'unlawful' and 'illegal' activities (see inter alia Essemlali and Watson 2013, p. 74; Watson 2015). The SSCS 'patrol vessels' are painted with navy gray or naval camouflage, and given bow numbers.[2] In preparation for engaging with Japanese whalers in the Southern Ocean, the *Sam Simon* had the words 'ANTI-POACHING' and its Japanese translation 密漁反対, emblazoned on its side (Correl 2014).

A distinction between DE and DA is evident in issue and target selection. Activists wishing to employ DE will select issues around an existing legal framework, where they can make a credible claim to be enforcing a law. By extension DE works effectively on targets which can be credibly framed as violating (international) law or laws. There is considerable flexibility in issue selection, and activists can engage in jurisdictional arbitrage in order to identify the laws they wish to enforce. The resonance of claims to enforcement is linked with the perceived legitimacy of the laws that practitioners are claiming to enforce. Thus, while in the international arena we tend to see activists making claims to enforce international laws, the strategy can and is applied to national, sub-national and even tribal laws. Claims need not be entirely technically accurate, as in the case of Japanese Antarctic 'research' whaling (JARPA I and II), where the SSCS asserted the illegality of the JARPA programme before it was adjudicated at the International Court of Justice (ICJ 2019).

In exploring the mechanisms of DE, Eilstrup-Sangiovanni and I found that enforcement was achieved through two means:

> (i) prevention or denial, whereby activists physically prevent perpetrators from carrying out illegal actions (for example, by standing as human shields between hunters and prey, or by disabling, damaging or confiscating equipment used in illegal activities); and (ii) coercion, whereby activists raise the material costs of non-compliance by, for example, damaging machinery or obstructing economic activities. Importantly, property damage or obstruction is not carried out indiscriminately in an effort to call attention to environmental problems, but is narrowly targeted against actors that violate environmental laws in order to change their payoffs (Eilstrup-Sangiovanni and Phelps Bondaroff 2014, p. 351).

In this way, DE functions in a manner similar to traditional DA, however, by framing actions as enforcement, activists are also able to circumvent 'normative debates and to mitigate backlash resulting from the controversial intersection of environmental problems with wider socioeconomic or cultural issues' (Eilstrup-Sangiovanni and Phelps Bondaroff 2014, p. 359).

DE seeks to avoid the challenges inherent in norm entrepreneurship and the complex and sometimes intractable stalemates that can develop when there are contested moral claims on multiple sides of an issue, a feature of many environmental issues (Eilstrup-Sangiovanni and Phelps Bondaroff 2014). It does so by appealing to the commonly held value of the rule of law, and by framing issues with truncated causal chains associated with the language of law enforcement, rather than by the

---

[2]For marine and dazzle camouflage, see images at Vessel Finder n.d. "MY Bob Barker"; and Marine Traffic n.d. "Sam Simon"; and for navy gray see Marine Traffic n.d. "Ocean Warrior,"; for bow numbers see images of the *Steve Irwin*, former SSCS flagship, at Whale Spotter (2012).

lengthy, meandering, and technical causal chains typically associated with complex environmental issues (Eilstrup-Sangiovanni and Phelps Bondaroff 2014, p. 352). Laws are perceived as powerful and legal arguments as persuasive, because they are grounded in the norms of a society—the shared understandings of that society (Finnemore and Sikkink 1998, p. 914–915).

DE exerts legal leverage, challenging governments' 'commitment to the rule of law' (Eilstrup-Sangiovanni and Phelps Bondaroff 2014, p. 359). To borrow from Mégret, DE is 'fundamentally challenging the state's behavior [sic] from the point of view of its own international obligations' (2008, p. 19–20). States may feel 'shame' when they are named as norm violators, but DE calls into question the very legitimacy of a state as an international actor—one capable of upholding its international agreements or enforcing laws within its jurisdiction. This is more than merely a rhetorical claim, it is one backed up with action. As Peter Hammarstedt, former first mate and now captain of the SSCS vessel *Bob Barker*, stated when discussing one of the goals of the SSCS's Antarctic anti-whaling campaign: 'We hope to encourage and inspire the governments of the world to actually do their job, the job that we are doing for them, enforcing the law down here' (*Whale Wars* 2011, at 10:16).

While DE practitioners may reject the idea that states have exclusive jurisdiction to enforce laws, and have attempted to exert legal leverage on states, this strategy is otherwise conservative in nature (Burbach 2012). DA practitioners circumvent existing power structures, and by so doing challenge, deny or reject them; they posit new norms and invoke moral arguments. The world envisioned by DE is one where existing (international) laws are followed and enforced by states. In enforcing existing laws and by encouraging states to do so, DE seeks to increase compliance with these structures. By speaking the language of law and sovereignty, DE addresses states directly, and ensures that this message is heard through the use of DA. By actively engaging in enforcement, sometimes using confrontational means, activists put down the soft power tool of persuasion and pick up a form of hard power, by so doing step increasingly into a realm otherwise reserved for state actors. For example, the fact that activists can effectively enforce compliance trumps objections from states that enforcement is impractical or impossible, further removing hurdles which would otherwise provide justification for non-intervention by states.

The use of DE also addresses one of the challenges of increased visibility and vulnerability to arrest and seizure resulting from operating at sea, by shielding activists from political and legal prosecution. By selecting targets that are engaged in illegal activity, activists invoke the principle of 'unclean hands', in as much as targets do not wish to draw attention to their own wrongdoings by indicting activists (Eilstrip-Sangiovanni and Phelps Bondaroff 2014, p. 357). This mechanism does not offer substantive protection to activists who are only targeting *norm* violators, especially when such norms are contested, as they often are with environmental issues.

Types of Direct Enforcement

This initial conceptualization of DE was formulated through an examination of the SSCS's Antarctic anti-whaling and Mediterranean bluefin tuna campaigns. Eilstrup-Sangiovanni and I point to a number of other organizations employing forms of DE, as an indication that the strategy was not idiosyncratic to the SSCS (2014, p. 359). In /examining the strategies of these various organizations, as well as subsequent iterations and modifications of that of the SSCS it is apparent that the application of DE has become more varied beyond denial and coercion. Three types of DE can be developed, which can then be added to the overall typology of DA at sea.

In consideration of the following three types of DE, it is valuable to note that there is overlap between them. For example, to identify targets upon which to conduct compellence, an activist needs some degree of monitoring and surveillance. Having a visible presence whilst conducting surveillance/monitoring operations can also likely have a deterrent effect. This is captured in Fig. 1 by the overlapping of the various DE categories.

Monitoring/Surveillance

While these two types of activities can be grouped together, it is useful to first distinguish between the terms. Generally speaking, 'monitoring' refers to the 'the systematic, continual, and active or passive observation of persons, places, things or processes', while surveillance refers to 'targeted monitoring of activities by police or security officials for specific evidence of crimes or other wrongdoing' (Encyclopedia of Science, Technology, and Ethics 2005). The former is routine and often overt, while the latter is targeted and often covert. Given the nature of the maritime environment, stationary surveillance is often limited to shore-based operations, and many actions combine elements of both.

Monitoring and surveillance are in a gray area when it comes to their inclusion in the typology, in so far as there are a number of NGOs that monitor compliance with laws and regulations and who could not be said to engage in DE. For example, the Environmental Investigation Agency (EIA) is well known for conducting undercover investigations to expose wildlife crime, and then releasing its 'findings in hard-hitting reports to campaign for improved governance and more effective enforcement' (EIA n.d. 'About Us'). Such an approach is an archetypal form of accountability politics, with reports being used primarily to 'name and shame' and to raise awareness.

The types of monitoring and surveillance which can be classified as DE are those which are principally focused around enforcement rather than advocacy; activists become an extension of law enforcement by gathering evidence necessary to catch offenders. The audience of the intelligence products they generate is the authorities rather than the public. This is not to suggest that organizations such as the EIA do not share their findings with the authorities. Parsing the distinction between these various strategies is fodder for future discussion, and Fig. 1 has been adjusted to reflect this distinction.

Examples of organizations engaging in monitoring and surveillance are plentiful. The FishAct, (formerly TBF) is a European-based marine conservation group launched by former SSCS crew member, which engages in what it describes as 'citizen led enforcement'. The organization purchased a Rigid Inflatable Boat (RIB) and operates several aerial drones, which it has deployed to patrol 'areas of the Mediterranean Sea to monitor the illegal use of driftnets and where possible taking action to challenge illegality' (TBF n.d. 'Driftnets'). It operates a Citizen Inspector Network (CIN), which engages ordinary citizens to participate in shore-based monitoring of fishing activities, whereby volunteers covertly inspect catches and gear, and monitor activity in fishing ports (Green 2016). The goal of the programme is to increase 'levels of independent surveillance of fishing operations in European seas by carrying out inspections in fishing ports, markets and coastal areas. Collected evidence is then used to aid policy work of other organisations and to initiate prosecution' (TBF n.d. 'Citizen Inspector Network/About').

On 27 November 2014, TBF signed a cooperation agreement with the Italian Coast Guard, detailing 'how the organizations will share evidence and intelligence in an effort to end illegal fishing operations… [and] work together to further enforcement of fisheries regulations in the area' (Tripp 2014). DE framing is clearly evident in Wietse van der Werf, founder of TBF's messaging, when he makes statements such as

> the Mediterranean Sea is a hotbed of illegal fishing activity, often with involvement from organised crime organisations, [and that]… we have investigated and exposed fishing crime in this region over the last years and are very pleased to announce that thanks to this new cooperation with the Italian Coastguard, the evidence collected by our Citizen Inspectors can now aid enforcement directly (Tripp 2014).

More recently, FishAct identified and gathered evidence of the use of illegal fishing gear in the Gulf of Gabès in Tunisia (FishAct 2018).

Other organizations with a monitoring and surveillance focus include groups such as the Coastal Guardian Watchmen (CGW). Founded in 2005, the CGW works to 'monitor, protect and restore the cultural and natural resources in our territories', including the Central Coast to the North Coast and Haida Gwaii, British Columbia, Canada. The clear DE approach of the CGW is apparent in selected elements of their vision statement, which notes that they:

- Monitor commercial and recreational activities on the North and Central Coast;
- Have the authority to carry out compliance and enforcement;
- Have a strong presence throughout our territories so that resource users regularly encounter and interact with us;
- Work with the federal and provincial government (through management agreements that respect the title and rights of First Nations) to ensure coordinated and robust monitoring and enforcement throughout our territories. (Coastal First Nations, Great Bear Initiative n.d. 'Coastal Guardian Watchmen Vision').

The organization further notes that 'we derive our authority and jurisdiction from our traditional laws to manage and safeguard the lands and waters of our territories for the health of future generations', demonstrating how DE can be used to enforce a variety of laws (Coastal First Nations, Great Bear Initiative n.d. 'Coastal Guardian Watchmen Vision').

Members patrol in RIBs looking for trophy hunters poaching grizzlies and other activities violating provincial and traditional laws. One coordinator of the Watchmen noted that sometimes 'it gets tense… [but] usually just having a presence is enough' (William Housty, chair of the Heiltsuk Nation resource management office in Bella, quoted by Prystupa 2015). In this way, these overt monitoring efforts also serve as a deterrent, with these two types of DE often walking hand in hand. This example also demonstrates that just as land-based operatives can monitor illegal activities at sea, sea-based operatives can monitor illegal activity on land. It can be concluded that many other organizations noted throughout this chapter engage in monitoring and surveillance, and that these kinds of activities are necessary to identify actors who are breaking the law and can serve as targets for further actions, such as compellence.

Deterrence

Unlike monitoring, which typically involves observing actors, deterrence and compellence seek to change their behaviour. Deterrence can be differentiated from compellence in as much as deterrence 'demands that the adversary refrain from acting, whereas compellence demands that the adversary undertake action' (Schaub 2004, p. 389). Deterrence functions by influencing the behaviour of others; notably by inducing them to refrain from undertaking a specific action. It does so by presenting them with undesirable consequences to carry out an undesirable action. Image and reputation play a significant role when an organization has its presence in an area that serves to deter would-be lawbreakers. For deterrence to function properly, the actor must have a sufficient threat capacity such that the target believes that the consequences of non-compliance will be unbearable and likely to be triggered (Schaub 2004, p. 400). This threat capacity is often communicated through image and reputation—those who appear menacing and capable of enacting undesirable consequences, or have a reputation for aggressive action, more effectively convey deterrence than those appearing weak or those who are unknown. Like most NGOs, those relying on DE depend on 'reputational capital', only theirs is oriented around conveying a high threat capacity (Sasser et al. 2006, p. 5).

The SSCS has gone to considerable lengths to cultivate its image as 'the most aggressive, determined, and effective marine wildlife conservation organization' (bold heading on SSCS leaflet n.d.). In addition to borrowing naval imagery, vessels often incorporate other iconography and structural elements that could easily be classified as 'aggressive'. For example, the 2010–2011 Antarctic anti-whaling campaign, which the SSCS named 'Operation No Compromise', several RIBs were decorated with 'shark mouth' nose art (popular with World War II airplanes) and SSCS's vessels were painted black and decorated with 'kill flags' of the vessels 'sunk' by the SSCS.[3] The rails of the *Bob Barker* were covered in prominent yellow anti-boarding spikes, which conveyed a 'hardcore look' and contributed to the SSCS's pirate brand. With respect to the organization's 'Jolly Roger' flag, Watson observed that 'some people say it's intimidating but usually they are the people we are trying to intimidate' (Watson 2014).

The use of this particular imagery certainly conveys a high threat capacity, but also seems to contradict the organizations efforts to portray itself as a law enforcement organization, given the criminal nature of pirates. The organization seems to vary its use of the Jolly Roger based on the nature of its campaign, adjusting its image and, as a result, the message it sends to its targets. When confronting the Japanese whaling fleet in the Southern Ocean, the SSCS prominently flew its Jolly Roger, however, when the SSCS partnered with West African states to combat illegal fishing (explored below), the organization reverted to the use of its original logo, the image of a humpback whale with a dolphin above it, both swimming to the left, in front of a circle.

The adjustment of its image to suit various circumstances demonstrates the SSCS's understanding of the role that image and reputation play in deterrence. The message conveyed by SSCS's reputation is that targets can expect confrontational action which will interfere with their operations, and those wishing to avoid such interference should desist. Such is the reputation of the SSCS that it has been able to win victories through reputation alone: In 1980, the Spanish whaler *Astrid* was quietly retired by her owners after the SSCS offered a \$25,000 USD 'bounty' for anyone who would sink the vessel. This threat was credible as earlier that year the SSCS had sunk two Spanish whalers—*Ibsa I* and *II*—at dockside (Darby 2007, p. 122–123).

Establishing an effective threat capacity need not be as elaborate as the approach of the SSCS's, as evidenced by the work of the Mendocino Abalone Watch (MAW). Founded in 2009 by a number of concerned residents of Mendocino Country, California, USA, this organization seeks 'to enhance regulatory enforcement and protection of the abalone resource along the Mendocino Coast' (MAW n.d. 'About'). MAW fills a valuable role on a coastline with over 100 beach accesses but only 5–7 official California Department of Fish and Game (CDFG) officers. Volunteers wearing prominent yellow jackets and hats patrol local beaches and monitor the activities of abalone harvesters to ensure compliance with very strict state limits and other regulations. MAW volunteers are trained by the CDFG in how to identify suspicious harvesting methods, applicable laws, as well as in methods of observing, recording and reporting. In addition to supporting the CDFG by recording and reporting suspicious activities to CDFG wardens, MAW volunteers have also been known to initiate citizen arrests of abalone poachers, and to confront and pursue poachers (see for example Jones 2011, p. A3). The high visibility of MAW volunteers, along with their capacity to call law enforcement and reputation for confrontations with poachers has a deterrent effect.

---

[3] More accurately representing the vessels the SSCS claims responsibility for taking out of operation, as not all the flags and names of vessels listed in the kill flags were in fact sunk.

Compellence

While deterrence seeks to prevent an illegal action from occurring, compellence undertakes to stop illegal activities after they are already underway. A portion of the literature on coercive diplomacy suggests that compellence can be more difficult to effectuate as compared with deterrence as

> the cost of compliance is partially determined by the degree to which it involves altering the status quo. It is presumed that the status quo possesses an inertia that makes altering it more costly than maintaining it, and that compellence is aligned with the former and deterrence with the latter (Schaub 2004, p. 397).

Although there is some debate about this, compellence does require maximal levels of intervention to physically compel a target to change behaviour already underway (Schaub 2004, p. 397). This type of DE is the 'prevention and denial' element identified by Eilstrup-Sangiovanni and Phelps Bondaroff (2014, p. 351).

A key example of this type of DE is the SSCS's Antarctic anti-whaling campaign. In the 2002/2003 austral summer, and annually from 2005 to 2017, the SSCS sent vessels to the Southern Ocean in an attempt to stop Japanese whaling operations (SSCS 2017, November 30). The specific goal of these operations was 'to enforce the global moratorium against commercial whaling and to enforce the protection regulations granted to the whales inside the official Southern Ocean [Whale] Sanctuary' (SSCS n.d. 'Operation Devine Wind'). In order to accomplish this, the SSCS actively prevented whaling from occurring by blockading the slipway of the whaling fleet's factory ship the *Nisshin Maru*, thereby preventing whales from being loaded and as a result, halting whaling operations (Phelps Bondaroff 2011). The SSCS also interfered in the hunt by harassing the harpoon vessels; attempting to foul their propellers with lines, throwing butyric acid stink bombs to contaminate whale meat and make decks unworkable, and a host of other confrontational tactics (Bethune 2010, p. 108).

This approach was effective: prior to interventions by the SSCS, the Japanese consistently met and exceeded their self-allotted quotas of minke and fin whales (see Annual Cruise Reports, ICR n.d. 'Publication list'). However, following the intervention of the SSCS, the Japanese saw their catches decline significantly. For example, in the 2010–2011 whaling season the whalers landed only 19% of their quota, and were forced to end their season early 'due to repeated harassment from activists at sea' (see for example Sky News 2011). In this way, the SSCS was able to use DE to change the behaviour of the Japanese whaling fleet, actively preventing it from whaling for a significant portion of its season.

Another example of compellence entails the confiscation or disabling of fishing gear used in illegal fishing operations. These cases differ from simple ghost gear recovery operations in so far as they target gear which is in use. Examples from SSCS operations abound. During the Operation Driftnet (January 2016), the SSCS vessel *Steve Irwin* encountered a fleet of Chinese vessels fishing with illegal driftnets in the South Indian Ocean. The SSCS described the objective of the campaign as being 'to locate the *Fu Yuan Yu* fleet, document and gather evidence of their illegal activity, disrupt their operations and confiscate their illegal driftnets' (SSCS 2016a). The SSCS claims to have recovered around 5 km of driftnet abandoned from the fleet after it fled (SSCS 2016a). Similarly, on 17 June 2010, SSCS divers entered bluefin tuna cages and cut the nets, releasing 800 tons of fish. This occurred after the *Steve Irwin* encountered the Italian bluefin tuna fishing vessel *Cesare Rustico* in the waters between Malta and Libya, and found it towing two cages containing bluefin tuna, after the 2010 annual tuna quota had been met (Vidal 2010).

In addition to physically stopping an action that violates the law and has negative ecological impacts, compellence DE also raises the material costs of non-compliance, thereby reducing the capacity of the target to act in the future (Eilstrup-Sangiovanni and Phelps Bondaroff 2014, p. 351). This effect is secondary, as increases in material costs have different impacts on varied targets, depending on a number of factors. While the principle of unclean hands has shielded the SSCS from many legal cases, the financial impact of SSCS actions can be roughly gauged by examining the rare cases in which the organization has been sued for damages. Following the SSCS's actions against the *Cesare Rustico*, the owners of this vessel, Fish and Fish Ltd. sued the SSCS for over €760,000 in damages, the case was eventually dismissed from the UK court in which it was filed (Scottish Courts and Tribunals 2011). In 2015 the Institute of Cetacean Research (ICR) and Kyodo Senpaku, the private company contracted by the ICR to conduct its Antarctic whaling operations, sued the SSCS in US court for damages for violating a temporary injunction for its vessels to stay 500 yards away from Japanese whaling vessels in the Southern Ocean. The SSCS agreed to pay $2.55 million USD, which included $2 million USD in damages (National Post 2015). The economic impact of the SSCS's actions on these targets varies significantly. The loss of €760,000 to a small private Maltese fishing company is difficult to compare to the impact of $2 million USD by the ICR, which has an annual operating cost of $65–$76 million USD, and which has received support from the Japanese government to cover past shortfalls (see inter alia Hoek 2010, p. 191; Kanehara 2009, p. 576–577). These economic impacts are a secondary mechanism of DE. The economic costs of compellence DE are much easier to calculate than that of deterrence, given the difficulty in calculating hypothetical profits or losses resulting from actions that were never undertaken.

Recent years have witnessed the emergence of a number of local fisher organizations in Mexico created with the purpose of protecting local stocks from overexploitation by outsiders and with the enforcement of local protected areas. The organization Eco-Alianza de Loreto A.C. operates the 'Observatorios Ciudadanos' (Citizen Observers), who 'work with park managers and local fishermen to coordinate, monitor and observe activities within the park, with a mission to promote conservation through education and training' (Eco-Alianza de Loreto n.d. 'Marine and Coastal Conservation'). Similar groups of *Pescadores Vigilantes* (vigilant or vigilante fishermen) have sprung up throughout Central and South America, with, for example, programmes in six of the ten provinces of Panama (Córdova 2015). Participants in these programmes are offered training, and consistent with the benefits of adopting enforcement imagery and symbolism, many of these programmes also provide participants with numbers and badges (Nonato López 2015).

These types of organizations and programmes demonstrate the interrelated nature of the three components of DE. While most of these organizations work essentially as monitoring groups, sharing intelligence on illegal fishing and drug trafficking with the authorities, their presence also serves as a deterrent. There have also been reports of them confronting poachers, and preventing them from continuing their activities. Conversations with an official with the Mexican Comisión Nacional de Acuacultura y Pesca (CONAPESCA) suggest rare occasions when these confrontations have been aggressive and resulted in violence (Global Marine Protected Areas Enforcement Conference, November 25–29, 2012, San Francisco, USA). When a Pescadores Vigilantes confronts someone stealing fish from their waters and prevents them from poaching, this is an example of compellence.

Activist-State Relations

In theorizing the mechanisms behind DE, Eilstrup-Sangiovanni and I (2014, p. 359) posited that in addition to direct outcomes, activists also apply legal leverage to governments, calling into question their ability to fulfil their international obligations by actively enforcing the law when governments are unable or unwilling to do so themselves. This, combined with the confrontational tactics often involved in executing DE-based operations, suggests that the use of DE would likely lead to states developing animus towards activists using DE. Given the nature of the 'legal shame' levelled at states by activists through the use of DE, one might expect that the relationship between activists and states would be equally tense. However, surveying other organizations using DE, as well as the evolution of the SSCS's strategy in recent years, reveals that the relationships between activists employing DE and states are considerably more varied. There is an emerging trend of activists collaborating and cooperating with states and government officials, and even launching joint operations. A brief examination of examples of the different kinds of relationships that can develop between activists and governments reveals the ongoing evolution of the strategy of DE. It also reveals that the strategy can be effective in a range of activist-government relationship scenarios.

This section examines a number of relationships that DE practitioners can have with states (and sub-state agencies): acquiescence, antipathy, concurrence and collaboration. The international arena is a complex theatre of operation, with numerous moving parts. Relationships between activists and states can vary between issues, governments, and sometimes between different agencies and units within a state, and can change over time. Given the capacity and standing of states in international relations, the nature of the relationship between activists and governments can significantly impact outcomes. However, this does not suggest that DE cannot produce results even in the face of a hostile relationship between activists and governments.

Acquiescence and Antipathy

Looking at the relationships between the SSCS and states involved in the Antarctic anti-whaling issue serves to highlight examples of antipathy and acquiescence. Prior to SSCS anti-whaling campaigns, the Australian government had done little to oppose Japanese whaling, a large portion of which takes place in the EEZ of the Australian Antarctic Territory (AAT). This position was logical, in so far as Australia was unlikely to want to upset one of its largest trading partners (Hutchinson 2006, p. 1). Japanese whaling in the Southern Ocean has a long history, dating back to the end of the Second World War, and since the global moratorium on commercial whaling, Japan has conducted whaling under the guise of 'research' (Morikawa 2009, p. 5). Over this time, the Australian government did not alter its position from one of inaction until several high profile compellence DE operations by the SSCS. The Australian government was forced to send naval assets to the Southern Ocean in order to monitor SSCS and ICR vessel engagements as a result of confrontational SSCS actions.[4] SSCS actions increased attention on the whaling issue and exerted considerable pressure on the Australian government (Epstein and Barclay 2013, p. 110–111), such that Australia would eventually take Japan to the ICJ over the issue in 2010 (Rothwell 2010, p. 4). Australia ultimately won the case, with the ICJ ruling that Japan's JARPA II whaling programme did not qualify as scientific research, and as such fell outside of the provisions of the 1946 International Convention for the Regulation of Whaling and regulations of the International Whaling Commission (IWC) which permit whaling for the purposes of science as an exception to the global moratorium on commercial whaling (ICJ 2019).

It took numerous SSCS operations before the Australian government could be leveraged into acting, during which Australia had ample opportunity and means to seriously obstruct the SSCS. The SSCS fleet operated out of Hobart, Tasmania and within the AAT-EEZ for a considerable portion of its Antarctic anti-whaling campaign; however, Australia did not overly frustrate SSCS operations. The Australian government did take some actions against SSCS, including putting up barriers to crew members acquiring visas and conducting vessel searches by the Australian Federal Police at the conclusion of campaigns.[5] However, these actions were largely symbolic: crew members eventually received visas, and the search of the vessels perfunctory, and apparently to satisfy a request by the Japanese government (Southwell 2011). Though it changed over time and was impacted by individual incidents, it seems that the best description of the relationship between the SSCS and Australia can be described as acquiescence; the reluctant acceptance of activists' use of DE by states without overt objections.

---

[4] For example, the Australian government sent the *Oceanic Viking* in 2008 and the *Oceanic Protector* in 2012. See Government of Australia (2010), Anton (2009, p. 339), The Sydney Morning Herald (2012).

[5] Paul Watson and Peter Hammarstedt both encountered problems getting Australian visas in 2009, and in 2011 a SSCS helicopter pilot was denied a visa (see SSCS 2009 and 2011).

The relationship between Japan and the SSCS is another matter and one characterized by antipathy. Tensions appeared to increase with each annual Antarctic operation. For example, when two SSCS crew members boarded the harpoon vessel *Yushin Maru #2* in January 2008, they were eventually released to the Australian government after considerable diplomatic wrangling (Heller 2007). The following year, the SSCS high tech trimaran, *Ady Gil*, was struck during an engagement and cut nearly in half by the *Shonan Maru No. 2*, a Japanese security vessel. Subsequent to this, the captain of the *Ady Gil*, Pete Bethune was detained by the Japanese when he boarded the *Shonan Maru No. 2*, in an attempt to make a citizen's arrest of the whaling vessel's captain. Bethune was returned to Japan at considerable expense, where he was charged with several minor offences and given a two year suspended sentence (Bethune 2010). In addition to the inclusion of a dedicated security vessel to accompany their fleet, the Japanese also placed armed guards on its vessels (McLeod 2008), and as noted above, pursued the SSCS through the American court system (National Post 2015). From these examples, it is clear that the relationship between the SSCS and Japan is antipathetic.

Despite this antipathy, the SSCS appears to have successfully leveraged Japan. Following the ICJ's ruling, Japan did not conduct whaling operations for the 2014/2015 whaling season. For the 2015/2016 Japan launched the New Scientific Whale Research Program in the Antarctic Ocean (NEWREP-A), which it operated for two seasons. This programme was given a quota of 333 minke whales by the Japanese government, a significant reduction to the 850 under JARPA II, which had been a sticking point with the ICJ (ICR n.d., 'Scientific Contribution'). In December 2018 Japan announced that it would be leaving the IWC, resuming commercial whaling in its waters, and ending its Southern Ocean whaling operations (McCurry and Weaver 2018). The exact role that leverage exerted by the SSCS played in altering the behaviour of the Japan and Australian governments is difficult to isolate, but is clearly evident. Both of these states preferred the status quo but were forced to change their behaviour as a result of the SSCS's use of DE.

Concurrence

Concurrence refers to activists and governments being concerned with an issue, and conducting parallel operations. In these situations, there may be some unidirectional intelligence sharing, either overt or through signalling, with information typically flowing from activists to governments. Any example of monitoring and surveillance efforts whereby unsolicited information is provided to the authorities could be classified as an example of concurrence. An examination of the SSCS's Operation Icefish (2014–2016), provides an example of this activist-state relationship, and the use of compellence DE.

Operation Icefish was a two year SSCS campaign to us 'innovative direct action tactics to fill a law enforcement void exploited by illegal toothfish operators' in the Southern Ocean (SSCS 2014c; and see Berube 2015; Sæter 2018; Ross and

Arnold 2018). On 17 December 2014 the SSCS vessel *Bob Barker* intercepted the Nigerian-flagged *Thunder*, fishing illegally within the Commission for the Conservation of Antarctic Marine Living Resources (CCAMLR) management region, and attempted to effectuate a 'citizen's arrest', ordering the fishing vessel to Fremantle, Australia (Sæter 2018). The SSCS informed CCAMLR and the *Bob Barker* pursued the *Thunder* as it fled west. Meanwhile SSCS crew from the *Sam Simon* gathered the gillnets abandoned by the *Thunder* as evidence to prevent further illegal fishing (SSCS 2014a, b).

This was the start of a pursuit that continued for the next 110 days and that covered over 10,000 miles. Over the course of the pursuit, attempts at fishing by the *Thunder* were prevented by SSCS vessels and nets were confiscated (SSCS 2015c). The pursuit, one of the longest in maritime history, ended on April 6, 2015, off the coast of Sao Tome when the *Thunder* sunk, apparently scuttled by her officers (SSCS 2015a; Milman 2015). As the *Thunder* sunk, SSCS crew members boarded it to retrieve evidence, which, along with all the confiscated nets, were handed over to government and Interpol authorities (SSCS 2015b, e). Over the course of the operation, the SSCS regularly shared updates with Interpol, and given that the SSCS is a non-state actor, this flow of information that was necessarily unidirectional (conversation with Interpol official, September 24, 2017).

The *Thunder* was one of a fleet of six vessels, branded the 'Bandit 6', notorious for toothfish poaching in the Southern Ocean (Gaworecki 2015). The remaining vessels were concurrently intercepted, arrested and detained by government authorities from New Zealand, Senegal, Cabo Verde, Malaysia, Thailand and Indonesia (see inter alia Pala 2015; Kearns 2015; SSCS 2015d, f). The final of the six, the *Viking*, was sunk by Indonesian authorities in March 2016 after it was detained by the Indonesian Navy earlier that year (Mollman 2016; SSCS 2016b). Meanwhile, Operation Sparrow, the Spanish government's investigations into illegal fishing, culminated in raids on premises owned by Vidal Armadores, a Spanish fishing company with convoluted ownership ties to a number of the 'Bandit 6' vessels. Vidal Armadores was eventually fined €17.84 million, disqualified from conducting fishing activities and banned from receiving public subsidies (Garat 2015).

Collaboration

Direct enforcement does not exclusively seeks to coerce those who are breaking the law and to leverage states which are otherwise unwilling to act, as it is not always the case that states are unwilling to act. Concurrence occurs when states are willing to act, but unwilling to share intelligence and otherwise cooperate with activists. In some cases, government inaction is not due to an unwillingness, but rather a lack or diminished capacity to act. In these cases, active collaboration between activists and states may occur. Joint operations, with activists and states working together cooperatively can be launched to tackle issues of shared concern. These relationships typically emerge when states have a diminished capacity to act relative to the capabilities of activists.

The ocean is large and so effective monitoring and enforcement are difficult, even for the best-equipped states. As a result, it is often of value for states to increase their monitoring, surveillance and deterrence capacities by cooperating with non-state actors. TBF/FishAct signing an agreement with the Italian Coast Guard, MAW volunteers being trained and working with the CDFG, and *Observatorios Ciudadanos,* and *Pescadores Vigilantes* being trained by local authorities are all examples of collaboration. In all these cases, state authorities are supplementing their capacity by harnessing civilian human intelligence assets. Governments will also often support cleanup efforts through grants (see for example Fishing For Litter n.d., 'What is Fishing for Litter'?; NOAA n.d., 'Fishing for Energy').

As noted, operating a vessel at sea is expensive, which applies equally to states and to activists. In some cases, activists are better equipped than state navies and coast guards, something which to an extent is unsurprising given the SSCS's claims to operate the world's largest private navy. Increasingly, a number of collaborative ventures between activists and states have emerged. The SSCS has been at the forefront of this evolution in DE, and has launched a number of collaborative ventures with West African states. An example is Operation Jodari, which the SSCS describes as

> a partnership with the Tanzanian government to stop IUU [illegal, unreported, unregulated] fishing by training and assisting Tanzanian officers in monitoring, control and surveillance of all fishing activity in Tanzanian waters, including fishing vessel inspections and boarding procedures. To do this, Sea Shepherd is providing the *Ocean Warrior* civilian patrol vessel with experienced crew, while the Tanzanian government is providing 10 law enforcement agents…(SSCS n.d. "Operation Jodari 2018").

These operations typically entail the SSCS providing vessels, crew for those vessels, fuel and other necessary supplies, with state partners providing armed coast guard and naval personnel and the authority to board, inspect and arrest vessels found to be violating national laws. In this way, activists are able to harness the power and authority of states, and states are able to benefit from the resources and experience of activists. This mutually beneficial relationship obviates the need for activists to exert legal leverage on states, dramatically streamlining the application of DE.

A number of SSCS's recent collaborative operations have produced results. For example, the first three months of Operation Sola Stella (2017–2019) resulted in the arrest of five illegal fishing vessels in the waters of Liberia (SSCS n.d. 'Operation Sola Stella'). The operations as part of Operation Albacore (2016–2018) conducted 'over 80 at-sea inspections and boardings, resulting in the arrest of eight IUU Congolese fishing trawlers and one Spanish long-liner', among other arrests in the waters of Gabon and São Tomé and Principe (SSCS n.d., "Operation Albacore"). On Operation Apex Harmony (2017), the SSCS's *Ocean Warrior* was used to conduct 'a dawn raid in collaboration with the Policia Nationale Timor Leste and arrested 15 vessels engaged in shark poaching' (interview Gary Stokes, SSCS Campaign Leader and now Director of OceansAsia, March 10, 2019; and see SSCS n.d., 'Operation Apex Harmony').

Another organization, Earthrace Conservation, founded by former SSCS captain Pete Bethune, has been focused on capacity building and training, and employs a number of former armed forces personnel to train enforcement officers from partner states, to conduct joint maritime patrols, and to carry out other specialist operations (Earthrace Conservation n.d. 'Latin America Maritime Program'). The organization, whose exploits are featured on the television program 'The Operatives', has a strong paramilitary image, which seems to obscure the line between conservation organization and private military contractor (PMC) (Earthrace Conservation n.d. 'The Operatives').

The PMC sector is a global industry with activity valued at $100 billion USD (Ivanciu 2015). The 'role of the private sector in the protection of seas and oceans continues to increase' (Marin et al. 2017, p. 193). An example of PMCs contracting with states to conduct maritime patrols is that of the PMC Hart-Nimrod, which signed an agreement with the Somali Transitional National Government in 1999 'to conduct fisheries enforcement with compensation derived from securing fishing licenses' (Berube 2015). While this effort failed, PMCs have been expanding operations at sea, notably offering personnel for anti-piracy measures (see inter alia Affi et al. 2016; Struwe 2012). What differentiates these efforts and others, from those of the organizations like Earthrace and the SSCS is that activist organizations are not commercial enterprises, offering their services, assets and personnel for free.

Conclusion

Activist strategy is constantly evolving, and nowhere is that more apparent than with the application of DA at sea. What first began as dramatic image events has gradually developed into a multiplicity of strategies, ranging from confrontational compellence DE which physically prevents targets from breaking the law; to high visibility monitoring operations which gather vital intelligence and serve to deter criminal activity. The relationships between activists and states can vary considerably: from overt animosity, where states use various means to lash out at activists groups using DE to exert legal leverage against them, to full-on collaboration, where activists and government personnel work hand in hand to combat illegal fishing.

Organizing these strategies and relationships into a typology allows us to gain an enriched understanding of the full range of use of DA at sea, and to explore some of the ways in which marine conservation activist strategy has been diversifying. However, the use of DA and DE at sea is not restricted to marine conservation, and there are a number of organizations employing versions of DA and DE for other causes at sea. For example, Women on Waves (WoW), an organization that seeks to prevent unsafe abortions and unwanted pregnancies, is notable for deploying sexual health and abortion clinics on board vessels just outside of the territorial waters of countries with restrictive access to abortion (Women on Waves 2017). Exploring how the strategy of WoW fits with this typology would be a useful way to test the applicability of this typology to areas other than marine conservation, and is a fruitful area for future inquiry.

In developing this typology, several areas for future discussion and investigation have been identified. For example, distinguishing the boundary between image events and DA will help further our understanding of both of these strategies, and has the potential of providing insight to practitioners seeking to increase their effectiveness. Understanding the nature of the power one is attempting to wield is important if one wishes to wield that power effectively. Similarly, a number of organizations, such as the EIA, gather intelligence and information about criminal activity and use this information in a variety of ways. Exploring how this information is used, and contrasting this with the use of information by groups engaged in monitoring and surveillance DE will further our understanding of both of these approaches.

The number activist-state collaborative operations have rapidly expanded in recent years. Organizations such as the SSCS, who previously presented a 'pirate' image and whose confrontational use of DE put them into conflict with states, are now working in close collaboration with law enforcement. This approach has only emerged in the past couple of years, and did so rapidly, suggesting that additional refinement of the strategy is on the horizon. As a result, two very promising areas for future research can be identified: exploring the blurring line between PMCs and environmental activists, and studying the evolving relationship between activists and states, specifically the development of collaborative operations between these actors.

References

Affi, L., Elmi, A. A., Knight, W. A., & Mohamed, S. (2016). Countering piracy through private security in the horn of Africa: Prospects and pitfalls. *Third World Quarterly, 37*, 934–950. https://doi.org/10.1080/01436597.2015.1114882.

Anton, D. K. (2009). Antarctic whaling: Australia's attempt to protect whales in the Southern Ocean. *Boston College Environmental Affairs Law Review*, *36*, 319–351. https://lawdigitalcommons.bc.edu/ealr/vol36/iss2/3.

Anton, D. K. (2010, April). Protecting whales by hue and cry–is there a role for non-state actors in the enforcement of international law? *The Australian National University College of Law Research Paper*, No. 10-08. https://dx.doi.org/10.2139/ssrn.1593305.

Anton, D. K. (2010, July 8). Dispute concerning Japan's JARPA II Program of 'scientific whaling' (Australia v. Japan). *The American Society of International Law*, *14*(20).

BBC News (2013, September 23). Russia 'seizes' Greenpeace ship after Arctic rig protest. Retrieved March 31, 2019, from https://www.bbc.com/news/world-europe-24170129.

Berube, C. (2015, October 28). Thunder road: Outsourcing the fight against illegal fishing. *War on the Rocks*. Retrieved March 31, 2019, from https://warontherocks.com/2015/10/thunder-road-outsourcing-the-fight-against-illegal-fishing/.

Bethune, P. (2010). *Whale warrior*. Auckland, New Zealand: Hodder Moa.

Bonnie A. (2011, November 14). Despite a helicopter pad, Greenpeace's new $23 million *Rainbow Warrior* ship is greener than predecessors. *Treehugger.com*. Retrieved March 31, 2019, from http://www.treehugger.com/aviation/greenpeace-new-rainbow-warrior-ship-is-greener-than-predecessors.html.

Burbach, D. T. (2012, April 1–4). Whale wars and future wars: Assessing the Sea Shepherd campaign against Japanese Antarctic whaling, presented at International Studies Association (ISA) 53rd Annual Convention, San Diego, USA.

Burnett, J. S. (2003). *Dangerous waters: Modern piracy and terror in the high seas.* London, UK: Plume Books.

Cameron, S. D. (2009, July). Paul Watson: The journey of Sea Shepherd. *The Green Interview.* Retrieved March 31, 2019, from https://thegreeninterview.com/interview/watson-paul/.

Caprari, A. M. (2010, July). Lovable pirates? The legal implications of the battle between environmentalist and whalers in the Southern Ocean. *Connecticut Law Review, 42*(5), 1493–1526.

Coastal First Nations, Great Bear Initiative (n.d.). Coastal Guardian Watchmen vision. Retrieved March 31, 2019, from https://coastalfirstnations.ca/coastal-guardian-watchmen-vision/.

Córdova, M. G. (2015, October 5). Los pescadores ayudan al Servicio Nacional Aeronaval de Panamá a combatir el crimen. *Diálogo: Revista Militar Digital.* Retrieved March 31, 2019, from https://dialogo-americas.com/es/articles/los-pescadores-ayudan-al-servicio-nacional-aeronaval-de-panama-combatir-el-crimen.

Correl, A. (2014, March 19). Photo of *Sam Simon*, from its Facebook page. Retrieved March 31, 2019, from https://bit.ly/2OBE7G0.

Darby, A. (2007). *Harpoon: Into the heart of whaling.* New South Whales, Australia: Allen & Unwin.

Doherty, B., Paterson, M., & Seel, B. (2000). Direct action in British environmentalism. In M. P. Seel & B. Doherty (Eds.), *Direct action in British environmentalism* (pp. 1–24). London, UK: Routledge.

Durland, S. (1987). Witness: The guerrilla theatre of Greenpeace. *Performance, 40,* 67–73.

Earthrace Conservation (n.d.). Latin America maritime program. Retrieved March 31, 2019, from https://www.earthrace.net/assets/.

Earthrace Conservation (n.d.). The Operatives. Retrieved March 31, 2019, from https://www.earthrace.net/operatives/.

Eco-Alianza de Loreto, A. C. (n.d.). Marine and coastal conservation - Observatorios Ciudadanos. Retrieved March 31, 2019, from http://ecoalianzaloreto.org/our-programs/marine-and-coastal-conservation/.

Eilstrup-Sangiovanni, M., & Phelps Bondaroff, T. (2014). From advocacy to confrontation: Direct enforcement by environmental NGOs. *International Studies Quarterly, 58*(2), 348–361. https://doi.org/10.1111/isqu.12132.

Encyclopedia of Science, Technology, and Ethics (2005). Monitoring and surveillance. Retrieved March 31, 2019, from https://www.encyclopedia.com/science/encyclopedias-almanacs-transcripts-and-maps/monitoring-and-surveillance.

Environmental Investigation Agency (n.d). About us. Retrieved March 31, 2019, from https://eia-international.org/about-eia/.

Epstein, C., & Barclay, K. (2013). Shaming to 'green': Australia-Japan relations and whales and tuna compared. *International Relations of Asia-Pacific, 13*(1), 95–124. https://doi.org/10.1093/irap/lcs019.

Essemlali, L., & Watson, P. (2013). *Captain Paul Watson: Interview with a pirate.* Ontario, Canada: Firefly Books.

FAO (2009, May 6). Ghost nets hurting marine environment. Retrieved March 31, 2019, from http://www.fao.org/news/story/en/item/19353/icode/.

Finnemore, M., & Sikkink, K. (1998, Autumn). International norm dynamics and political change. *International Organizations, 52*(4), 887–917. https://www.jstor.org/stable/2601361.

FishAct (2018, October 14). Facebook post. Retrieved March 31, 2019, from https://bit.ly/2UaBss7.

Fishing For Litter (n.d.). What is fishing for litter? Retrieved March 31, 2019, from http://www.fishingforlitter.org.uk/what-is-fishing-for-litter.

Garat, J. (2015, October 21). Operation sparrow: Catching pirates in Antarctica. *Huffington Post.* Retrieved March 31, 2019, from https://www.huffpost.com/entry/operation-sparrow-catching-pirates-in-antarctica_b_8155722.

Gaworecki, M. (2015, June 1). Have the 'Bandit 6' poached their last toothfish in the Southern Ocean? *Mongabay*. Retrieved March 31, 2019, from https://news.mongabay.com/2015/06/have-the-bandit-6-poached-their-last-toothfish-in-the-southern-ocean/.

Ghost Fishing (n.d.). Introduction. Retrieved March 31, 2019, from http://www.ghostfishing.org/introduction/.

GhostNets Australia (2012). 2012 annual report. Retrieved March 31, 2019, from http://www.ghostnets.com.au/process/download/docs/3/.

GhostNets Australia (n.d.). Indigenous rangers. Retrieved March 31, 2019, from https://www.ghostnets.com.au/ranger-activities/.

Government of Australia (2010, June 16). Question time brief: Whaling–Australian government role, including boarding of Sea Shepherd. Documents obtained under Freedom of Information Act 1982 request, FOI-135.

Graeber, D. (2009). *Direct action: An ethnography*. Oakland, USA: AK Press.

Green, M. (2016, February 24). The Black Fish: Undercover with the vigilantes fighting organised crime at sea. *The Guardian*. Retrieved March 31, 2019, from https://www.theguardian.com/environment/2016/feb/24/black-fish-undercover-with-vigilantes-fighting-organised-crime-at-sea.

Gunn, R., Hardesty, B. D., & Butler, J. (2010, August). Tackling 'ghost nets' local solutions to a global issue in Northern Australia. *Ecological Management and Restoration, 11*(2), 88–98. https://doi.org/10.1111/j.1442-8903.2010.00525.x.

Heller, P. (2007). *The whale warriors: The battle at the bottom of the world to save the planets largest mammals*. New York, USA: Free Press.

Hoek, A. (2010). Sea Shepherd Conservation Society v. Japanese whalers, the showdown: Who is the real villain? *Stanford Journal of Animal Law and Policy, 3,* 159–193.

Honniball, A. N. (2015). Private political activists and the international law definition of piracy: Acting for 'private ends.' *Adelaide Law Review, 36*, 279–328. https://dspace.library.uu.nl/handle/1874/328999.

Hopper, T. (2015, June 25). Farley Mowat's final act of defiance: Eco-activists' former flagship goes down leaking oil in N.S. harbour. *National Post*, Canada. Retrieved March 31, 2019, from https://nationalpost.com/news/canada/farley-mowats-final-act-of-defiance-eco-activists-former-flagship-goes-down-leaking-oil-in-n-s-harbour.

Hunter, R. (2004). *The Greenpeace to Amchitka: An environmental odyssey*. Vancouver, Canada: Arsenal Press.

Hutchinson, A. (2006). Baleen out the IWC: Is international litigation an effective strategy for halting the Japanese scientific whaling program. *Macquarie Journal of International and Comparative Environmental Law, 3,* 1–33.

ICJ (2019). Whaling in the Antarctic (Australia v. Japan: New Zealand intervening–overview of the case. Retrieved March 31, 2019, from https://www.icj-cij.org/en/case/148.

ICR (n.d.). Publications list. Retrieved March 31, 2019, from https://www.icrwhale.org/CruiseReportJARPN.html.

ICR (n.d.). Scientific contribution–JARPA/JARPAII/NEWREP-A. Retrieved March 31, 2019, from https://www.icrwhale.org/scJARPA.html.

Ivanciu, C. (2015). Guns for hire: Private military companies and their status under international humanitarian law, presented at International Conference of Scientific Paper, AFASES, Basov, Romania, 2015, May 28–30. Retrieved March 31, 2019, from http://www.afahc.ro/ro/afases/2015/afases_2015/socio/Ivanciu%20Cosmin.pdf.

Jabour, J., & Iliff, M. (2009). Theatre sports in the Southern Ocean: Engagement options for Australia in whale research protest action. *Australian Journal of International Affairs, 63*(2), 269–289. https://doi.org/10.1080/10357710802389488.

Jones, R. (2011, May 26). Abalone Watch volunteers assist F&G wardens. *The Mendocino Beacon*, A3.

Jordan, T. (2002). *Activism! Direct action, hacktivism, and the future of society*. London, UK: Reaktion Books.

Kanehara, A. (2009). Legal responses of Japan to the impediments and harassments by foreign vessels against Japanese vessels during research whaling in the Antarctic Sea. *Japanese Yearbook of International Law, 52,* 553–582.
Kearns, M. (2015, September 14). Toothfish poaching vessel, listed by Interpol, escapes Thailand. *Seafood Source.* Retrieved March 31, 2019, from https://www.seafoodsource.com/news/supply-trade/toothfish-poaching-vessel-listed-by-interpol-escapes-thailand.
Keck, M. E., & Sikkink, K. (1998). *Activists beyond borders: Advocacy networks in international politics.* Ithaca, USA: Cornell University Press.
Keck, M. E., & Sikkink, K. (1999). Transnational advocacy networks in international and regional politics. *International Social Science Journal, 159,* 89–101. https://doi.org/10.1111/1468-2451.00179.
Khatchadourian, R. (2007, November 5). Neptune's navy: Paul Watson's wild crusade to save the oceans. *The New Yorker.* Retrieved March 31, 2019, from https://www.newyorker.com/magazine/2007/11/05/neptunes-navy.
Lee, M. F. (1995). *Earth First! Environmental apocalypse.* Syracuse, USA: Syracuse University Press.
Marin, J., Mudrić, M., & Mikac, R. (2017). Private maritime security contractors and use of lethal force in maritime domain. In G. Andreone (Ed.), *The future of the law of the sea: Bridging gaps between national, individual and common interests* (pp. 191–212). Switzerland: Springer Open.
Marine Traffic (n.d.). Ocean Warrior. Retrieved March 31, 2019, from https://www.marinetraffic.com/en/ais/details/ships/shipid:4297098/mmsi:244890807/imo:9791262/vessel:OCEAN_WARRIOR.
Marine Traffic (n.d.). Sam Simon. Retrieved March 31, 2019, from https://www.marinetraffic.com/en/ais/details/ships/shipid:252503/mmsi:244810415/imo:9053452/vessel:SAM_SIMON.
McCurry, J., & Weaver, M. (2018, December 26). Japan confirms it will quit the IWC to resume commercial whaling. *The Guardian.* Retrieved March 31, 2019, from https://www.theguardian.com/environment/2018/dec/26/japan-confirms-it-will-quit-iwc-to-resume-commercial-whaling.
McLeod, S. (2008, March 7). Armed guards will protect whalers: Japan. *ABC News*, Australia. Retrieved March 31, 2019, from https://www.abc.net.au/news/2008-03-08/armed-guards-will-protect-whalers-japan/1065972.
Mégret, F. (2008). July 19). *Civil disobedience and international law: Sketch for a theoretical argument.* https://doi.org/10.2139/ssrn.1163270.
Mendocino Abalone Watch (n.d.) About. Retrieved May 13, 2013, from http://www.mendoabwatch.com/about-maw/.
Milman, O. (2015, April 7). Captain deliberately sank illegal fishing vessel, claim Sea Shepherd rescuers. *The Guardian.* Retrieved March 31, 2019, from https://www.theguardian.com/environment/2015/apr/07/captain-deliberately-sank-illegal-fishing-vessel-claim-sea-shepherd-rescuers.
Moffa, A. L. I. (2012). Two competing models of activism, one goal: A case study of anti-whaling campaigns in the Southern Ocean. *Yale Journal of International Law, 37*, 201–214. https://digitalcommons.law.yale.edu/yjil/vol37/iss1/6.
Mollman, S. (2016, March 15). An illegal fishing vessel wanted by 13 nations has finally been destroyed by Indonesia. *Quartz.* Retrieved March 31, 2019, from https://qz.com/639408/an-illegal-fishing-vessel-wanted-by-13-nations-has-finally-been-destroyed-by-indonesia/.
Morikawa, J. (2009). *Whaling in Japan: Power, politics and diplomacy.* London, UK: C. Hurts & Co., Ltd.
Murphy, D. (2010, January 7). How Bob Barker joined Sea Shepherd Paul Watson and the whale wars. *Christian Science Monitor.* Retrieved March 31, 2019, from http://www.csmonitor.com/World/Global-News/2010/0107/How-Bob-Barker-joined-Sea-Shepherd-Paul-Watson-and-the-whale-wars.
National Post (2015, June 10). Sea Shepherd and Paul Watson agree to pay Japanese whale killers millions to settle lawsuit. Retrieved March 31, 2019, from https://nationalpost.com/news/world/sea-shepherd-and-paul-watson-agree-to-pay-japanese-whale-killers-millions-to-settle-lawsuit.

NOAA (n.d.). Fishing for energy supports new derelict gear innovation projects. Retrieved March 31, 2019, from https://marinedebris.noaa.gov/partnerships/fishing-energy-supports-new-derelict-gear-innovation-projects.

Nonato López, R. (2015, October 23). SENAN do Panamá luta sem trégua contra o narcotráfico. *Diálogo: Revista Militar Digital*. Retrieved March 31, 2019, from https://dialogo-americas.com/pt/articles/senan-do-panama-luta-sem-tregua-contra-o-narcotrafic.

Nucci, B. (2013). From comity to comedy: The ninth circuit's blanket injunction on the Sea Shepherd's Southern Ocean activities may be laughing in the face of established international law. *Arizona Journal of Environmental Law and Policy, 3*, 1051–1059.

Ocean Conservancy (2014). Turning the tide on trash: 2014 report. Retrieved March 31, 2019, from https://oceanconservancy.org/wp-content/uploads/2017/04/2014-Ocean-Conservancy-ICC-Report.pdf.

Pala, C. (2015, June 17). The hunt for the last Chilean sea bass poachers. *Foreign Policy*. Retrieved March 31, 2019, from https://foreignpolicy.com/2015/06/17/chilean-sea-bass-toothfish-thunder-sea-shepherd/.

Parmentier, R. (1999). Greenpeace and the dumping of wastes at sea: A case of non-state actor's intervention in international affairs. *International Negotiation, 4*(3). https://doi.org/10.1163/15718069920848561.

Phelps Bondaroff, T. (2008, Summer). Throwing a wrench into things: The strategy of radical environmentalism. *Journal of Military and Strategic Studies, 10*(4), 1–23.

Phelps Bondaroff, T. (2011, Spring). Sailing with the Sea Shepherds. *Journal of Military and Strategic Studies, 13*(3), 1–55.

Poon, L. (2018, March 1). Pick up trash while you exercise. It's called plogging. *CityLab*. Retrieved March 31, 2019, from https://www.citylab.com/life/2018/03/plogging-exercise-trash-running/554456/.

Prystupa, M. (2015, February 24). Coastal Guardian Watchmen confront armed trophy hunters to save grizzlies. *The National Observer*, Canada. Retrieved March 31, 2019, from https://www.nationalobserver.com/2015/02/24/news/coastal-guardian-watchmen-confront-armed-trophy-hunters-save-grizzlies.

Rawley, C., & Berube, C. (n.d.). The non-state navy: Sea Shepherd as a case study for 21st Century maritime non-state actors. *Small Wars Journal*. Retrieved March 31, 2019, from https://smallwarsjournal.com/jrnl/art/the-non-state-navy-sea-shepherd-as-a-case-study-for-21st-century-maritime-non-state-actors.

Reynolds, R. (2008). *On guerilla gardening: A handbook for gardening without boundaries*. London, UK: Bloomsbury Publishing.

Roeschke, J. E. (2009). Eco-terrorism and piracy on the high seas: Japanese whaling and the rights of private groups to enforce international conservation law in neutral waters. *Villanova Environmental Law Journal, 20*, 99–137. https://digitalcommons.law.villanova.edu/elj/vol20/iss1/6.

Ross, M., & Arnold, A. (2018, August 18). Thunder versus Sea Shepherd: The true story of the world's longest ship chase. *ABC News*, Australia. Retrieved March 31, 2019, from https://www.abc.net.au/news/2018–08-19/catching-thunder-illegal-fishing-worlds-longest-ship-chase/10114450.

Rothwell, D. R. (2010, May 31). Australia v. Japan: JARPA II whaling case before the international court of justice. Proceedings instituted by Australia against Japan: Application instituting proceedings. *The Hague Justice Portal*.

Sæter, K. (2018). *Catching Thunder: The true story of the world's longest sea chase* (D. Oatley, Trans.). London, UK: Zed Books Ltd.

Sasser, E. N., Parkash, A., Cashore, B., & Auld, G. (2006). Direct targeting as an NGO political strategy: Examining private authority regimes in the forestry sector. *Business and Politics, 8*(3), 1–32. https://doi.org/10.2202/1469-3569.1163.

Scarce, R. (2006). *Eco-warriors: Understanding the radical environmental movement* (Updated ed.). California, USA: Left Coast Press Inc.

Schaub, G. (2004, June). Deterrence, compellence, and prospect theory. *Political Psychology, 25*(3), 389–411. https://www.jstor.org/stable/3792549.
Scottish Courts and Tribunals (2011). Fish and Fish Limited for an order under section 27 of the Civil Jurisdiction and Judgments Act 1982. [2011] CSOH 122, P783.11. Retrieved March 31, 2019, from https://www.scotcourts.gov.uk/search-judgments/judgment?id=146186a6-8980-69d2-b500-ff0000d74aa7.
SeaDoc Society (n.d.). California lost fishing gear recovery project. Retrieved March 31, 2019, from https://www.seadocsociety.org/california-lost-fishing-gear-removal-project.
Shaiko, R. G. (1993, July). Greenpeace U.S.A: Something old, new, borrowed. *Annals of the American Academy of Political and Social Science,* 528, 93. https://doi.org/10.1177/000271629352 8001007.
Sky News (2011, February 18). Japan announces end to whaling season. Retrieved April 20, 2011, from http://www.skynews.com.au/businessnews/article.aspx?id=579284.
Southwell, P. (2011 March 3). *Commonwealth of Australia Crimes Act 1914: Section 3E: Search Warrant for Premises or a Conveyance.* Signed over to T. Dawe. Australia Federal Police, 18828, 15:11.
SSCS Australia (2011, December 5). Australian government attempts to sabotage Sea Shepherd's campaign to defend whales. Retrieved March 31, 2019, from https://bit.ly/2JRRaVl.
SSCS Australia (2016, July 4). Sea Shepherd Global launches new patrol vessel, Ocean Warrior. Retrieved March 31, 2019, from https://www.seashepherd.org.au/news-and-commentary/news/sea-shepherd-global-launches-new-patrol-vessel-ocean-warrior.html.
SSCS (2008). 990: Return of organization exempt from income tax. *Charity Navigator*. Retrieved March 31, 2019, from http://www.charitynavigator.org/index.cfm?bay=search.summary&orgid=5784.
SSCS (2009). 990: Return of organization exempt from income tax. *Charity Navigator*. Retrieved March 31, 2019, from http://www.charitynavigator.org/index.cfm?bay=search.summary&orgid=5784.
SSCS (2010, June 17). Sea Shepherd frees 800 bluefin from floating cages below Libya's infamous line of death. Retrieved May 16, 2012, from http://www.seashepherd.org/news-and-media/2010/06/17/sea-shepherd-frees-800-bluefin-fromfloating-cages-below-libyas-infamous-line-of-dea th-174.
SSCS (2014a, December 17). Sea Shepherd intercepts toothfish poachers in the Southern Ocean. Retrieved March 31, 2019, from https://www.seashepherdglobal.org/latest-news/sea_shepherd_intercepts_toothfish_poachers/.
SSCS (2014b, December 30). Sea Shepherd retrieves 25 kilometers of illegal gillnet from the Southern Ocean. Retrieved March 31, 2019, from https://www.seashepherdglobal.org/latest-news/sea_shepherd_retrieves_25k_illegal_gillnet/.
SSCS (2014c, September 16). Operation Icefish–Sea Shepherd launches new Southern Ocean campaign. Retrieved March 31, 2019, from https://www.seashepherdglobal.org/latest-news/ope ration_icefish_new_southern_ocean_campaign/.
SSCS (2015a, April 6). Poaching vessel, Thunder, sinks in suspicious circumstances. Retrieved March 31, 2019, from https://www.seashepherdglobal.org/latest-news/thunder_sinks_in_suspici ous_circumstances/.
SSCS (2015b, February 25). Sea Shepherd hands-over evidence of Thunder's illegal fishing to authorities in Mauritius. Retrieved March 31, 2019, from https://www.seashepherdglobal.org/lat est-news/sea_shepherd_hands_over_evidence_thunder/.
SSCS (2015c, February 7). Sea Shepherd shuts down operations of the Thunder and confiscates illegal gear. Retrieved March 31, 2019, from https://www.seashepherdglobal.org/latest-news/sea_shepherd_shuts_down_operations_of_thunder/.
SSCS (2015d, March 29). Second internationally wanted toothfish poaching vessel, Viking, detained in Malaysia. Retrieved March 31, 2019, from https://www.seashepherdglobal.org/latest-news/vik ing_arrested/.

SSCS (2015e, March 5). Interpol takes custody of evidence against Thunder from Sea Shepherd. Retrieved March 31, 2019, from https://www.seashepherdglobal.org/latest-news/interpol_takes_custody_of_evidence_against_thunder/.

SSCS (2015f, May 22). Military and police board last two toothfish poaching vessels in Cabo Verde. Retrieved March 31, 2019, from https://www.seashepherdglobal.org/latest-news/military_and_police_board_last_two_vessels/.

SSCS (2016a). Operation Driftnet: A campaign to stop an illegal driftnet fleet. Retrieved March 31, 2019, from https://www.seashepherdglobal.org/our-campaigns/opdriftnet/learn-more/.

SSCS (2016b, February 26). The last toothfish poacher, Viking, arrested in Indonesia. Retrieved March 31, 2019, from https://www.seashepherdglobal.org/latest-news/the_last_toothfish_poacher_viking_arrested/.

SSCS (2016c, February 29). Coral propagation courses. Retrieved March 31, 2019, from https://www.seashepherdglobal.org/latest-news/coral-propagation-courses/.

SSCS (2017, November 30). Whale defense campaign history. Retrieved March 31, 2019, from https://www.seashepherdglobal.org/latest-news/whale-defense-campaign-history/.

SSCS (n.d.). Operation Sola Stella: Sea Shepherd combats illegal fishing in Liberia, West Africa. Retrieved March 31, 2019, from https://www.seashepherdglobal.org/our-campaigns/operation-sola-stella/.

SSCS (n.d.). Mission statement. Retrieved March 31, 2019, from https://seashepherd.org/mission-statement/.

SSCS (n.d.). Operation Albacore: Our third campaign to stop illegalf in Gabon, West Africa. Retrieved March 31, 2019, from https://www.seashepherdglobal.org/our-campaigns/operation-albacore/.

SSCS (n.d.). Operation Apex Harmony–Timor Leste: Defending sharks from illegal fishing in East Timor. Retrieved March 31, 2019, from https://www.seashepherdglobal.org/our-campaigns/operation-apex-harmony-timor-leste/.

SSCS (n.d.). Operation Devine Wind–campaign history. http://www.seashepherd.org/campaigns/operation-divine-wind/operation-divine-wind/campaign-history-15. Accessed 22 January 2013.

SSCS (n.d.). Our flagship: Thank you & farewell to the M/Y Steve Irwin. Retrieved March 31, 2019, from https://www.seashepherdglobal.org/our-campaigns/my-steve-irwin/.

Struwe, L. B. (2012). Private security companies (PSCs) as a piracy countermeasure. *Studies in Conflict & Terrorism, 35,* 588–596. https://doi.org/10.1080/1057610X.2012.684660.

Stuart, A., Thomas, E. F., Donaghue, N., & Russell, A. (2013). 'We may be pirates, but we are not protesters': Identity in the Sea Shepherd Conservation Society. *Political Psychology, 20*(20), 1–25. https://doi.org/10.1111/pops.12016.

Surfrider Pacific Rim (n.d.). Remote cleans. Retrieved March 31, 2019, from https://pacificrim.surfrider.org/remote-cleans/.

TBF (n.d.). Citizen Inspector Network/about. Retrieved January 11, 2016, from http://theblackfish.org/cin/about/.

TBF (n.d.). Driftnets. Retrieved August 22, 2013, from http://www.theblackfish.org/driftnets/.

The Sydney Morning Herald (2012, January 13). Three whaling activists transferred to customs boat. Retrieved March 31, 2019, from http://www.smh.com.au/environment/whale-watch/three-whaling-activists-transferred-to-customs-boat-20120113-1pyyb.html.

Tracy, J. (1996). *Direct action: Radical pacifism from the Union Eight to the Chicago Seven.* Chicago, USA: University of Chicago Press.

Trentmann, F. (2019). Consumer boycotts in modern history: States, moral boundaries, and political action." In D. Feldman (Ed.), *Boycotts past and present: From the American Revolution to the campaign to boycott Israel.* Palgrave MacMillan. https://doi.org/10.1007/978-3-319-94872-0_2.

Tripp, E. (2014, December 8). Italian Coast Guard and NGOs partnership to combat illegal fishing. *Marine Science Today.* Retrieved March 31, 2019, from http://marinesciencetoday.com/2014/12/08/coast-guard-black-fish-partnership/.

Vanderheiden, S. (2005, September). Eco-terrorism or justified resistance? Radical environmentalism and the 'war on terror'. *Politics and Society, 33*(3), 425–447. https://doi.org/10.1177/0032329205278462.

Vessel Finder (n.d.). MY Bob Barker–research Vessel. Retrieved March 31, 2019, from https://www.vesselfinder.com/vessels/MY-BOB-BARKER-IMO-5280540-MMSI-246847000.

Vidal, J. (2010, June 18). Sea Shepherd activists free hundreds of threatened bluefin tuna off Libya. *The Guardian*. Retrieved March 31, 2019, from https://www.theguardian.com/environment/2010/jun/18/sea-shepherd-release-bluefin-tuna-libya.

Wapner, P. (1995, April). Politics beyond the state: Environmental activism and world civic politics. *World Politics, 47*(3), 311–340. https://www.jstor.org/stable/2950691.

Watson, P. (2014, October 4). Why the Jolly Roger. Facebook post. Retrieved March 31, 2019, from https://www.facebook.com/captpaulwatson/photos/a.443115070931/10152698232595932/?type=3&theater.

Watson, P. (2015, December 4). Opposing the Japanese pirate whalers in the Southern Ocean whale sanctuary. *SSCS, Australia*. Retrieved March 31, 2019, from https://www.seashepherd.org.au/news-and-commentary/commentary/opposing-the-japanese-pirate-whalers-in-the-southern-ocean-whale-sanctuary.html.

Whale Spotter (2012, December 3). Sea Shepherd Conservation Society fleet. Retrieved March 31, 2019, from http://whalespotter.com.au/?p=762.

Whale Wars (2011). Delivering the final blow. Season 4, episode 10 (aired 2011, August 12).

Whale Wars (2012). Operation bluefin, 14:40 (aired 2012, April 21).

Women on Waves (2017, February). Abortion ship in Guatemala, February 2017. Retrieved March 31, 2019, from https://www.womenonwaves.org/en/page/6686/in-collection/517/abortion-ship-in-guatemala–february-2017.

World Animal Protection (2014). Fishing's phantom menace: How ghost fishing gear is endangering our sea life. Retrieved March 31, 2019, from http://www.worldanimalprotection.org/sites/default/files/int_files/sea-change-tackling-ghost-fishing-gear-summary.pdf.

Weyler, R. (2004). *Greenpeace: How a group of ecologists, journalists, and visionaries changed the world*. Vancouver, Canada: Raincoast Books.

Zelko, F. (2013). *Make it a green peace! The rise of countercultural environmentalism*. New York, USA: Oxford University Press.

The IPBES Conceptual Framework: Enhancing the Space for Plurality of Knowledge Systems and Paradigms

Lavínia Pereira and Olivia Bina

Abstract Conceptions of nature and of humanity's place within it, we argue, are the best arena to finally confront and resolve the fundamental epistemological and ontological divide that stands in the way of transformative change. We develop this idea focusing on the Intergovernmental Science-Policy Platform on Biodiversity and Ecosystem Services (IPBES), whose goal is 'strengthening the science-policy interface for biodiversity and ecosystem services'. In 2015, the IPBES created a *Conceptual Framework* to support policy-makers, and different stakeholders, in their assessment of complex interactions between the natural world and human societies. One of the main goals of the *Framework* is to allow plurality, bringing together different knowledge systems, and its particular representations of humans-nature relations. In this chapter, we explore the epistemological and ontological assumptions that lay behind the Framework's endeavour, aiming to contribute to strengthen its ability to achieve such a goal. First, we set out the theoretical context and the analytical tools for a critical review the IPBES *Framework*; second, we expose limitations and hidden assumptions that risk undermining the *Framework's* pursuit of plurality of knowledge; finally, we identify and discuss the conditions to create the space that allows for effective dialogue across paradigms and knowledge systems. We conclude arguing for the need to transcend paradigms and to embrace integral approaches (re)connecting biodiversity conservation and humans and nature.

L. Pereira (✉) · O. Bina
Instituto de Ciências Sociais, Universidade de Lisboa, Av. Prof. Aníbal de Bettencourt, 9, 1600-189 Lisboa, Portugal
e-mail: lavinia.pereira@ics.ulisboa.pt

O. Bina
Chinese University of Hong Kong, Hong Kong, China

World Academy of Art and Science, Minneapolis, MN, USA

J. C. Pereira and A. Saramago (eds.), *Non-Human Nature in World Politics*, Frontiers in International Relations, https://doi.org/10.1007/978-3-030-49496-4_15

Introduction

> When a system is far from equilibrium, small islands of coherence have the capacity to shift the entire system. - Ilya Prigogine, Nobel Laureate in Chemistry.[1]

> The loss of species, ecosystems and genetic diversity is already a global and generational threat to human well-being. Protecting the invaluable contributions of nature to people will be the defining challenge of decades to come. Policies, efforts and actions—at every level—will only succeed, however, when based on the best knowledge and evidence. This is what the IPBES Global Assessment provides. - Sir Robert Watson, IPBES Chair.[2]

The multiple crises and instability of our times have led to repeated calls for systems change and transformative change, argued against the background of the double narrative of an Anthropocene (Steffen et al. 2007) and a Capitalocene (Moore 2017; Haraway 2015). The first is an umbrella concept, arising from the field of global environmental change, seeking to explain how humanity (especially rich nations) has become a force of nature capable of affecting planetary boundaries and a safe operating space for humanity (Rockström et al. 2009). The second looks at the same phenomenon by bringing together history and theory, politics and ecology, economy and culture, to offer an interpretation of why we are in the Anthropocene, potentially heading for collapse. The 'what' and the 'why' have much to do with our understanding of the relation, connectedness, interdependence and belonging of the humans-nature nexus (e.g. Bennett et al. 2016; Dalby 2016; Dibley 2012; Ellis 2011).

Combined, these narratives set a clear agenda for global systems change and sustainability transformation, most recently set out in the UN Agenda '*Transforming our world: the 2030 Agenda for Sustainable Development*' (UN 2015), and reaffirmed with redoubled urgency in the '*Global assessment report on biodiversity and ecosystem services*' of the Intergovernmental Science-Policy Platform on Biodiversity and Ecosystem Services (IPBES 2019b). The recent call for 'system change not climate change' triggered by climate strikes initiated by the young Greta Thunberg is a societal expression of such agenda. Central to systems change and transformation are the type and role of knowledge, including policy-relevant knowledge, and mindsets and paradigms (EEA 2016). These crucial levers of systems change, applied to the specific case of IPBES, are the focus of this chapter.

IPBES has already made a significant contribution to such levers by defining a *Conceptual Framework* (Díaz et al. 2015). This encourages diverse understandings of humans-nature connectedness, by integrating disciplines, perspectives and

[1]Prigogine was a pioneer of self-organizing systems theory. His work is fundamental to the theory of Regenerative Economics, and to the theory of change of the Regenerative Communities Network (cited in the Newsletter Capital Institute Summer Update 2019, https://capitalinstitute.org/about-us/newsletter-archives/).

[2]https://www.ipbes.net/news/ipbes-global-assessment-preview.

knowledge systems including Indigenous and Local Knowledge (ILK), for biodiversity conservation.[3] The underlying driver of IPBES's *Framework* is the need for plurality and diversity of representations of humans-nature relations in a context that many have argued is dominated by globalized capitalism and the economization of values (Göpel 2016; Sandel 2013), which, in turn, are built on a view of humans and nature as separate (Latour 2004). In this chapter, we speak to this broader context by looking at the specific case of IPBES's *Conceptual Framework*.

The IPBES was established in 2012, by its Member States as an independent organization, under the initiative of the United Nations Environmental Program (UNEP 2010). The goal of the IPBES is to assess the state of biodiversity and ecosystem services, and help decision-making by 'strengthening the science-policy interface… for the conservation and sustainable use of biodiversity, long-term human well-being and sustainable development' (http://www.ipbes.net). The creation of its *Framework* is the result of a participatory and multidisciplinary work aiming at facilitating collaboration and communication across scientific disciplines, and diverse stakeholders, providing a common tool to inform policy-makers and different stakeholders on the elements involved in assessing socio-ecological dynamics (Díaz et al. 2015).

In this chapter, we aim to contribute to strengthening the *Framework's* capacity for plurality in two steps: through an overview of the existing critique of its achievements and weaknesses, and through the subsequent proposal for a more effective space of dialogue that can tackle the deep-seated reductionist approach linked to the current dominant socio-economic system.

Our inquiry builds on a purposefully interdisciplinary analytical framework that brings two broad arenas into dialogue. First, the epistemic foundations of global environmental knowledge that is politically relevant, and of the systems change discourse for sustainability transitions, which is implicit in agendas such as that of UN 2030 Agenda on Sustainable Development (UN 2015). This will broaden the meaning of the existing critique of the IPBES *Framework*, and help propose more effective ways of ensuring dialogue between epistemologies and ontologies of multiple environmental knowledges. Second, we combine these perspectives with those of transitions research and literature, which maintain that 'thinking', 'policies' and 'practices' are key pillars of change (EEA 2016). We will focus on 'thinking'—the act of producing and using knowledge. 'Thinking' as the umbrella dimension for the mindsets and paradigms that characterize existing knowledge systems, and which according to Meadows (1999) are the deepest levers for transformative change; and 'thinking' to question unacknowledged assumptions, which might be working as institutionalized epistemologies (Jasanoff 2005; Miller 2008), thus representing 'barriers to desired transformations towards sustainability' (Abson et al. 2017, p. 35).[4]

[3]Biodiversity conservation is understood here as a science that addresses 'the dynamics of coupled human-natural systems' (Kareiva and Marvier 2012, p. 962).

[4]Through this chapter, we will refer to knowledge mainly in two different ways: its production and negotiation within the scope of environmental change, and the broader scope of mindsets and paradigms that characterizes the discourse of systems change for sustainability transformation. Other expressions such as knowledge systems, ideas and ways of knowing will also be used within the context of the IPBES *Framework* and the cross-cultural work that the Platform aims to build.

We do this in three main steps: first, we set out the theoretical context and the analytical tools that will allow us to critically review the IPBES *Framework*; second, we expose existing limitations and hidden assumptions behind the *Framework* that risk undermining its own pursuit of a plurality of knowledge; finally, we identify and discuss the conditions to create the space for co-generation of knowledge that allows for the contribution of multiple rationalities and effective dialogue across paradigms and knowledge systems. We conclude exploring the opportunity to transcend paradigms and benefit from integral approaches connecting biodiversity conservation and humans and nature.

Environmental Knowledge, Production and Change: Setting the Context for IPBES

IPBES Conceptual Framework: An Overview

The initial concept for an *IPBES Framework* was captured by a diagram, resulting from a workshop held in Paris in 2012, built around three key elements: (a) Biodiversity, (b) Ecosystem goods and services and (c) Human well-being; though it was not considered satisfactory, perceived as too utilitarian by a wide range of participant countries. The Bolivian delegates emphasized the need to include a plurality of values concerning biodiversity and humans-nature relations, namely Indigenous and Local Knowledge (ILK), in the conceptual tool of the Platform (Borie and Hulme 2015).

As a result, the 2015 *Framework* integrates the contributions from ILK: (a) Mother Earth/Systems of Life, (b) Nature's gifts and (c) Living-well in balance and harmony with Mother Earth. ILK contributions are framed under 'inclusive categories' and headlines ('Nature', 'Nature's benefits to people' and 'Good quality of life', respectively). For instance, the headline—*Nature*—is supposed to include both its meaning in Western science concerning categories such as biodiversity, ecosystems, the biosphere and living natural resources; and their equivalents in non-Western knowledge systems—mainly of indigenous peoples from South American Andes—and their understanding of 'Mother Earth' and systems of life.

Nature's benefits to people, the second element, 'refers to all the benefits that humanity…—in rural and urban settings—obtains from nature'. It includes 'Ecosystem goods and services', provisioning, regulating and cultural services, concepts originating within the Western paradigm, including functions such as the production of oxygen; and 'Nature's Gifts', which originates in non-Western knowledge systems, indicating values such as those attributed to pollinator species 'two of the most important totemic species' of Mayan people (Díaz et al. 2015, p. 6). In 2017, the term 'benefits' was replaced by 'contributions', hence 'Nature's Contributions to

People'[5] (NCP), in an effort to make it more 'comprehensive and neutral' (Pascual et al. 2017, p. 9). The use of 'Nature's contributions to people' aims at 'providing space for context-specific perspectives recogniz[ing] that there are multiple ways of understanding and categorizing relationships between people and nature' (Díaz et al. 2018, p. 272).

The third key element is *Good quality of Life*: 'highly value-based' and hence considered to be contextual (Díaz et al. 2015, p. 7), it is conceived as an inclusive category for 'Human well-being', 'Living in harmony with nature' and 'Living in balance and harmony with Mother Earth', comprising spiritual and material dimensions, physical and mental health of individuals, and harmonious relationship with nature. Such a diversity of indicators places the concept of 'Quality of Life' well beyond mainstream measure of 'Gross Domestic Product' (GDP) or 'Human Development Index' (HDI). We return to this subject in the next section.

Next, we wish to better understand the existing critiques of this *Framework.* We do so by placing these critiques in conversation with the closely related contexts of knowledge production for Global Environmental Change (GEC) and of systems change, aiming to illuminate three aspects: the claim that the *Framework* helps integrate the diversity of knowledge systems, the reasons that make it relevant to the GEC context, and the difficulties in making it effective.

Plurality and Values Underlying Environmental Knowledge

Deep-seated tacit and unquestioned views that constitute the epistemic foundations of public life (Jasanoff 2005; Miller 2008) have been barriers to effective plurality integration in the making of politically relevant environmental knowledge: either through the association between environmental knowledge production and ecosystem services and econometric logics (Hulme 2010; Turnhout et al. 2013), or through the invisibility and ineffectiveness of alternative paradigms of humans-nature relations within environmental agendas and international governance platforms (O'Brien 2019).

The literature on global environmental knowledge has been keen to show the major risks of using econometric logics in environmental thinking precisely because it reproduces the view that dissociates humans and nature, which has contributed to environmental disruption in the first place (Turnhout et al. 2013; Yusoff 2010). It reproduces the modernity project that studies ecosystems 'as if humans are not part of it' (Turnhout et al. 2013, p. 158). Scholars also denounce the difficulties of letting alternative paradigms enter the environmental change agendas. Reductionist, deterministic, mechanistic approaches have been dominating public discourses concerning, for instance, climate change (e.g. O'Brien 2019). Conditioned by 'civic epistemologies' (Jasanoff 2005), politically relevant global environmental

[5]The updated version of the IPBES *Conceptual Framework* including the expression 'Nature's contributions to people' was published in 2019 (see IPBES 2019b).

knowledge marginalizes alternative paradigms, as they do not fit socially accepted tacit patterns of knowledge validity and objectivity.

As O'Riordan and Jordan (1999) state, the way we actually understand biodiversity and nature will always be framing assessment and practices for global environmental issues. If tacit 'knowledge-ways' (Jasanoff 2005, p. 255) are framed by the general perception of humans and nature as separated, and conditioned by the actual 'extinction of experience' (e.g. Zylstra et al. 2014; Soga and Gaston 2016)—then this perception will be adopted by institutionalized actors and stakeholders looking into the existing frameworks and tools for knowledge production.

This matters, as these perceptions—and the knowledge that follows—are performative: as Turnhout (2018, p. 368–369) states, 'the conceptualization of environmental knowledge as not only partial, situated and selective but also performative, has enabled deeper insight into how environmental knowledge produces effects'. Knowledge shapes the way we experience and represent reality; it frames the way we set goals and value orientations that justify action (see also O'Riordan and Jordan 1999).

Moreover, as Hulme (2010, p. 559) argues, the aim to produce 'global kinds of knowledge' about environmental change is a problem in itself: its outcomes risk being 'insensitive to the peculiarities of place and context', which is why the *Framework* gives ILK a central role in environmental knowledge-making. Failure to acknowledge the performative dimension of knowledge can lead to being unaware that what we claim to be true within a particular knowledge system is partial and contextualized. Exposing this limitation holds the key to unlocking system change.

Expose to Transform: Deep Leverage

There have been growing calls for 'a forum for multiple scientific and nonscientific understandings of biodiversity' (Turnhout et al. 2013, p. 156) and thus for the inclusion of diverse disciplinary insights on environmental knowledge (Castree et al. 2014; Lahsen 2016), especially from the social sciences and humanities (Jetzkowitz et al. 2018) still barely visible compared to the predominant role of natural sciences (e.g. Timpte et al. 2018; Vadrot et al. 2018). Yet, they remain inadequately addressed in the literature on conservation sciences and on the IPBES *Framework*.

We argue that the nexus between calls for greater plurality in knowledge systems, and for the need to acknowledge and expose value orientations and the normative dimension underpinning environmental change knowledge, practices and agendas—is central to the potential of IPBES *Framework*. We, therefore, bring these literature and perspectives in conversation with the broader theme of systems change for sustainability transformation. Following Göpel (2016) and Meadows (1999): to expose assumptions and blind spots is to enable change and make progress towards sustainability transformation, since the way we conceptualize environmental knowledge and humans-nature relations will frame solutions for socio-ecological problems (Leichenko and O'Brien 2019).

Transition theory literature echoes the above debates in global environmental change domains, when discussing **deep leverage points** for sustainability transformations and the need to go beyond sectorial, incremental and techno-managerial transitions (Hölscher et al. 2018, p. 2), by addressing the sources of the system—mindsets and paradigms (Meadows 1999). Adopting transition and incremental measures through short-term approaches which 'typically favor techno-managerial and behavioral solutions' (Leichenko and O'Brien 2019, p. 185), means to address only shallow levers of change, easy to tackle but without broader systematic impact. Moreover, these approaches tend to support 'prevailing social, economic, or gender relations of power and inequality' (Leichenko and O'Brien 2019, p. 187).

The 2019 IPBES Report reinforces the urgency of transformative (deep) change, claiming that without it, neither the UN Sustainable Development Goals, nor the goal of reverting dramatic biodiversity loss (IPBES 2019a, p. 14–16), can be achieved. It thus echoes Meadow's appeal to tackle the sources of the system itself:

> Paradigms are the sources of systems. From them, from shared social agreements about the nature of reality, come system goals and information flows, feedbacks, stocks, flows and everything else about systems (1999: 18).

Instead of adopting approaches to environmental change that operate within the existing system without questioning it, we need to expose and deal with the sources of the system, creating the space and opportunity for paradigms to be shifted or even, and desirably, transcended (Meadows 1999).

Enhancing Plurality: Four Concepts

The literature on politically relevant global environmental knowledge and on systems change are essential to locate the *Framework* in its wider context, and to help understand the significance of its claim of delivering an effective plurality of knowledge. We explore such potential through four concepts originated in related fields, including environmental social sciences, education for sustainability and transdisciplinary research: 'boundary objects' and 'multiple rationalities' (Dunkley et al. 2018), 'third space' (Turnbull 2003) and the concept of 'transdisciplinarity' according to Nicolescu's view (2010).

Dunkley et al. (2018, p. 784) propose the 'boundary object' as a tool to articulate the relationship between scientific and non-scientific discourses, and to allow collaboration between different communities, letting them expose and discuss worldviews, without necessarily implying consensus. With 'multiple rationalities', the authors intend to refer to a situation in which 'the idea of a single rationality is abandoned' and existing power imbalances between different ways of knowing are acknowledged (Dunkley et al. 2018, p. 789).

'Third space' is an expression explored by Turnbull (2003, p. 233–234), referring to a shared space allowing different rationalities to express themselves 'without power and epistemological constraints'; a space for co-evolution and reflexivity about hidden assumptions. Nicolescu's ambitious approach to 'transdiciplinarity' allows us to think beyond what is commonly understood as working across and beyond disciplines, building on Gödelian's idea of levels of reality: 'signifying the coexistence between complex plurality [of levels of reality] and open unity', which 'implies the impossibility of a self-enclosed complete theory' and that 'knowledge is forever open' (Nicolescu 2010, p. 19).

By allowing a more effective dialogue between knowledge systems and paradigms, these concepts will be central both to our understanding of the existing critique of successes and limitations of IPBES *Framework* (section 'The Framework: Achievements and Persistent Weaknesses'), and to contribute to its potential (section 'The 'Third Space': Radical Reflection and the Transcending of Paradigms') to serve its original aim.

The Framework: Achievements and Persistent Weaknesses

Having described the most relevant dimensions of the IPBES *Framework*, and located its contribution in the wider context of GEC and systems change, we will now review the existing critique in academic literature in order to understand what are the main lines of thinking about the *Framework* since 2015. We performed a search through the engines of 'ScienceDirect' and 'Google Scholar' using a range of keywords: 'IPBES', 'IPBES Conceptual Framework', which was also combined with qualifiers arising from the literature itself, including: 'ecosystem services', 'biodiversity', 'nature's contributions to people'. We identified and selected texts according to their relevance to the overall theme, mainly peer-reviewed articles, and public documents produced within the institutional scope of the IPBES. Additional articles were also identified through the bibliographical references within the original sample, which included links with themes such as GEC, knowledge production to policy-making, Climate Change and interdisciplinarity. The selected texts were analysed and categorized, through the process of reviewing data, coding it and analysing it at the same time, following Glaser and Strauss (2017).

Sections 'Achievements' and 'Persistent Weaknesses' explore the main achievements and weaknesses of the *Framework,* identified within the revised literature.

Achievements

The dominant representation of humans-nature relations, which is deeply intertwined with the dominant socio-economic system that still frames 'global kinds of knowledge' (Hulme 2010) needs to be challenged. Diversity and plurality of paradigms concerning such relations need to be effectively included, informing and shaping politically relevant knowledge for global environmental governance. The Framework promotes this in four ways.

Human Well-being and Harmony with Nature as Ultimate Ends

The IPBES *Framework*, places 'Quality of Life' (QoL) and notions of well-being, at its core. In doing so, it distinguishes itself in two important ways compared to other proposals on science-policy interfaces concerning biodiversity conservation and assessment, ecosystem services approaches and concepts of sustainability (e.g. UN 2015; MEA 2005). First, it places 'Human well-being' and 'Living in balance and harmony with nature' as ultimate ends within the overall framework (in line, for instance, with Costanza et al. 2016). Second, it expands its understanding beyond common material indicators of wealth and economic growth, viewing human well-being from outside of the market/consumption and needs/scarcity logic (Göpel 2016). By suggesting that QoL and human well-being are ultimate ends of policy-making, IPBES steps away from the default focus on the growth of Gross Domestic Product (GDP), linking to the potentially transformative agenda of 'Beyond GDP' (Stiglitz et al. 2018).

However, its main transformative potential lies in its explicit connection and acknowledged interdependence between humans and nature's well-being: the *Framework* enriches the above definitions and indicators by adding 'the ethical and ecologically sustainable utilization of nature as key components of the concept of human well-being' (Díaz et al. 2015, p. 7).

Diversity of Values and Knowledge Systems Concerning Humans and Nature

To ensure inclusiveness, IPBES goes beyond the notion of ecosystems services (as in the Millennium Ecosystems Assessment—MEA (2005)), explicitly integrating contributions from non-Western worldviews (Montana 2017), and redressing 'the balance between disciplines, gender and geographical regions' (Morin et al. 2017, p. 553). Miller (2008, p. 1913) considers it 'a very interesting experiment in epistemological democracy in international governance', which 'systematically sought out participants from among developing countries, researchers and indigenous groups… [and] granted them considerable epistemological flexibility in their contributions', including ILK from Andean, East Asian and South Pacific cultures (Díaz et al. 2015). It is the ILK contribution to the conception of QoL that has introduced in the *Framework* an often neglected, yet fundamental, leverage point for sustainability transformation (Abson et al. 2017): the harmonious relationship with nature as central to human well-being.

The pursuit of diversity and plurality in value representation through the *Framework* (Brooks et al. 2014; Peterson et al. 2018), and the resulting extension of the spectrum of values and perspectives on humans-nature relations, is thus a major achievement. Acknowledging worldviews conveying different values related to 'Nature's Contributions to People' (NCP) is crucial to transformative practices, but 'rarely taken into account in decision making' (Pascual et al. 2017, p. 9): the IPBES *Framework* has given visibility to systems of knowledge too often marginalized by mainstream approaches to biodiversity conservation and by the dominant view of humans and nature as separated. This can be considered one of the main achievements of the *Framework*.

Humans and Nature: A Relational Approach

One of the fundamental aspects of the *Framework* is precisely the relational dimension that underpins much of what was argued above. According to Chan et al. (2016), defining relational values offers an opportunity to bring to the discussion the extra-material benefits (e.g. social, cultural, spiritual) of humans-ecosystems interactions. As Pascual and colleagues argue '[r]elational values reflect elements of cultural identity, social cohesion, social responsibility and moral responsibility towards nature' (Pascual et al. 2017, p. 2012); through the relational lenses, NCP aims to better accommodate pluralistic valuation of humans-nature relationship, able to embody 'symbolic relationships with natural entities' deeply intertwined with a sense of identity and spiritual fulfilment (Pascual et al. 2017, p. 11). Thus, by thinking the relational dimension of values concerning humans and nature beyond the dichotomic debates between intrinsic or instrumental value, the *Framework* takes a critical step for environmental protection (Chan et al. 2016, p. 1462).

Beyond Current Trajectories, Towards Transformative Change

The 2019 report by the IPBES warns against the dangers and risks of insisting on 'current trajectories' concerning biodiversity conservation, and its failures. The persistence of indirect drivers underpinned by societal values and behaviours that include 'rapid human population growth, unsustainable production and consumption and associated technological development' won't allow the achievement of a sustainable use of nature, necessary to reverse the dramatic trend of biodiversity loss (IPBES 2019a, p. 15). The Report thus concludes that only a 'system-wide reorganization across technological, economic and social factors, including paradigms, goals and values' (IPBES 2019a, p. 33) would prevent the negative trends projected 'to continue to 2050 and beyond' (IPBES 2019a, p. 16).

The Report's clear call for transformative change stands out in the context of international environmental governance, for its efforts to connect what keeps being divided (something Miller (2008) acknowledges also to the MEA Report): the destiny of humanity and all other forms of life on Earth, and to question the dominant forms of knowledge and worldviews that reproduce such divide. Its appeal to engage in such transformative change, echoes theories of system change for sustainability transformations, and in particular to Meadows' (1999) deepest leverage points for sustainability change: values, goals and mindsets that underpin paradigms.

The above-mentioned achievements are, therefore, significant, and emphasize not only the potential contribution of the *Framework* in pushing the debate and agenda-setting into a new, diverse and plural arena, but also the Platform's capacity to improve itself building on criticism, over the past few years. The *Framework's* achievements are also central to transformative systems change. The clarity it offers in terms of interdependence of all forms of life on Earth helps to explain why a far more central consideration of biodiversity is needed in the politics and policy of implementation of the UN 2030 Agenda for Sustainable Development. In this sense, the 2019 Report aligns with calls for transformative change (e.g. Leichenko and O'Brien 2019), by stressing the urgency of countering the inertia of current trajectories of institutionalized structures and epistemologies (Jasanoff 2005; Miller 2008).

Persistent Weaknesses

The IPBES's far-reaching mission and the goal set for its *Framework* expose it to criticism as well as praise. The intention of the *Framework* is 'to support communication across disciplines and knowledge systems and between knowledge and policy' (Díaz et al. 2015, p. 3). Scholars from different disciplinary backgrounds, concerned with the production of global kinds of knowledge for environmental governance, question its ability to allow effective and equitable dialogues (Borie and Hulme 2015; Turnhout et al. 2016; Dunkley et al. 2018), calling upon the Platform to do a 'constant reflexive scrutiny' (Turnhout et al. 2016, p. 70), questioning its intents, implementation and tools for knowledge production.

In section 'Achievements', we have shown why the IPBES intention is laudable, given its potential to challenge the persistent views on humans-nature relations beyond dominant socio-economic systems (Costanza et al. 2016; Göpel 2016; Moore 2016; Raworth 2017). In this section, we argue that significant obstacles to the effective production and sharing of plural policy-relevant knowledge remain, and that the way in which IPBES attempts to implement its aims is not without criticism. Consequently, we concur with scholars and commentators who argue that the full potential of the IPBES *Framework* has yet to be attained.

The main critique around which we focus suggests that the architects behind the *Framework* have insufficiently unpacked the valuation processes that underpin the different knowledge systems.[6] The *Framework* is premised on the conflation of different knowledge systems, and their heterogeneous, often oppositional worldviews (Dunkley et al. 2018, p. 789) under the same framework. However, this premise is developed without fully exposing and critically examining the different meanings, values and ontologies that those systems represent. The limitations and partial views

[6] A knowledge system is 'a body of propositions that are adhered to, whether formally or informally, and are routinely used to claim truth' (Díaz et al. 2015, p. 13). A knowledge system is produced within a specific cultural context, translating its particular ways of knowing and representing reality.

concerning cross-cultural approaches and 'global kinds of knowledge' (Hulme 2010) are core challenges for the *Framework*—and a failure to question them may constitute a flaw that ultimately undermines the original intention of IPBES. Such a flaw may be aggravated by predefined power relations between knowledge systems that generally characterize environmental knowledge production, and global environmental agendas (Hulme 2010).

Having in mind the broader context outlined in section 'Environmental Knowledge, Production and Change: Setting the Context for IPBES', we now critically review and categorize three tensions, which—despite significant efforts referred to earlier—might still undermine the *Framework* effectiveness: (A) epistemic primacy and power balances, (B) normative and descriptive dimensions and (C) qualitative and quantitative approaches.

The Risk of Epistemic Primacy and Unbalanced Distribution of Power

The IPBES *Framework* is a rare opportunity to make ILK visible and relevant to policy in international governance forums. However, putting different systems of knowledge together within a visual composition does not imply an effective understanding of their meaning within their original context, nor does it ensure that they can be successfully translated without losing their original sense. The 2015 *Framework* claim of equivalence between ILK and the Western scientific paradigm is not adequately conceptualized: there is the risk of undermining the very diversity that the *Framework* was meant to enable by imposing a pre-existing grid of understanding, and reducing the contribution and novelty of minority knowledge systems to what is already known within the dominant paradigm.

We suggest that the concept of 'boundary object', introduced in section 'Enhancing Plurality: Four Concepts', helps understand the risks entailed in working across knowledge systems:

> Boundary objects can support cross-boundary communication and knowledge translation by connecting those operating in different social worlds, whilst preserving their autonomy (Rödder 2017)... The development of a boundary object is important in cross-cultural work... where processes of knowledge generation are based on different epistemologies (Dunkley et al. 2018, p. 782).

To preserve the diversity of epistemologies is one of the main goals of 'boundary objects', such as the IPBES *Framework*. However, the risk of epistemic primacy of a dominant knowledge system puts into question the effectiveness of such diversity: it might cancel the opportunity to co-generate knowledge and the advantages that would arise from having contributions from different worldviews, ways of knowing and valuations of humans-nature relations.

According to Löfmarck and Lidskog (2017, p. 28), the risk of subordinating ILK to the Western Scientific knowledge system, or to turn it into a complement of it, is significant, since the framework does not provide 'forms for dealing with contrasting rationalities, diverging ontological claims and different criteria for knowledge validation'. Incommensurability between knowledge systems is both expected and accepted

(Dunkley et al. 2018), hence putting together an inclusive framework should not contribute to obliterate the valuable contributions of ILK concerning humans-nature relations, by making them *scientific* (Löfmarck and Lidskog 2017, p. 28), but rather: to enable the integration of plurality of meaning and values, thus strengthening the science-policy interface.

Scholars have identified the following interrelated problems: the assumption of 'epistemic primacy' by those who hold the interpretation and transfer of non-Western knowledge systems into the IPBES *Framework* is still not fully acknowledged, leading to the risk of marginalization, and persistent power inequalities (Löfmarck and Lidskog 2017; Tengö et al. 2017; Turnhout et al. 2016). These debates and critiques were heeded, and lead authors of the 2015 proposal took steps: acknowledging the risk of power imbalances 'a lack of capacity in interpreting these knowledge systems could easily lead them to being disregarded…' (Díaz et al. 2018, p. 272), as well the 'cultural incommensurability and resistance to universal perspectives on human-nature relations', introducing a more subtle approach when dealing with NCP, making the distinction between generalizable, contextual and mixed perspectives, and considering them as equally useful to rebalance power distribution (Díaz et al. 2018, p. 272).

The question then becomes whether these efforts are sufficient to enable the full potential of IPBES *Framework* in moving beyond the persistent limitations of international governance forums and the production of 'global kinds of knowledge' (Hulme 2010) to deal with multiple crises. The risk of erasing cultural and contextual differences in meaning-making through the production of globalized knowledge that aims to offer universal and de-contextualized views is dissected by different scholars working within the context of climate change (e.g. Hulme 2010), and discussed as a transversal risk of GEC agendas. Without acknowledging the performative dimension of all knowledge systems and how they frame in partial and contextualized ways goals, beliefs and value systems that justify actions (O'Riordan and Jordan 1999), one cannot overcome the risk of 'epistemological violence' (Radcliffe et al. 2010).

Unfortunately, the 2019 IPBES Global Assessment Report does not dissipate these concerns when reporting on the progress towards the Aichi Targets[7]: there is insufficient information concerning the integration of ILK and the effective participation of Indigenous Peoples and Local Communities (IPLCs) (IPBES 2019a, p. 34).

The Risk of Conflating Normative and Descriptive Dimensions

The obstacles arising from unexamined and, or, unacknowledged epistemic primacy are compounded by the use of the *Framework* to artificially establish borders between knowledge systems: 'the CF fails to capture the different perspectives, beliefs and values of knowledge holders *at finer scales*, which is what it set out to achieve' (Dunkley et al. 2018, p. 790). The laudable attempt to work across knowledge

[7]Aichi Targets are part of the 'Strategic Plan for Biodiversity 2011–2020' of the Convention on Biological Diversity (https://www.cbd.int/sp/targets/).

systems, leads to the attempt to capture them in a stabilized framework, which in turn risks an artificial reification and dichotomization of knowledge itself, denying the necessary flexibility for effective debates. We explore this risk by engaging with the notion of *value* within the *Framework,* arguing that the difficulty in clarifying its meaning in the context of IPBES leads to further difficulties and misconceptions.

The *Framework* refers to 'Nature's contributions to people—NCP' (Pascual et al. 2017). It is within this framing that the different perspectives on humans-nature relations become central, and that the risk of conflating intrinsically different perspectives and approaches becomes tangible (e.g. Borie and Hulme 2015; Löfmarck and Lidskog 2017; Maier and Feest 2016). First, we find the mainstream Western scientific approach: *descriptive* by definition, and concerned with the 'discover[y of] facts about things in the spatio-temporal world, as well as their... non-normative properties and relationships' (Maier and Feest 2016, p. 337); its focus is on ecosystem goods and services through the lens of a functionalist, utilitarian perspective on nature (e.g. Kenter 2018; Vadrot et al. 2016).[8] Second, we have the ILK approach: essentially *normative*, consisting in 'a body of beliefs about why certain things do matter and what this entails for how people ought to behave with respect to these things' (Maier and Feest 2016, p. 337), referring to concepts such as 'Nature's Gifts' (Díaz et al. 2015).

Thus, on the one hand, 'Nature's Gifts' dimensions are explicit about the ways and reasons why humans-nature relations have meaning and value to local and/or indigenous people, and why their protection is central. On the other, Western scientific claims of neutrality, together with ecosystems services functionalist approaches, make it harder to understand why biodiversity matters. In fact, goods and services describe functions of desire and/or preference, which have 'marginal normative importance' (Löfmarck and Lidskog 2017, p. 323).

Despite these radically different ways to value humans-nature relations, the *Framework* suggests an equivalence between 'Nature's Gifts' normative claims, and 'Ecosystem Goods and services' scientific and functionalist claims. Both dimensions are subsumed under NCP, understood in the IPBES context as 'economic facts represented by valuations and indexes of neoclassical and happiness economics' (Maier and Feest 2016, p. 337): this entails displacing the normative dimension of ILK's worldview from its original context of meaning-making to include it in an econometric de-contextualized and potentially global standpoint.

As discussed in section 'Environmental Knowledge, Production and Change: Setting the Context for IPBES', this constitutes a common trait of global kinds of knowledge production typically linked to GEC, which tends to be 'insensitive to the peculiarities of place and context' aiming to produce 'globalised explanations and predictions of environmental change' (Hulme 2010, p. 559). In Jasanoff's words (2010): 'a view from everywhere'. The problem with these de-contextualized

[8]Without explicitly including in the IPBES *Framework* references to the co-existing alternative (though invisible) Western paradigms concerning Life, Biodiversity and humans relations to it, there is the risk of a reductionist reading of (Western) concept of 'Nature', for instance, in the assessment practices. As O'Brien (2019) states concerning Climate Change, the paradigm used to frame Policy Agendas is still deterministic, atomistic and mechanistic.

ways of knowledge production is that they block co-generation of meaning-making, hence cancelling the main purpose of science-policy and cross-cultural platforms such as IPBES. Without being exposed and acknowledged, this 'globalized instinct' erases the diversity of ways of knowing and valuing 'embedded in local practices and knowledge-making traditions' (Hulme 2010, p. 560).

We argue that ambiguities concerning the meaning of *value* within the *Framework* reflect difficulties in integrating locally and contextualized knowledge-making, and underpins many of the problems discussed above. The additions and the clarifications made in 2017 to the IPBES initial *Framework* (2015) included an attempt to address this problem:

> the word 'value' can refer to a *principle* associated with a given worldview or cultural context, a *preference* someone has for a particular state of the world, the importance of something for itself or for others, or simply a *measure*[9] (Pascual et al. 2017, p. 9).

This commendable attempt nonetheless falls into the same difficulty that invalidates the overall attempt, by dealing with intrinsically different valuation processes as if they were equivalent. The quote above includes three alternative ways of understanding the notion of 'value' that combine in a way that is suggestive of interchangeability. Yet, one should not conflate values with individual preferences or interests, which characterize particular practices and lifestyles (Maier and Feest 2016). The difficulty in clarifying the notion of 'value' thus persists, and it is an example of the difficulty in allowing the space for effective plural and contextualized knowledge generation, which risks undermining the *Framework* further.

The Risk of Conflating Qualitative and Quantitative Approaches

Building on the previous overview of epistemic dimensions, questions of power distribution and the tension between normative and descriptive understandings, we now turn to the question of quantification and economization (Hulme 2010; Göpel 2016).

The focus on ecosystem services is central to this third category of risks in relation to the *Framework*. The scientific community that contributed to the IPBES *Framework* sees ecosystem services as a pragmatic notion that enables the translation of scientific knowledge about biodiversity into a tangible (and quantifiable) dimension for policy-makers and stakeholders involved in the decision processes. Borie and Hulme (2015, p. 493) describe the way people involved in the process of discussing and ratifying the *Framework* at the Paris meeting in 2012 adopted this pragmatic approach, arguing that 'ecosystem services' is an 'ideologically neutral notion' and the most effective way to communicate biodiversity-issues to decision-makers, hence the best way to enable science-policy dialogue.

The argument is that the supposed neutrality surrounding the notion of ecosystem services facilitates pragmatism, providing an easy way for those involved to grasp the

[9]Our emphasis.

value of biodiversity and of humans-nature relations, thus enabling the dialogue and inclusion of different perspectives. But, can 'ecosystem services'—an expression created under the realm of service economy—be considered neutral? And is it able to 'capture what centrally matters about biodiversity or nature' (Maier and Feest 2016, p. 328)?

The limitations and advantages of the use of the Ecosystem Services Framework for biodiversity conservation have been widely discussed in the literature beyond the specific debate on IPBES. Gunton and colleagues, for example, identify in ecosystem services two of the problems that we see being replicated in the debate around IPBES: incoherence in definitions and a 'narrow approach to valuation' (2017, p. 249). We have partly explored the first in section 'Persistent Weaknesses' (A) and (B). Here we look at the narrow approach to valuation in two ways: the reductionist use of economic language and frames, and the tension between the functionalist logics versus the notion of intrinsic value of biodiversity.

The *Framework's* 'econometric perspective' risks displacing (or distorting) other forms of—non-quantitative—valuation, despite the efforts to make it plural and inclusive. Although arguments have been made to the contrary (Díaz et al. 2015; Pascual et al. 2017), the *Framework* is steeped in economic and econometric language: '"Goods" are either "assets" or "services". An asset, in economic language, is something *owned* that's expected to provide *benefits*' (Maier and Feest 2016, p. 334). A benefit thus becomes a quantifiable dimension within a predefined framework of profits and costs, and yet it can be argued that something can be *a good* independently of providing any kind of quantifiable benefit or prejudice; moreover, something can provide us with benefits even if that is not what provides it with *value*. The fact that these two dimensions—value and benefit—can be attributed to the same thing does not make them the same.

We suggest that by conflating qualitative and quantitative dimensions the *Framework* ends up generating more misunderstandings than pragmatic clarification. Given the starting point of IPBES, whereby plural epistemologies and worldviews must be engaged, we argue that the *reason why biodiversity matters and why it is worth protecting* is made *less* clear through the use of economic language.

The problem with the persistent use of ecosystems services and econometric logics within global environmental knowledge platforms is both the risk of becoming a default view—which depends only on those involved in the *Framework*'s implementation and their deep-seated epistemologies (Jasanoff 2005; Miller 2008)—and the fact that it reproduces a harmful reductionist and dualistic view of humans and nature as separated (Turnhout et al. 2013; O'Brien 2019). That persistence prevents alternative paradigms of humans and nature connectedness to be effectively integrated (Borie and Hulme 2015), and the urgent need for effective current socio-economy systems to change and enable sustainability transformation (Göpel 2016; Meadows 1999). This last obstacle is captured well in the words of Weber and Kurt (2015, p. 58):

> Civilization still operates as if reality is about organizing inert, dead matter in efficient ways. It is impossible to achieve sustainablity with our prevailing "operating system" for

economics, politics, and culture if the underlying "bios"—our unconscious assumption about reality—remains tied to an ideology of dead matter.

Ecosystems services discourse risks depriving biodiversity of its value as something worthy of dignity in itself, independently of any exterior measure, or measurable contribution to our preferences and lifestyles, hence alienating our responsibility towards its conservation (independently from our specific interests). The benefit can be understood as a secondary outcome of the existence and value of biodiversity and nature and of our relation to them: being worthy of respect and care *independently* of the benefit we can take from them.[10]

We do not intend to argue here that there is necessarily an incompatibility between intrinsic value and relational value, or even that IPBES is effectively contributing to a commodification of nature (Hahn et al. 2015). But we concur with scholars who emphasize that we need to be aware of the risk of using quantitative, econometric scales which, under the 'filter effect' of dominant socio-economic systems and 'civic epistemologies', might effectively create distortions rather than clarity in communication: it risks restricting the very essence of the *Framework's* purpose, ultimately worsening rather than solving the biodiversity crisis (Borie and Hulme 2015). We argue that the need to quantify something not quantifiable with the aim to provide a more pragmatic framing, results in a paradox: to include something not quantifiable in an econometric measure in order to demonstrate its value is, in fact, simultaneously, to withdraw value from it.

The 'Third Space': Radical Reflection and the Transcending of Paradigms

Starting from the discourse on systems change for sustainability transformation, we focused on the IPBES *Framework* as a 'case' in order to deepen our understanding of the epistemological and ontological assumptions that underpin its potential to change systems. We did so building on Göpel (2016) and Meadows' (1999) calls to expose unexamined assumptions and blind spots underpinning mindsets and paradigms, whose implications populate the debate around IPBES and, more generally, global kinds of environmental knowledge.

In section 'Persistent Weaknesses', with the help of the literature reviewed, we identified three main interconnected concerns that are emblematic of those contexts of global environmental knowledge, where: a 'globalising instinct' (Hulme 2010), econometric views (Turnhout et al. 2013) and civic epistemologies (Jasanoff 2005),

[10] We believe that a teleological reading—understood according to the acknowledgement of an *autopoietic* dimension in all living organisms (Weber 2013) as intrinsically purposeful (van der Schyff 2010)—can help us understand the notion of intrinsic worth in a much richer and consequent way than the one offered by the IPBES *Framework*: 'the values inherent to nature, independent of human judgment, and therefore beyond the scope of anthropocentric valuation approaches' (Díaz et al. 2015, p. 14).

shape de-contextualized and reductionist approaches to knowledge generation, preventing effective integration of other ways of knowing and valuing humans-nature relations.

In this section, we reflect on *how* to strengthen the Framework's capacity for plurality, hence presenting our own contribution to this field of knowledge, having set out the reasoning as to *why* plural ways of knowing environment and humans-nature relations matter in contributing to more informed and effective practices for biodiversity conservations, and *what* is still not working about the IPBES *Framework*'s in its attempt to promote such plurality.

The above-mentioned blind spots needed to be exposed to enable an effective plurality of alternative paradigms and knowledge systems and, ultimately, to allow to transcend paradigms and promote integral approaches to sustainability transformation. Once exposed the assumptions that underpin both global environmental knowledge debates and the IPBES context, we intend to contribute to strengthen the Framework in order to overcome the problems reviewed in section 'Persistent Weaknesses', building in notions such as 'third space' (Turnbull 2003), 'boundary objects', 'multiple rationalities' (Dunkley et al. 2018) and 'transdisciplinary objects' (Nicolescu 2010).[11]

We propose that in order to work as a tool for 'integrative, cross-paradigm, coproduced knowledge' (Díaz et al. 2015) (i.e. as a 'boundary object'), the *Framework* should enable the creation of a 'third space' (Turnbull 2003). This is a sharing space, in which '[t]he power and antagonism of different rationalities are acknowledged, in contrast to denying, erasing or suppressing other ways of knowing' (Dunkley et al. 2018, p. 789). Despite the need for structure and a certain level of standardization (to enable data collection and validation across different contexts),[12] the *Framework* should allow for flexibility, and mutual dissention (Dunkley et al. 2018), leaving open 'an additional *third space*' (Turnbull 2003) for future developments, negotiation and co-evolution, triggered by cross-fertilization between different cultural or knowledge systems, without forcing consensus.

Here we return to the idea of 'multiple rationalities' (section 'Persistent Weaknesses') to activate the contribution of less dominant and visible epistemologies and ways of knowing, provided not only by ILK, but also by other paradigms within the Western tradition which are not sufficiently represented—if not excluded—in global kinds of knowledge, such as global environmental science-policy interfaces (e.g. O'Brien 2019). In fact, it is not only the effective integration of ILK into the processes of knowledge generation that is at risk: in the context of IPBES, Western science is framed as representing a single view, which does not explicitly refer to the multiplicity of co-existing paradigms concerning humans-nature relations that exist within the Western tradition: including, for example, the 'Enlivenment' proposal by

[11] See section 'Enhancing Plurality: Four Concepts' for definitions of these concepts.

[12] The need for such collection and processing of data is, of course, itself, the product of a worldview and practice of governance intrinsically linked to sectorial and techno-managerial approaches within the Western scientific paradigm, its link to the dominant socio-economic systems, and the typical way of global production of knowledge to inform science-policy platforms.

Andreas Weber (2013). As a result, mainstream Western science risks promoting reductionist interpretations not only due to dominant econometric approaches but also to the hidden and institutionalized lenses of 'civic epistemologies' (Jasanoff 2005), which will ultimately condition the interpretation and implementation of the *Framework* itself.

In order to avoid both reductionist and/or marginalizing readings of less dominant ways of knowing, the creation of a space where multiple rationalities, and ways of knowing and valuing are visible and allowed to exist and co-evolve, is of pivotal relevance. As Dunkley et al. argue (2018, p. 792) '[i]t is not necessary to merge heterogeneous concepts and viewpoints to advance knowledge': the most important feature of a 'boundary object' is to enable a 'sharing space' to allow cooperation and co-evolution despite dissention. The creation of such a place, as we've argued in section 'The Framework: Achievements and Persistent Weaknesses', referring to Meadows' understanding of assumptions and blind-spots, and to Jasanoff's (2005) view of civic epistemologies depends on each knowledge system being capable of reflexivity.

As Turnbull (2003) also stresses, all knowledge systems have ontological assumptions about whom to trust, what counts as evidence, what criteria of validity are generally accepted, etc. And only through the acceptance of this heterogeneity will we be able to interact and co-evolve:

> It is in the revelation of these hidden messy and actively constructed components that the possibility lies of working together with people in other spaces. Herein lies the real work of putting on the motley. If we do not actively celebrate the messiness of all our knowledge making we will in the long run condemn ourselves to an inevitable death brought on by the inflexibility and sterility of a monoculture (Turnbull 2003, p. 232).

There is an effective performative dimension of each knowledge system which defines goals, values and power distribution that needs to be accepted and exposed so to enable the creation of an equitable place where heterogeneity and plurality are conditions for knowledge generation. Messines and dissention can be a generative dimension 'without the shadow of a single transcendent rationality' (Turnbull 2003, p. 234). But we need to embrace the heterogeneity within plurality.

Nicolescu (2010) goes further, proposing a process of co-generation of knowledge to work across and *beyond* different knowledge systems, and not only disciplines. He promotes a reading of transdisciplinary objects (such as biodiversity and humans-nature relations) that allows us to acknowledge the plurality of levels of reality, and the limits of each particular paradigm to embrace the whole reality, thus advocating the need to further transcend them. As a quantum physicist, Nicolescu's approach to transdisciplinarity is enriched by his insight on the limits of disciplines, and on the limits of specific scientific paradigms. Through the example of the classic modern Western scientific paradigm, he shows how an ideal of freedom (portrayed in the separation between subject and reality), which resulted in the creation of a scientific method, can turn into a 'potential danger' when 'incarnated by the ideology of scientism' (Nicolescu 2010, p. 21), and considered as *the* explanatory paradigm.

One of the contributions of quantum physics principles has been calling into question both the idea of separability between subject and object, and the absolute exteriority of the material world. Quantum revolution puts an end to that misconception, namely through Heisenberg's 'Uncertainty Principle' associated with dimensions of indeterminacy and non-separability. Hence, Nicolescu's project for transdisciplinarity questions the alleged objectivity of the classical Western scientific model simultaneously seen as a *legitimate* and a *limited* way to represent reality.

This approach acknowledges the limits of mainstream Western modern Newtonian paradigm—which is still dominant in the public arena of GEC (see O'Brien 2019), and highlights the essential step of reflexivity for effectively enabling the purpose of a framework such as IPBES's.

In order to benefit from heterogeneity and plurality, the Framework needs to create a 'sharing space' where paradigms are questioned and transcended. Nicolescu (2010, p. 26) states that 'every level [of Reality] is characterized by its incompleteness: the laws governing this level are just a part of the totality of laws governing all levels', emphasizing the need to acknowledge the incompleteness of each and every paradigm. This leads us back to Meadow's claim on the ultimate need to transcend (and abandon) the paradigmatic way of thinking:

> There is yet one leverage point that is even higher than changing a paradigm. That is to keep oneself unattached in the arena of paradigms, to stay flexible, to realize that *no* paradigm is "true," that every one, including the one that sweetly shapes your own worldview, is a tremendously limited understanding of an immense and amazing universe that is far beyond human comprehension (Meadows 1999, p. 19).

Only this level of reflexivity and awareness will allow the IPBES *Framework* to create the space both for cross-cultural work, based on different epistemologies, and to transcend those epistemologies and the paradigms underpinning them. The *Framework* needs to create the conditions to benefit from a plurality of knowledge systems and paradigms, not from a relativist point of view, but from one in which we agree that all the resulting insights, despite representing a limited point of view, have a legitimacy to inform.

Thus, the ability to transcend the paradigmatic way of thinking can allow processes of 'mutual creative fertilization' towards transmodern (Turnbull 2003, p. 233), plural and integral approaches (see e.g. Leichenko and O'Brien 2019; Wilber 2007; Woiwode 2016). Our claim is that such approaches would help to strengthen the *Framework,* overcoming the limits of global kinds of knowledge, effectively benefitting from the inputs of diverse knowledge systems and marginalized ways of knowing and valuing the relations between human and non-human forms of life, thus contributing to a transformative change towards more sustainable futures.

Conclusions

In May 2019, the IPBES Global Assessment Report on Biodiversity and Ecosystem Services, the first global assessment on biodiversity since the Millennium Ecosystem Assessment in 2005 (MEA 2005), stated that

> Goals for conserving and sustainably using nature and achieving sustainability cannot be met by current trajectories, and goals for 2030 and beyond may only be achieved through transformative changes across economic, social, political and technological factors... Aichi Biodiversity Targets and the 2030 Agenda for Sustainable Development will not be achieved based on current trajectories (IPBES 2019a, p. 14).

Thus, transformative change is a condition for life on Earth to continue and thrive. As part of global environmental governance initiatives, the IPBES strives to overturn the decline of biodiversity (IPBES 2019a), and seeks to redefine the humans-nature relationship by contributing to a sustainable use of biodiversity, as well as long-term human well-being and global sustainable development. Our analysis focused on its *Conceptual Framework* as a means of improving science-policy interface, and crucially as an instrument that places the significance of multiple knowledge systems at the forefront of GEC. Conceptions of nature and of humanity's place within it are, we have argued, the best arena to finally confront and resolve the fundamental epistemological and ontological divide that stands in the way of transformative change.[13] In the words of the American writer James Baldwin (Baldwin and Peck 2016), influential voice of the civil rights movement: 'Not everything that is faced can be changed; but nothing can be changed until it is faced'.

The dramatic levels of biodiversity loss and the growing level of genetic homogenization can no longer be tackled through the same business-as-usual approach of techno-managerial, mitigation and incremental practices. A fundamental, system-wide reorganization across technological, economic and social factors, including paradigms, goals and values (precisely what Meadows (1999) calls *deepest* leverage points) is now explicitly demanded. Hence our focus on the untold blind spots and unacknowledged assumptions that we suggest act as deep barriers to the inclusion of diversity and plurality of knowledge systems and paradigms concerning humans-nature relations and biodiversity value. These barriers undermine attempts to disrupt the resilient and pervasive view of humans and nature as separated, which in turn underpins the reductionist, dominant socio-economic system (and hidden institutionalized views and practices), and its environmental change agendas.

Building on the discourse on systems change for sustainability transformation (see Göpel 2016; and Meadows 1999) and on the epistemic foundations of global environmental knowledge (see e.g. Hulme 2010; Leichenko and O'Brien 2019), we attempted to clarify the drivers of persistent problems that still block transformative

---

[13]The recent viral dissemination of Bendell's (2018) work in relation to deep adaptation to climate change is an indicator of the prescience of such debates. Göpel M. (2016) *The Great Mindshift. How a New Economic Paradigm and Sustainability Transformations go Hand in Hand*: Springer International Publishing.

change, preventing an effective integration of plural and alternative ways of valuing biodiversity and perceiving humans-nature relations.

We acknowledged the main achievements of the IPBES *Framework*, namely the consideration of human and nature well-being as ultimate ends, which—if applied—could trigger transformative change, as suggested by the 'beyond GDP' agenda; the need to include a diversity of values and knowledge systems concerning humans and nature, the adoption of a relational approach to humans and nature, and, crucially, IPBES's acknowledgment of a need to move away from current trajectories and adopt transformative changes.

We then focused on a detailed analysis of the meaning and consequences of three major risks noted by the existing literature on the *Framework*: the risk of epistemic primacy and resulting unbalances in power to shape and contribute knowledge which echoes the critique of global kinds of knowledge in GEC; of invalidating the *Framework* by conflating ILKs normative and Western sciences descriptive approaches, while failing to fully engage with heterogeneous notions of value underpinning 'Nature's gifts' and 'Nature's contributions'; and the risk of conflating qualitative and quantitative approaches, failing to contain the dominance of econometric language and perspectives and their reductionist appeals to pragmatism. We stressed the need to expose existing blind spots and to raise awareness about the assumptions underlying the *Framework,* including answers to the reasons why biodiversity matters and why it is worth protecting, as a key step in reflexivity.

Bringing a range of disciplinary perspectives on these well-rehearsed questions into dialogue allowed us to identify and propose a new, radical space of knowledge production, combining boundary objects, multiple rationalities, third space and transdisciplinarity. The resulting 'space' can help address persistent weaknesses, and give the *Framework's* purpose a reach beyond the merely operational, strengthening plurality and acceptance of heterogeneity and creating the opportunity to transcend paradigms and benefit from integral approaches (concerning biodiversity conservation and humans and nature connectedness).

For the IPBES *Framework* to fully embrace its role, as a device for allowing a much-needed dialogue between knowledge systems, dissention will have to be embraced and a space for trans-rationality created. Embracing dissention creates the opportunity to raise awareness about the limits of the dominant paradigm, revealing its powerful barriers towards more plural ways to know biodiversity and humans and nature's well-being, and its denial of the paradigm shift(s) needed for sustainability thinking to flourish.

References

Abson, D. J., Fischer, J., Leventon, J., et al. (2017). Leverage points for sustainability transformation. *Ambio, 46,* 30–39.

Baldwin, J., & Peck, R. (2016). *I am not your negro.* (Film) Magnolia Pictures.

Bennett, E. M., Solan, M., Biggs, R., McPhearson, T., Norström, A. V., Olsson, P., et al. (2016). Bright spots: Seeds of a good Anthropocene. *Frontiers in Ecology and the Environment, 14*(8), 441–448.
Borie, M., & Hulme, M. (2015). Framing global biodiversity: IPBES between mother earth and ecosystem services. *Environmental Science & Policy, 54,* 487–496.
Brooks, T. M., Lamoreux, J. F., & Soberón, J. (2014). Ipbes ≠ ipcc. *Trends in Ecology & Evolution, 29*(10), 543–545.
Castree, N., Adams, W. M., Barry, J., Brockington, D., Büscher, B., Corbera, E., et al. (2014). Changing the intellectual climate. *Nature Climate Change, 4*(9), 763.
Chan, K. M., Balvanera, P., Benessaiah, K., et al. (2016). Opinion: Why protect nature? Rethinking values and the environment. *Proceedings of the National Academy of Sciences, 113,* 1462–1465.
Costanza, R., Fioramonti, L., & Kubiszewski, I. (2016). The UN sustainable development goals and the dynamics of well-being. *Frontiers in Ecology and the Environment, 14,* 59.
Dalby, S. (2016). Framing the Anthropocene: The good, the bad and the ugly. *The Anthropocene Review, 3*(1), 33–51.
Díaz, S., Demissew, S., Carabias, J., et al. (2015). The IPBES conceptual framework—connecting nature and people. *Current Opinion in Environmental Sustainability, 14,* 1–16.
Díaz, S., Pascual, U., Stenseke, M., et al. (2018). Assessing nature's contributions to people. *Science, 359,* 270–272.
Dibley, B. (2012). 'The shape of things to come': Seven theses on the Anthropocene and attachment. *Australian Humanities Review, 52,* 139–153.
Dunkley, R., Baker, S., Constant, N., et al. (2018). Enabling the IPBES conceptual framework to work across knowledge boundaries. *International Environmental Agreements: Politics, Law and Economics, 18,* 779–799.
EEA (2016). *Sustainability transitions: Now for the long term.* Copenhagen: European Environment Agency (EEA), European Environment Information and Observation Network (EIONET).
Ellis, E. (2011). The planet of no return: Human resilience on an artificial earth. *The Breakthrough Journal, 2,* 37–44.
Glaser, B. G., & Strauss, A. L. (2017). *Discovery of grounded theory: Strategies for qualitative research.* Routledge.
Göpel, M. (2016). *The great mindshift. How a new economic paradigm and sustainability transformations go hand in hand.* Springer International Publishing.
Gunton, R. M., van Asperen, E. N., Basden, A., Bookless, D., Araya, Y., Hanson, D. R., et al. (2017). Beyond ecosystem services: Valuing the invaluable. *Trends in Ecology & Evolution, 32*(4), 249–257.
Hahn, T., McDermott, C., Ituarte-Lima, C., Schultz, M., Green, T., & Tuvendal, M. (2015). Purposes and degrees of commodification: Economic instruments for biodiversity and ecosystem services need not rely on markets or monetary valuation. *Ecosystem Services, 16,* 74–82.
Haraway, D. (2015). Anthropocene, capitalocene, plantationocene, chthulucene: Making kin. *Environmental Humanities, 6*(1), 159–165.
Hölscher, K., Wittmayer, J. M., & Loorbach, D. (2018). Transition versus transformation: What's the difference? *Environmental Innovation and Societal Transitions, 27,* 1–3.
Hulme, M. (2010). Problems with making and governing global kinds of knowledge. *Global Environmental Change, 20,* 558–564.
IPBES (2019a). Summary for policymakers of the global assessment report on biodiversity and ecosystem services of the Intergovernmental Science-Policy Platform on Biodiversity and Ecosystem Services. In S. Díaz, J. Settele, E. S. Brondízio E. S., et al. (Eds.), *IPBES secretariat.* Bonn, Germany. https://doi.org/10.5281/zenodo.3553579.
IPBES (2019b). Global assessment report on biodiversity and ecosystem services of the Intergovernmental Science-Policy Platform on Biodiversity and Ecosystem Services. In E. S. Brondizio, J. Settele, S. Díaz, & H. T. Ngo (Eds.), *IPBES secretariat.* Bonn, Germany.
Jasanoff, S. (2005). *Designs on nature: Science and democracy in Europe and the United States.* Princeton University Press.

Jasanoff, S. (2010). A new climate for society. *Theory, Culture & Society, 27,* 233–253.
Jetzkowitz, J., Van Koppen, C. S. A., Lidskog, R., Ott, K., Voget-Kleschin, L., & Wong, C. M. L. (2018). The significance of meaning. Why IPBES needs the social sciences and humanities. *Innovation: The European Journal of Social Science Research, 31*(1), S38–S60.
Kareiva, P., & Marvier, M. (2012). What is conservation science? *BioScience, 62*(11), 962–969.
Kenter, J. O. (2018). IPBES: Don't throw out the baby whilst keeping the bathwater. Put people's values central, not nature's contributions. *Ecosystem Services, 33,* 40–43.
Lahsen, M. (2016). Toward a sustainable future earth: Challenges for a research agenda. *Science, Technology and Human Values, 41*(5), 876–898.
Latour, B. (2004). *Politics of nature, how to bring the sciences into democracy.* Harvard University Press.
Leichenko, R., & O'Brien, K. (2019). *Climate and society: Transforming the future.* Cambridge, Medford, MA: Polity Press.
Löfmarck, E., & Lidskog, R. (2017). Bumping against the boundary: IPBES and the knowledge divide. *Environmental Science & Policy, 69,* 22–28.
Maier, D., & Feest, A. (2016). The IPBES conceptual framework: An unhelpful start. *Journal of Agricultural and Environmental Ethics, 29,* 327–347.
Meadows, D. (1999). Leverage points: Places to intervene in a system. *The Sustainability Institute, 3,* 19 pp.
Miller, C. A. (2008). Civic epistemologies: Constituting knowledge and order in political communities. *Sociology Compass, 2*(6), 1896–1919.
Montana, J. (2017). Accommodating consensus and diversity in environmental knowledge production: Achieving closure through typologies in IPBES. *Environmental Science & Policy, 68,* 20–27.
Moore, J. W. (Ed.). (2016). *Anthropocene or capitalocene?: Nature, history, and the crisis of capitalism.* Oakland: PM Press.
Moore, J. W. (2017). The capitalocene, part I: On the nature and origins of our ecological crisis. *The Journal of Peasant Studies, 44*(3), 594–630.
Morin, J.-F., Louafi, S., Orsini, A., & Oubenal, M. (2017). Boundary organizations in regime complexes: A social network profile of IPBES. *Journal of International Relations and Development, 20*(3), 543–577.
Nicolescu, B. (2010). Methodology of transdisciplinarity-levels of reality, logic of the included middle and complexity. *Transdisciplinary Journal of Engineering & Science, 1*(1), 19–38.
O'Brien, K. (2019). Quantum social change: You matter more than you think.... In Keynote-*Leverage points 2019: International conference on sustainability research and transformation.* 2019, February 8. Retrieved June 27, 2019, from https://www.youtube.com/watch?v=spMcXBdZBmM.
O'Riordan, T., & Jordan, A. (1999). Institutions, climate change and cultural theory: Towards a common analytical framework. *Global Environmental Change, 9,* 81–93.
Pascual, U., Balvanera, P., Díaz, S., et al. (2017). Valuing nature's contributions to people: The IPBES approach. *Current Opinion in Environmental Sustainability, 26,* 7–16.
Peterson, G., Harmáčková, Z., Meacham, M., Queiroz, C., Jiménez-Aceituno, A., Kuiper, J., et al. (2018). Welcoming different perspectives in IPBES: "Nature's contributions to people" and "Ecosystem services". *Ecology and Society, 23*(1).
Radcliffe, S. A., Watson, E. E., Simmons, I., et al. (2010). Environmentalist thinking and/in geography. *Progress in Human Geography, 34,* 98–116.
Raworth, K. (2017). *Doughnut economics: Seven ways to think like a 21st-century economist.* Chelsea Green Publishing.
Reid, W. V., Mooney, H. A., Cropper, A., Capistrano, D., Carpenter, S. R., Chopra, K., et al. (2005). *Ecosystems and human well-being: Synthesis.* Washington, DC: Island Press.
Rockström, J., Steffen, W., Noone, K., Persson, Å., Chapin, F. S., Lambin, E. F., et al. (2009). A safe operating space for humanity. *Nature, 461*(7263), 472–475.

Sandel, M. J. (2013). Market reasoning as moral reasoning: Why economists should re-engage with political philosophy. *Journal of Economic Perspectives, 27*(4), 121–140.
Soga, M., & Gaston, K. J. (2016). Extinction of experience: The loss of human–nature interactions. *Frontiers in Ecology and the Environment, 14,* 94–101.
Steffen, W., Crutzen, P. J., & McNeill, J. R. (2007). The Anthropocene: Are humans now overwhelming the great forces of nature. *AMBIO: A Journal of the Human Environment, 36*(8), 614–621.
Stiglitz, J., Fitoussi, J., & Durand, M. (2018). *Beyond GDP: Measuring what counts for economic and social performance.* Paris: OECD Publishing. Retrieved August 16, 2019, from https://doi.org/10.1787/9789264307292-en.
Tengö, M., Hill, R., Malmer, P., Raymond, C. M., Spierenburg, M., Danielsen, F., et al. (2017). Weaving knowledge systems in IPBES, CBD and beyond—lessons learned for sustainability. *Current Opinion in Environmental Sustainability, 26,* 17–25.
Timpte, M., Montana, J., Reuter, K., Borie, M., & Apkes, J. (2018). Engaging diverse experts in a global environmental assessment: Participation in the first work programme of IPBES and opportunities for improvement. *Innovation: The European Journal of Social Science Research, 31*(1), S15–S37.
Turnbull, D. (2003). *Masons, tricksters and cartographers: Comparative studies in the sociology of scientific and indigenous knowledge.* Taylor & Francis.
Turnhout, E. (2018). The politics of environmental knowledge. *Conservation and Society, 16,* 363–371.
Turnhout, E., Waterton, C., Neves, K., & Buizer, M. (2013). Rethinking biodiversity: From goods and services to "living with". *Conservation Letters, 6*(3), 154–161.
Turnhout, E., Dewulf, A., & Hulme, M. (2016). What does policy-relevant global environmental knowledge do? The cases of climate and biodiversity. *Current Opinion in Environmental Sustainability, 18,* 65–72.
UN (2015). Transforming our world: The 2030 Agenda for Sustainable Development. Retrieved August 19, 2019, from https://www.un.org/ga/search/view_doc.asp?symbol=A/RES/70/1&Lang=E.
UNEP (2010). *Report of the third ad hoc intergovernmental and multi-stakeholder meeting on an intergovernmental science-policy platform on biodiversity and ecosystem services.* Retrieved August 19, 2019, from https://www.ipbes.net/system/tdf/downloads/UNEP_IPBES_3_3_EN.pdf?file=1&type=node&id=14239.
Vadrot, A., Akhtar-Schuster, M., & Watson, T. (2018). The social sciences and the humanities in the intergovernmental science-policy platform on biodiversity and ecosystem services (IPBES). *Innovation: The European Journal of Social Science Research, 31*(1), S1–S9. https://doi.org/10.1080/13511610.2018.1424622.
Vadrot, A., Jetzkowitz, J., & Stringer, C. (2016). Social sciences: IPBES disciplinary gaps still gaping. *Nature, 530*(7589), 160.
van der Schyff, D. B. (2010). The ethical experience of nature: Aristotle and the roots of ecological phenomenology. *Phenomenology & Practice, 4*(1).
Weber, A. (2013). *Enlivenment: Towards a fundamental shift in the concepts of nature* (Vol. 31). Heinrich Böll Foundation.
Weber, A., & Kurt, H. (2015). Towards cultures of aliveness. Politics and poetics in a postdualistic age—an anthropocene manifesto. *Solutions, 5,* 58–65.
Wilber, K. (2007). *Integral spirituality: A startling new role for religion in the modern and postmodern world.* Shambhala Publications.
Woiwode, C. (2016). Off the beaten tracks: The neglected significance of interiority for sustainable urban development. *Futures, 84,* 82–97.
Yusoff, K. (2010). Biopolitical economies and the political aesthetics of climate change. *Theory, Culture & Society, 27*(2–3), 73–99.
Zylstra, M. J., Knight, A. T., Esler, K. J., et al. (2014). Connectedness as a core conservation concern: An interdisciplinary review of theory and a call for practice. *Springer Science Reviews, 2,* 119–143.

Concluding Discussion: The Planetary Is Not the End of the International

Olaf Corry

Abstract Drawing on chapters of this book as well as wider literatures, in this concluding chapter I first situate the relative invisibility of non-human nature in IR, pointing to the demise of geopolitics around the Second World War as part of a wider bifurcation of knowledge into "social" and "natural" sciences. Secondly, I argue that current attempts to take account of non-human nature have tended to bring with them globalist framings that underplay or even obscure the importance of the international. Thirdly, I outline an outlook that does not feature prominently in the rest of this book, but which might provide an additional way of further developing its goals, allowing a theorisation of society that has the non-human at its core to form the building block for a materialist theory of the international. The overall aim is to take stock of attempts to grasp how the metabolism between humans and non-human nature is itself multiple, intrinsically bound up with and marked by relations between societies—separate yet coexisting socio-ecological entities.

With the global ecological crisis becoming ever more acute, and a deepening sense of turmoil in world politics, it is timely to reassess, as this book does, whether existing paradigms and tools of analysis of non-human nature and world politics are fit for purpose. Each chapter of this volume puts forward critiques and potential ways forward towards a world politics more attuned to the mutual imbrication of human and non-human worlds. Anthropocentric and state-centric assumptions in International Relations (IR) led the discipline for decades to treat nature as little more than a set of resources or as a backdrop for competitive human dramas.

The distinction between "human" and "non-human nature" adopted by the editors provides a starting point for pushing back against anthropocentrism in IR, and for working out the implications of ever more global and dramatic entanglements between human societies and the rest of nature. Humans are also part of nature after all, although the social also remains distinct in its processes (though not in its

O. Corry (✉)
University of Copenhagen, Copenhagen, Denmark
e-mail: oc@ifs.ku.dk

J. C. Pereira and A. Saramago (eds.), *Non-Human Nature in World Politics*,
Frontiers in International Relations, https://doi.org/10.1007/978-3-030-49496-4_16

fundamental material substance). At the same time, this volume shows that the entanglement of human and non-human cannot be the end-point of analysis, particularly not if the specific aim of the book is to grasp the role of the non-human in *world* politics (and the key role of the international for non-human nature). The fundamental problem of the world being divided into multiple social units does not disappear with the recognition that humans are part of and dependent upon the rest of nature. On the contrary, international logics and challenges only become more important once it is fully realised that multiple, uneven societies relate very differently to non-human nature as well as to each other.

With this in mind, the gathering storm of accelerating climate change, precipitous biodiversity loss, soil degradation, ocean acidification and rapacious resource extraction cannot be understood by reference to some innate characteristic of "humans". Rather, it is a result of a particular economic-political system that at a particular historical juncture emerged in the interests of *some* humans, and in some societies. Moreover, it arose through a confluence of international pressures and opportunities upon and for those countries (Anievas and Nisancioglu 2015) and subsequently spread via international dynamics to take in most of the world, albeit highly unevenly. That unevenness is now part and parcel of a global crisis (as it is usually understood), as inequality and difference thwart collective international action (Burton-Chellew, May and West 2013). Recognition of the massive and irrevocable impact of human activity on non-human nature (conceived unhelpfully by some as the "end of Nature", McKibben 1989) must therefore be coupled to analysis of "the international": the many consequences of the world being divided into multiple, uneven, historically and geographically varied societies (Rosenberg 2016). In short, for all the talk of global and Earth Systems Crisis, the "planetary" is not the end of the "international". It is a radicalisation of it.

It may seem obvious that the planetary environmental crisis does not mean we are living in a singular global society. But just as IR as a discipline early on lost sight of non-human nature, the core idea of the international is today surprisingly weak in debates about ecological challenges, technological change and existential risks (Boström and Circovic 2011). The core subject matter of the modern discipline of IR emerged in essentially negative terms as "politics outside the state", rather than in positive terms as the profound general consequences of the fact that social life is not, and has never been, singular (Kurki and Rosenberg 2020). The "politics outside the state" definition of IR led to the false idea that IR has little or nothing to say about domestic politics (e.g. Kenneth Waltz's insistence that IR should abstract from domestic factors when theorising the international) (Waltz 1979). But it also has a corollary allowing "internalism" to proliferate in social analysis: the fallacy that developments inside a given society can be analysed as if they were determined by factors only within that society (Tenbruck 1994, p. 80, quoted in Rosenberg 2016). Most of the classical sociologists were deeply beholden to internalist assumptions (or unsystematic about the effects of the international on their object of analysis), and most Science and Technology Studies (STS) and Environmental Sociology that engage the question of the intermingling of human and non-human conceive of society basically in the singular (e.g. see Lockie 2015; Salter and Walters 2016).

On top of this disciplinary weakness, international thinking has been further weakened first by globalisation discourse and then, with the escalating environmental crisis, by the direct importation into IR of epistemologies, such as those underpinning climate and other Earth sciences, that treat the world as one single space or system. As I have suggested elsewhere, the core challenge for IR at this juncture is to allow "human-natural intermeshing and international dynamics to be finally grasped together" (Corry 2020). This book can be seen as a step further down that route.

Drawing on chapters of this book as well as wider literatures, in this concluding chapter, I *first* situate the relative invisibility of non-human nature in IR, pointing to the demise of geopolitics around the Second World War as part of a wider bifurcation of knowledge into "social" and "natural" sciences. *Secondly*, I argue that current attempts to take account of non-human nature have tended to bring with them globalist framings that underplay or even obscure the importance of the international. *Thirdly*, I outline an outlook that does not feature prominently in the rest of this book, but which might provide an additional way of further developing its goals: allowing a theorisation of society that has the non-human at its core to form the building block for a materialist theory of the international. The overall aim is to take stock of attempts to grasp how relations between societies are part of a double dialectic simultaneously involving metabolism between humans and non-human nature, in societies that are themselves in a constant exchange with each other in the international. Boundaries between 'human' and 'non-human' are themselves drawn within this multiplicitous socio-ecological interchange.

Situating Anthropocentric IR

A sharp division between the human and non-human is currently entrenched in the disciplinary mainstream of IR, just as it is in other social sciences. But as Kavalski and Zolkos (2016, p. 143) note, the distinction between "social" and "natural" factors in world politics is a relatively recent invention, and non-human nature "used to have an important place in the study and practice of international affairs". Geopolitical ideas of the nineteenth and early twentieth century put a version of non-human nature centre stage in their theories and analyses. Swedish geographer Rudolf Kjellén coined the term "geopolitics" in 1899, and his and Joseph Ratzel's ideas went on to inspire the likes of Harold J. MacKinder and Alfred Mahan in Britain, Klaus Haushofer in Germany and later Nicholas Spykman in the US. They held geography to be the most important factor in world politics, partly because of its stability (Spykman and Nicholl 1944). The topography of the Earth, its climates, landmasses and oceans set the basic frame for world politics, even as technology and politics changed how that frame affected the (leading) nations of the world. Their attention to geography, European *Machtpolitik* and in some cases demography led them to advocate strategies such as autarchy (Kjellén), domination of the Heartland (MacKinder) or the opposite Rimland theory (Spykman). Haushofer's idea of "Lebensraum" infamously fed ideas about geography and (pseudo)-biological race theory into Nazi geopolitical strategies of expansion and conquest.

After the Second World War, the excesses of *Geopolitik* and related ideas of sociobiology, including eugenics, fell somewhat out of favour. Among those keen to establish IR as an independent discipline, Hans Morgenthau condemned geopolitics as "a pseudoscience erecting the factor of geography into an absolute that is supposed to determine the power, and hence the fate, of nations" (Morgenthau 1985, [Morgenthau 1948], p. 174). Although realists retained a certain focus on resources, and variants of geopolitics have recently made a revival (Porter 2015; Kaplan 2013), the modern discipline of IR was shaped as a distinctly social science concerned with social structure, institutions, norms and latterly identities and discourse. Human nature was initially retained by classical realists as a key to understanding world politics, but non-human nature became either taken for granted or treated as a resource or backdrop to competitive state action. By the 1980s, non-human nature had made a return but only in the guise of "the environment", an issue for IR to consider but outside the international itself.

This process was part of a wider bifurcation of knowledge production into "social" and "natural" sciences. This was not a neat and orderly process. Multiple, interlinked disciplinary moves took place involving the natural sciences both as model and foil for social science. "Social science" emerged in the late nineteenth century as an umbrella term in English to describe all those disciplines, outside of physical science and art, which analyse human society as a discrete object (Porter and Ross 2003, p. 2). Sociology formed early as a discipline through a protracted argument about how a general science of society might be related to "the general science of life", namely biology (Renwick 2012, p. 2). But as Steve Fuller cautions, "it is worth remembering that even at the dawn of the 20th Century neither biology nor sociology was a clearly defined field anywhere" (2012, p. xiii). In addition, often hidden "double transference" where "human society comes to be rigidly determined by irrevocable natural laws, which were [themselves] conceptually modelled after human society" (Foster et al., York, Clark 2011, p. 311) complicated matters further. For example, social Darwinism involved the transference first of Victorian competitive individualist ideas to evolutionary theory, and then that was transferred back in efforts to understand and theorise society (ibid.). Similarly, biologists borrowed the idea of "community" from sociology after which biology was used to make sense of human groups.

The social/natural science division itself had emerged downstream from the demise of a unitary conception of natural philosophy. Thus, the so-called father of sociology, Auguste Comte, distinguished social science from biology only after biology had been separated from chemistry and physics (Heilbron 2003, p. 42). The animate/inanimate dividing line replaced triads such as animal/plant/mineral, and was used to demarcate new disciplinary boundaries, e.g. between biology and geology (Fuller 2012, p. xiii). Competing ideas of what made knowledge production "scientific" were tightly bound up with this process too, with natural science again playing a large role in determining what could be considered to be "scientific" study of society (Porter and Ross 2003). Focus on behaviour as the empirically observable expression of society added to the anthropocentric shift that happened throughout this process. Somewhat separately, in the materialist camp, Marx engaged deeply with natural sciences in his quest to theorise human society as a part of the natural, but also historical, world.

Nevertheless, by the end of the twentieth century, the study of the international had come to be theorised in terms of social factors and concepts including the structural position of units in a system, institutions, identity, laws, norms and social practices. The two leading models of international structure concerned solely the relative position of subjects (i.e. states): anarchy and hierarchy (Corry 2010). Non-human nature essentially posed collective action dilemmas for environmental governance, primarily understood through institutionalist and neorealist approaches (see Corry 2017a).

Is the Planet Obscuring the International?

While the environment has been studied using existing IR approaches and "Global Environmental Politics" (Corry and Stevenson 2017), the significance of the current debate among critical scholars is that non-human nature is again being explored as something *internal* to society and world politics. The boundaries between social and natural sciences are blurring more widely, recently through the "Anthropocene" concept (Hamilton 2017). The editors thus aim with this volume "to contribute to overcome IR's ecological blindness and anthropocentrism and enhance the field's capacity to both deal with the challenges posed by a rapidly changing planet and have a positive and constructive impact on policy and practice" (Pereira and Saramago, Chap. 1). They are building on "an emerging body of literature that is seeking to reorient research towards novel ontological, epistemological, methodological, ethical, moral, political, and governance approaches aligned with the planetary real" (Pereira and Saramago, Chap. 1). Volumes such as Eroukhmanoff and Harker (2017) and Cudworth, Hobden and Kavalski (2017) have pushed the conversation on posthumanist approaches to IR forwards. A growing literature on "new materialism" in IR brings entanglements between human and non-human matter, including technologies and devices or "humanised non-human nature" to the fore (Connolly 2013; Acuto and Curtis 2013; Amicelle et al. 2015; Dittmer 2017).

This book continues and extends that trend. Rafi Youatt (Chap. 5) offers a critique of mountains as global objects identified and classified by Western cartography and colonialism. He points instead to their local situatedness, introducing "routes" as pluralistic ways of viewing their multiple existences. Cameron Harrington (Chap. 4) challenges us to think of the Anthropocene in "quantum" terms instead of Newtonian mechanics, highlighting complexity, inter-relationality, nonlinearity and entanglement—rather than simply an accelerated version of mechanistic interference in non-human nature. Shofwan Al Banna Choiruzzad (Chap. 8) traces the creation and evolution of the modern state and how it brought with it an extractivist and instrumentalist stance towards the non-human, pointing out how this happened via colonisation (and then continued via decolonisation). Bob Carter and Oliver Harris (Chap. 2) seek to overcome "kinds thinking" that abstracts humans from non-human nature at multiple scales. Other chapters look at the representation of non-human nature in leading states (Heinzekehr, Chap. 9; Collins, Chap. 10), future global governance (Burke and Fishel, Chap. 3; Pereira and Bina, Chap. 15) and existing law and international institutions and how these take account of non-human nature. Readers are invited to consider how non-human nature might be represented by the EU (Schunz, De Botselier and López Piqueres, Chap. 12) or how rights of nature might be advanced at member state level (Schoukens, Chap. 11) or by NGOs in the Arctic and Antarctic (D. Burke, Chap. 13; Phelps Bondaroff, Chap. 14).

These interventions, thus, share a collective call to break down the human/non-human divide. However, their different approaches to the issue also raise the larger question of how taking non-human nature into account should affect the "international" itself? And how might the international—the simple but highly consequential fact that multiple uneven societies coexist (Rosenberg 2016)—affect how human/non-human intermingling happens and changes?

My argument is that an answer to these questions must build upon, but also in some respects move beyond, the various approaches found in this volume. Emphasis on entanglements between "human" and "non-human" constitutes a fundamental starting point, but also run the risk of pushing the idea of the international into the background and ignoring how these entanglements exist in the context of the uneven and combined features of the "international". Rather than geography and race, the new field of Earth Systems Science is increasingly becoming the major interlocutor for world politics and non-human nature, perhaps along with Ecology and STS, none of which are equipped for taking the international into account. A hard-hitting *Planet Politics Manifesto* from 2016 aimed to awaken IR as a discipline from its anthropocentric slumbers and to catalyse "a politics to nurture worlds for all humans and species co-living in the biosphere" (Burke et al. 2016, p. 500). It casts ecological stress and biodiversity collapse as a fundamental challenge to IR as a discipline. But epistemologies such as those of Earth Systems Sciences that the authors used to lampoon IR posit the world precisely as one single space and system rather than one of disparate, interacting societies. That has to then be added.

More generally, globalist framings also tend to dominate the diagnosis of the contemporary world crisis in a way that they did not just half a century ago. The idea of the climate as a global phenomenon is today almost overpowering in public discourse. Yet the idea of a singular global climate is a surprisingly recent development (Allan 2016) emerging with globalist infrastructures such as satellites and other measuring apparatus and global circulation models simulating the Earth. What began as local weather measurements and spread eventually to all corners of the world were merged to describe a single integrated climate system (Edwards 2010) setting up the climate as an object of governance (Corry 2013; Hamilton 2018). The emergence of the wider field of Earth Systems Science based on the idea that the integrated "spheres" (hydrosphere, atmosphere, cryosphere, etc.) combine and interact into a giant overall system of which the climate is just one element is even more recent. Since the formulation of James Lovelock's Gaia hypothesis in the 1960s, it has grown rapidly to wield profound influence over how environmental crisis, ethics and global governance are conceived (Lövbrand et al. 2009) even affecting sentiments in Pope Francis' encyclical on the environment (2015). Earth System thinking and political ecology fed on planetary imagery, including the early pictures taken from space by astronauts. The "Earthrise" picture taken from the moon lent its name to a so-called "Earthrise era" in which a new self-awareness of the world as a singular but vulnerable "spaceship Earth" emerged, helping to foster the modern environmental movement (Lazier 2011).

These scientific and political discourses, while supporting forms of global cosmopolitanism and planetary ethics, also tend to hide the persistence of the "international" or cast IR as something that originally had the global as its point of departure, only to lose its way in statist geopolitics (Burke et al. 2016, p. 501). This turns Justin Rosenberg's idea of IR on its head. He sees IR as the only discipline that has as its core that social existence "is not and has never been unitary: it has always comprised a multiplicity of social entities evolving in different ways, and yet interacting with each other in real time" (Kurki and Rosenberg 2020). But with the planetary real in focus, the international as a problem is easily marginalised as illusory or anachronistic.

Worries that social sciences risk being colonised by epistemologies and ontologies emanating from natural sciences are, as mentioned above, by no means new; the main threat of "false prophets" for politics usually come from biology (including genetics) (Bell 2006). The current urgency to foreground non-human nature, therefore, requires careful and genuine interdisciplinarity rather than a submission to or collapse back into other disciplines. For Cameron Harrington writing on IR and the Anthropocene (itself originally a geological category, see Baskin 2015), IR is no longer just a subdiscipline of political science and economics, but is also subordinate to the geophysical sciences (Harrington 2016, p. 481). Unless care is taken, the necessary task of bringing non-human nature and its limits to bear on world politics could unintentionally bolster the dominance of particular (Western) natural science paradigms, spilling inevitably over into knowledge production in relation to political and social issues, most of which are deeply marked by the international. Moreover, arguing that entities identified by Earth System Science such as biomes or ecosystems have rights and can act, or that IR could be a subdiscipline of geophysical sciences tees up those sciences (and potentially scientists) to be the authoritative voices of the new "real" of world politics, the "*signals* of the Earth" (Burke and Fishel, Chap. 3).

The policy framings designed to deal with climate change and other global natural (dis)equilibria may contribute to the occlusion of the international in a similar way. Earth Systems Governance (Biermann 2014) or the Planetary Boundaries framework, for example, foregrounds that the world consists of multiple systems with built-in tipping points framing a "safe operating space" for humanity (Wijkman and Rockström, 2013). Simon Schunz, Bram De Botselier and Sofía López Piqueres (Chap. 12) in their chapter measure the EU's environmental diplomacy precisely in relation to the concept of planetary boundaries. In his chapter, Rafi Youatt (Chap. 5) interestingly describes how globalist framings (perhaps Planetary Boundaries is among them) have abstracted the mountains of the world into a generic category. This tears them away from local (including indigenous) understandings and interests, rendering them amenable to global governance. The idea of geoengineering the Earth to limit the rise of global average temperature (itself a globalist measure) is a particularly globalist framing of what world politics is now about, as model experiments often assuming a global benevolent actor (Corry 2017b). Recent calls to use technology to somehow refreeze the Arctic to stem rapid melting have led to scientific simulations and studies to test how this might be done (Corry 2017c). Such research and advocacy contribute to viewing the Arctic as a "global embedded space" (Keil and Knecht 2017). Yet the international politics of the Arctic can make any such project essentially a pipe dream. Even if it were technically possible, even just the littoral states (the US, Russia, Canada, Norway and Denmark) disagree markedly on whether global warming and Arctic melting constitute an unacceptable threat or a strategic and economic opportunity.

The point is not that international dynamics never appear in analysis of non-human nature, but rather that a purely globalist or planetary framing of analysis risks hiding the inescapable international dimension of human/non-human nature relations. In this book, several starting points to think human/non-human relations are provided that disclose their international dimension. Choiruzzad (Chap. 8) interestingly links colonisation and state-building with the spread of anthropocentrism and extractivism in Indonesia. As he explains, it was in the process of colonisation that power was centralised in Jakarta (and Amsterdam), and control over local environments came more systematically into the hands of groups with little interest in local human or non-human natures. First, the colonisers brought systematic extractivist notions of nature as a resource to bear, but then decolonisation continued the disenfranchisement of local indigenous groups disempowering some who at least some of the time had different, more holistic views of human/non-human relations. Burke and Fishel (Chap. 3) in their chapter find that a politics of "*democratic holism* that takes in the entire Earth" must in the end deal with fragmentation of the world into multiple societal units: "regional member states would need to agree to yield or pool jurisdiction in environmental law-making, along the EU model, whilst ensuring their ability to make additional environmental laws or retain existing ones until the Ecoregion Assembly does so". Hendrik Schoukens (Chap. 11) notes the "remarkable surge in the number of jurisdictions in which nature has been granted explicit rights and personhood" and how the politics of statehood, science, postcolonial movements and other international undercurrents are evolving to gradually allow national and international law to take on board non-human nature in new ways. In Stephen Quilley's (Chap. 7) chapter, the international is identified as a key cause of the spread of a culture of individual restraint and group tolerance. He argues that the Treaty of Westphalia provided "an international guarantee for the emergence of small, cellular societies, internally homogenous and with fixed, semi-permeable but controllable borders" that were a precondition for the development of pacific relations and psychologies of self-restraint in relation to other humans and non-human nature. Similar themes are present in Rasmus Karlsson's (Chap. 6) chapter when he contrasts ecomodernist visions of a high-energy, globally integrated economy with those of deep ecology's arguments in favour of low-energy and re-localised modes of life. In the process, Karlsson implicitly points to a discussion of how competing discourses, visions on environmental sustainability and potential ecological futures are necessarily caught up with the uneven distribution of technology and the various levels of integration between human societies. The globalisation of mountains perceptively described by Youatt (Chap. 5) is a product of international processes: a liberal governmentality has been installed via Western-dominated empires and international institutions (see Joseph 2012). Equally, his preferred conception of mountains as "places with multiple entities and forces involved in steering and even shaping the very things that move through them" (Youatt, Chap. 5) owes its reality to the fundamental multiplicity of social life, divided into distinct and different social units that usually straddle mountain ranges adding to their plural existences. Heinzekehr (Chap. 9) writes about Chinese environmental regulation, which is not in itself an international subject. But the international does play a key role here because the Chinese evolving position on non-human nature is also deeply bound up with China's relationship to the West: "the idea of ecological civilization is meant as a direct alternative to the path of market-based capitalism taken in Western history" (Heinzekehr, Chap. 9). It is worth adding that the regime's environmental doctrine "ecological civilization", part of the regime's "Marxism with Chinese characteristics" (a mixture of indigenous and international philosophical influences), contributes "to a national self-identity based on environmental responsibility by projecting a fictionalized image of itself *as an alternative to Western capitalism*" (Heinzekehr, Chap. 9, emphasis added). Similarly, Lavínia Pereira and Olivia Bina (Chap. 15) examine how knowledge about non-human nature is plural beyond the Western scientific canon, but also how that plurality comes under pressure from "globalized capitalism and the economization of values". Both the plurality of knowledges about nature and the hegemonic Western epistemic regime (and the struggle between the two) are a consequence of the very fact that the world consists of multiple, uneven and at times competing societies. Cameron Harrington (Chap. 4), while calling for the recognition of interdependence and entanglement via quantum social theory, also recognises that modes of being and perceiving during the Holocene have been essentially characterised by a diversity of actors organised as sovereign states. In one way, the argued-for transition to a quantum perspective serves to further reinforce the recognition of the "diverse ways in which subjects and objects coexist" and to "observe a diverse set of worldly realities" (Harrington, Chap. 4), although such strong notions of entanglement potentially also strain the idea of discrete societies. Danita Burke's (Chap. 13) chapter raises the spectre of colonialism by putting suspicions against environmental NGOs among indigenous communities in the Arctic into the context of Western anti-seal hunting campaigns and wider histories of colonisation. In the process, she emphasises how the successful implementation of environmental programmes on the part of international organisations/NGOs depends also on the recognition of the uneven and diverse ways of relating to non-human nature that have historically developed in different regions of the globe. Teale Bondaroff's (Chap. 14) chapter implicitly shows how the “international” frames the activity of non-state environmental activists and NGOs engaged in direct action by highlighting how the pursuit of their agendas can both be supported by, and enter into conflict with, different states. Environmental activists’ agendas, even if frequently legitimised with planetary or globalist language, are thus necessarily embedded in the uneven character of world politics and their cooperative/competitive geopolitical dynamics. Finally, Simon Schunz, Bram de Botselier and Sofia López Piqueres’s (Chap. 12) chapter implicitly demonstrates the impact of the “international” in the EU’s incapacity to make its reformist environmental discourse match its practice, which is rather focused on the maintenance of the *status quo*. The fact that the EU exists in a “complex global context”, where it competes with other actors with uneven and diverse capacities, pressures it to consider non-human nature protection as less important than socio-economic goals such as promotion of jobs and economic growth. On the other hand, the international is also in the character of the EU in terms of the diversity of its member states, which itself is a product of uneven processes of development, making it particularly difficult to arrive at the required consensus for bolder non-human nature protection measures.

The various contributions of the present volume, thus, constitute important starting points to rethink human/non-human nature relations in world politics. But, simultaneously, their awareness of the international character of these relations also tends to be combined with predominantly globalist or planetary conceptual categories of analysis that, in some instances, dilute the insights into the international that might otherwise be foregrounded more. In Burke and Fishel’s (Chap. 3) proposed Earth System Council and the Global Ecoregion Assembly, it is not always clear exactly how such bodies would assert institutional power, or achieve the pooling of jurisdictions required, as long as there are multiple uneven (often competing) societies. While the UN Security Council with its permanent veto-wielding members to some extent takes geopolitical realities (or those that existed at the end of the Second World War at least) into account, how will power politics be handled in an Earth System Council?

Quilley (Chap. 7) arguably deploys some globalist (internalist) assumptions, despite the weight given to Westphalian sovereignty in explaining civilisational processes. His assumption that degrowth or a contraction in consumption and extraction of resources overall would translate into a fairly uniform spread of communitarian and nativist sentiments, a process of cellularisation of the world, resulting in lower levels of interaction and psychological restraint, suggests a more limited notion of the international. Countries are deemed to go through the same basic patterns of civilisation and de-civilisation, despite their varied position in the global economic and political landscape.

Similarly, Rasmus Karlsson (Chap. 6) describes ecomodernist thinking that also takes a planetary view of a “high-energy planet” and even entertains “the possibility of humanity becoming a multi-planetary species and thereby securing not only its own long-term survival but that of life more generally”. His discussion of nuclear power makes a passing reference to “the long shadow of the Cold War” and the unusually high levels of trust in Sweden, but does not elaborate much on dynamics born of the international more widely. Unevenness of the international and distributional questions are subsumed in a species-on-a-globe discourse: e.g. "traditional environmentalism comes with the assumption that humanity can somehow collectively lessen its appetite" (Karlsson, Chap. 6).

While the international is thus present, there is a tension between the analytical gaze of societal multiplicity on the one hand and global systems and circulations associated with planetary notions of non-human nature on the other, along with other globalist framings such as civilisation, global society, world economy and "humanity"/human species characteristics.

A Materialist International

While some begin from theories of entanglement with non-human nature, another way of theorising non-human nature in world politics instead takes a strong theory of the "international" as its point of departure—and then considers human/non-human interaction within that frame. This strategy has its own challenges. Most IR analysis of climate change has hitherto reproduced the focus and theory of the IR "isms" and applied them to it: realism exploring climate conflict and neoliberal institutionalism fixating on institutional designs. Furthermore, Rosenberg's idea of positing the consequences of there being a multiplicity of human societies as the "deep ontology" of IR does not in itself overcome anthropocentrism (Corry 2018). However, it could still provide a way to bring in insights from geology, ecology, biology and Earth Systems Sciences without smuggling the globalist, planetary or internalist perspectives of other disciplines in smothering analysis of societal multiplicity. I have argued elsewhere (Corry 2020) that if a broad notion of the international as societal multiplicity is married to a materialist theorisation of "society" that puts relations between human and non-human nature at its core, this could get us beyond the anthropocentrism while exposing latent globalism in analysis of human/non-human interaction.

While some posit that the current age or epoch—the Anthropocene (Crutzen 2006) or global risk society (Beck 2006)—involves a *novel* intermingling between humans and non-human nature that dissolves the international, a materialist notion of the "societal" includes relations with the non-human from the very beginning. Materialist analysis is based, fundamentally, on the idea that humans as physical, practical beings must transform non-human nature via their (collective) labour in order to survive—and that this process is highly consequential for social organisation and hence historical. There is a "metabolism" between social and natural, but it is not universal or constant. This is what John Bellamy Foster calls "a dialectical notion of natural history" (2000, p. 19)—one that he finds in Marx's deep engagement with both Darwin and Malthus and Marx's interest in capitalist agriculture's destruction of soil ecology. The exploitation of non-human nature is intimately linked, not simply to anthropocentrism or Western science but to different local modes of organisation and power including ownership and also to modes of making security and protection (Deudney 2000). Foster's view forms a key part of an ongoing rehabilitation of historical materialism as a collected theory of society *and* ecology (Foster 2000;

Saito 2017). The "metabolic rift" that Marx sees emerging between capitalist society and non-human nature as the cycle of nutrients is broken down is inherently connected to a rift emerging between city and country (later developing fully between different classes). But this is dialectically related, in turn, to inter-societal relations, e.g. via imperialism and (neo)colonialism.

Rather than pitting vitalism and ecocentrism as the cure for mechanical and anthropocentric views of non-human nature, this tradition aims instead to transcend those categories "by recovering the deeper critique of the alienation of humanity from nature" (Foster 2000, p. 19–20). Whereas in popular Anthropocene discourse, it is the (purportedly novel) *unity* of society and nature that needs explaining; it is for the materialist tradition, the *separation* of the two, as a result of an historical process that is remarkable.

Except it is precisely not "society" and "nature" in the singular that are in a metabolism. Societies are multiple, and this fact has material consequences that go far beyond the security dilemma. A focus on societal multiplicity in the context of a materialist understanding of the "societal" might thus serve "both to reveal the ecological substratum of the international and to integrate an international dimension into our understanding of the Anthropocene as a historical conjuncture" (Kurki and Rosenberg 2020, forthcoming). The uneven development of different societies means that different parts of the world have different ways in which humans and the non-human intermingle, and these ways vary and change historically and thereby nonrandomly. While each society is in a metabolism with its local ecological niche, it will also be in an interaction with the rest of non-human nature, partly via the international, e.g. via trade, imperialist forays from the centre to the periphery (Clark and Foster 2009) or via the spread or transfer of scientific ideas, technologies and cultural forms, mutating as they combine with local conditions.

In this perspective, there is not simply a general "entanglement" between human and non-human, but a patterned set of logics of exchange and exploitation between social and natural systems as well as between different societal units. There is a logical connection between how different social groups relate to (and exploit) *each other* on the one hand and how a given society is "entangled" with non-human nature, e.g. through empires or other forms of hierarchy. This should ideally allow non-human nature in world politics to be taken seriously on board *without* throwing out or marginalising analysis of intra-human relations. In this respect, several of the contributions in the present volume are particularly relevant. Choiruzzad (Chap. 8) maintains an admirable focus on the links between exploitation of one set of people by another (colonisation) on the one hand and the exploitation of non-human natures on the other (the archipelago of Indonesia). Pereira and Bina's (Chap. 15) analysis of plural knowledge of non-human nature illustrates the complex links between epistemologies and the politics of non-human nature, and hints at how knowledges are connected to wider struggles between groups, often amid the realities and legacies of colonialism and extractivist practices. Schoukens (Chap. 12) also points to the interlinkages between social struggles and the shifting status of non-human nature noting that "the eagerness with which the Colombian judges make a bridge between a human rights-based litigation strategy and rights of nature is remarkable". Carter and

Harris (Chap. 2) say that their analysis of car driving as a "practice of freedom in a world where more substantive freedoms are no longer possible" is not based directly on IR. Yet the chapter does indirectly highlight the importance of the international automobile industry and the spread of neoliberal individualism as a constitutive part of the human/non-human assemblage of transportational devices (cars) and social practices (driving) that they interestingly evoke in their chapter. In Collins' (Chap. 10) chapter (not unlike Heinzekehr's on China), it becomes clear that the US's position on climate change—ostensibly a "domestic" question—evolves through international pressures from other states, a globalised oil and gas industry, media structures, sciences of climate change (e.g. James Hansen's testimony in 1988), class interests and domestic political ideologies of individual freedom. As often in the US environmentalist debates, the idea of Manifest Destiny seems to lurk in the background which itself is the result of European migration and colonisation of North America and the expropriation of native lands and ecologies. The metabolism in the US between humans and non-human nature is clearly in another simultaneous dialectic with other societies.

Final Remarks

With wildfires raging, oceans acidifying and glaciers melting, and gathering storms in world politics as institutions and norms of cooperation come under fire, the efforts to rethink the role of non-human nature in world politics in this book are extremely timely and serve as important critiques and starting points. Of course, further research that builds upon its insights remains to be done. This is urgent both because IR as a discipline was theoretically weakly founded in terms of offering a general contribution to all social analysis (not just politics beyond states), but also because the sciences that currently dominate diagnosis of global ecological crisis do not leave much space for acknowledging societal multiplicity. Theoretical solutions sought in sociology, STS or posthuman ethics are also often of an internalist disposition and less attuned to the implications of social life not being singular.

There are, thus, important traps to be avoided in future research that seeks to grasp world politics and ecology simultaneously. The framing of "non-human nature" in world politics in this book is laudable in its mission to recognise that humans are too part of nature. At the same time, the very notion of "non-human" paradoxically promotes the category of "human" and potentially thereby the notion of an undifferentiated "humanity" as a historical agent in world history. Bringing in concepts from natural sciences such as the Anthropocene and planetary boundaries may allow crucial aspects of the non-human world to be represented, but also risk obscuring the unevenness of the world, and thereby one category of social dynamics that is driving or exacerbating the problems disappears from view. Moreover, no sooner are attempts made to break down conceptual barriers between human and non-human, before old problems of whether or how to transfer methods and concepts between social science and natural science reappear. The founders of sociology insisted that social phenomena had emergent features that differed from those of biology, necessitating different concepts, theories and methods. The human and non-human, despite their interactions and mutual conditioning, should not be collapsed into each other in one 'hybrid' category. If they are, the materiality of non-human nature could effectively be denied, as could the emergent social characteristics of human-infused systems. The importance of social structures may disappear in "global emissions" and "global temperatures" while the very idea of "anthropogenic climate change" would cease to make sense. If the relations between human and non-human nature are to be analysed, the different emergent characteristics of each must still be recognised, just as is the distinction between them and always historical notions of the "human" emerge. Maintaining a (non-Cartesian) dualism may, therefore, still be useful: all of life—human or not—is part of nature, but not all of it behaves according to the same logic (Andreas Malm refers to this as substance monism with process dualism, Malm 2018). As Kavalski and Zolkos (2016, p. 141) stress, the "Anthropocene" "does not deny the qualitative distinctions between these two systems".

But it helps to remember that the impact on natural habitats from societal activity is not a new or recent development born of modern science, a "new modernity" (Beck 1992) or the Anthropocene epoch. The mutual transformation and metabolism between human and non-human is rather a fundamental aspect of the social in the first place. Crucially, as this chapter has sought to emphasise, those two systems are themselves fragmented by multiplicity. An international politics that takes non-human nature seriously will therefore ultimately also be taking relations and power between societies seriously. The exploitation of human groups is deeply linked to patterns of exploitation of non-human nature. Some human groups are recognised as grievable and fully human, while others are not, landing on the alienated, non-human side of social-ecological divides (Butler 2009). Such fragmentary logics will persist as long as there is more than one society coexisting. And despite the planetary character of ecological challenges and existential risks, multiplicity looks set to continue for the foreseeable future.

References

Acuto, M., & Curtis, S. (2013). *Reassembling international relations theory*. Houndmills: Palgrave Pivot, Palgrave.

Anievas, A., & Nişancıoğlu, K. (2015). *How the west came to rule*. London: Pluto Press.

Allan, B. B. (2016). Producing the climate: States, scientists, and the constitution of global governance objects. *International Organization, 71*, 1–32.

Amicelle, A., Aradau, C., & Jeandesboz, J. (2015). Questioning security devices: Performativity, resistance, politics. *Security Dialogue, 46*(4), 293–306.

Baskin, J. (2015). Paradigm dressed as epoch: The ideology of the Anthropocene. *Environmental Values, 24*(1), 9–29.

Beck, U. (1992). *Risk society. Towards a new modernity*. London: Sage.

Beck, U. (2006). Reflexive governance: politics in the global risk society. In J. P. Voss, D. Bauknecht & R. Kemp, (Eds.), *Reflexive governance for sustainable development* (pp. 31–56). Cheltenham: Edward Elgar Publishing.

Bell, D. (2006). Beware of false prophets: Biology, human nature and the future of international relations theory. *International Affairs, 82*(3), 493–510.
Biermann, F. (2014). *Earth system governance: World politics in the Anthropocene.* Cambridge, MA: MIT Press.
Bostrom, N., & Cirkovic, M. M. (Eds.). (2011). *Global catastrophic risks.* Oxford: Oxford University Press.
Burke, A., Fishel, S., Mitchell, A., Dalby, S., & Levine, D. J. (2016). Planet politics: A manifesto from the end of IR. *Millennium: Journal of International Studies, 44*(3), 499–523.
Burton-Chellew, M. N., May, R. M., & West, S. A. (2013). Combined inequality in wealth and risk leads to disaster in the climate change game. *Climatic Change, 120*(4), 815–830.
Butler, J. (2009). *Frames of war. When is life grievable?.* London: Verso.
Clark, B., & Foster, J. B. (2009). Ecological imperialism and the global metabolic rift: Unequal exchange and the guano/nitrates trade. *International Journal of Comparative Sociology, 50*(3–4), 311–334.
Connolly, W. (2013). The 'new materialism' and the fragility of things. *Millennium: Journal of International Studies, 41*(3), 399–412.
Corry, O. (2010). What is a (global) polity?. *Review of International Studies, 36*(S1), 157–180.
Corry, O. (2013). The rise and fall of the global climate polity. In H. Bulkeley & J. Stripple (Eds.), *Governing the climate: New approaches to rationality, power and politics* (pp. 219–234). Cambridge: Cambridge University Press.
Corry, O. (2017a). The 'nature' of international relations: From geopolitics to the Anthropocene. Ch. 9. In C. Eroukhmanoff & M. Harker (Eds.), *Reflections on the posthuman in international relations–the Anthropocene, security and ecology* (pp. 102–118). Bristol: E-International Relations Publishing.
Corry, O. (2017b). The international politics of geoengineering: The feasibility of Plan B for tackling climate change. *Security Dialogue, 48*(4), 297–315.
Corry, O. (2017c). Globalising the arctic climate: Geoengineering and the Emerging Global Polity. In *Governing Arctic Change* (pp. 59–78). Palgrave Macmillan, London.
Corry, O. (2018). Societies are not the only source of multiplicity. *International Relations, 32*(2), 243–245.
Corry, O. (2020). Nature and the international: Towards a materialist understanding of societal multiplicity. *Globalizations*, 1–17.
Corry, O., & Stevenson, H. (2017). *Traditions and trends in global environmental politics: International Relations and the Earth.* London: Routledge.
Crutzen, P. J. (2006). The "anthropocene". In *Earth system science in the anthropocene* (pp. 13–18). Springer, Berlin, Heidelberg.
Cudworth, E., Hobden, S., & Kavalski, E. (Eds.). (2017). *Posthuman dialogues in international relations.* Routledge.
Deudney, D. (2000). Geopolitics as theory: Historical security materialism. *European Journal of International Relations, 6*(1), 77–107.
Dittmer, J. (2017). *Diplomatic material: Affect, assemblage, and foreign policy.* Duke University Press.
Edwards, P. N. (2010). *A vast machine: Computer models, climate data, and the politics of global warming.* Cambridge, MA: MIT Press.
Eroukhmanoff, C., & Harker, M. (Eds.). (2017), *Reflections on the posthuman in international relations–the Anthropocene, security and ecology.* Bristol: E-International Relations Publishing.
Foster, J. B. (2000). *Marx's ecology: Materialism and nature.* New York: Monthly Review Press.
Foster, J. B., Clark, B., & York, R. (2011). *The ecological rift: Capitalism's war on the earth.* New York: Monthly Review Press.
Francis, P. (2015). *Encyclical on climate change and inequality: On care for our common home.* Melville House.
Fuller, S. (2012). Foreword. In C. Renwick (Ed.), *British sociology's lost biological roots: A history of futures past* (pp. xii–xvii). New York: Palgrave MacMillan.

Hamilton, C. (2017). *Defiant earth: The fate of humans in the Anthropocene*. Cambridge: Polity Press.
Hamilton, S. (2018). The measure of all things? The Anthropocene as a global biopolitics of carbon. *European Journal of International Relations, 24*(1), 33–57.
Harrington, C. (2016). The ends of the world: International relations and the Anthropocene. *Millennium, 44*(3), 478–498.
Heilbron, J. (2003). Social thought and natural science. In T. M. Porter, & D. Ross (Eds.), *The Cambridge history of science: The modern social sciences*.
Hornborg, A., & Crumley, C. L. (Eds.). (2016). *The world system and the Earth system: Global socioenvironmental change and sustainability since the neolithic*. London: Routledge.
Joseph, J. (2012). *The social in the global: Social theory, governmentality and global politics* (Vol. 122). Cambridge University Press.
Kaplan, R. (2013). *The revenge of geography. What the map tells us about coming conflicts and the battle against fate*. New York, NY: Random House.
Kavalski, E., & Zolkos, M. (2016). The recognition of Nature in International Relations. In P. Hayden & K. Schick (Eds.), *Recognition and global politics: Critical encounters between state and world* (pp. 139–155). Manchester University Press.
Keil, K., & Knecht, S. (2017). Introduction: The arctic as a globally embedded space. In *Governing arctic change* (pp. 1–18). London: Palgrave Macmillan.
Kurki, M., & Rosenberg, J. (2020). Multiplicity–A new common ground for international theory? In *Introduction to Special Issue of Globalizations on Multiplicity*. Forthcoming.
Lazier, B. (2011). Earthrise; or, the globalization of the world picture. *The American Historical Review, 116*(3), 602–630.
Lockie, S. (2015). What is environmental sociology? *Environmental Sociology, 1*(3), 139–142.
Lövbrand, E., Stripple, J., & Wiman, B. (2009). Earth system governmentality reflections on science in the Anthropocene. *Global Environmental Change, 19*(1), 7–13.
Malm, A. (2018). *The progress of this storm: Nature and society in a warming world*. London: Verso Books.
McKibben, B. (1989). *The end of nature*. New York: Random House.
Morgenthau, H. (1948/1985). *Politics among nations*. (Revised by Kenneth W. Thompson). New York, NY: Knopf.
Porter, P. (2015). *The global village myth: Distance, war, and the limits of power*. Washington, DC: Georgetown University Press.
Porter, T. M., & Ross, D. (2003). Introduction: Writing the history of social science. In T. M. Porter, & D. Ross, (Eds.), *The Cambridge history of science: The modern social sciences* (Vol. 7, pp. 1–10).
Renwick, C. (2012). *British sociology's lost biological roots: A history of futures past*. Springer.
Rosenberg, J. (2016). International relations in the prison of political science. *International Relations, 30*(2), 127–153.
Saito, K. (2017). Marx in the Anthropocene: Value, metabolic rift, and the non-Cartesian dualism. *Zeitschrift für kritische Sozialtheorie und Philosophie, 4*(1–2), 276–295.
Salter, M. B., & Walters, W. (2016). Bruno Latour encounters international relations: An interview. *Millennium: Journal of International Studies, 44*(3), 524–546.
Spykman, N. J., & Nicholl, H. R. (1944). *Geography of the Peace*, Chicago.
Tenbruck, Friedrich. (1994). Internal history of society or universal history? *Theory, Culture and Society, 11*(1), 80.
Waltz, K. (1979). *Theory of international politics*. New York: McGraw Hill.
Wijkman, A., & Rockström, J. (2013). *Bankrupting nature: Denying our planetary boundaries*. London: Routledge.

The manufacturer's authorised representative in the EU is Springer Nature Customer Service Centre GmbH, Europaplatz 3, 69115 Heidelberg, Germany. If you have any concerns regarding our products, please contact ProductSafety@springernature.com

Printed and bound by CPI Group (UK) Ltd, Croydon, CR0 4YY
30/09/2026
02225504-0012